Heat Transfer - Fundamentals, Enhancement and Applications

Edited by Salim Newaz Kazi

Published in London, United Kingdom

Heat Transfer - Fundamentals, Enhancement and Applications
http://dx.doi.org/10.5772/intechopen.97951
Edited by Salim Newaz Kazi

Contributors
R. Yuvaraj, Obaid Ur Rehman, Nor Erniza Mohammad Rozali, Marappa Gounder Ramasamy, Sudarmadji Sudarmadji, Sugeng Hadi Susilo, Asrori Asrori, Miha Bobič, Bijoy Kumar Purohit, Zakir Hussain, PVR Sai Prasad, Rania Jradi, Christophe Marvillet, Mohamed Razak Jeday, Tatsuhiko Aizawa, Naoki Ono, Hiroki Nakata, Naga Ramesh Korasikha, Thopudurthi Karthikeya Sharma, Gadale Amba Prasad Rao, Kotha Madhu Murthy, Thanh Nhan Phan, Moises Meza Pariona, Seung Jin Oh, Yeongmin Kim, Yong-yoo Yang, Yoon Jung Ko, Nickan H. Hossein Ghahramani, Martin Sharp, Michael Morgan, Mehdi Seddighi, Nguyen Minh Phu, Nguyen Van Hap, Tahir Naseem

First published in London, United Kingdom, 2023 by IntechOpen
IntechOpen is the global imprint of INTECHOPEN LIMITED, registered in England and Wales, registration number: 11086078, 5 Princes Gate Court, London, SW7 2QJ, United Kingdom

British Library Cataloguing-in-Publication Data
A catalogue record for this book is available from the British Library

Additional hard and PDF copies can be obtained from orders@intechopen.com

Heat Transfer - Fundamentals, Enhancement and Applications
Edited by Salim Newaz Kazi
p. cm.
Print ISBN 978-1-80355-939-1
Online ISBN 978-1-80355-940-7
eBook (PDF) ISBN 978-1-80355-941-4

For EU product safety concerns:
IN TECH d.o.o., Prolaz Marije Krucifikse Kozulić 3, 51000 Rijeka, Croatia,
info@intechopen.com or visit our website at intechopen.com.

Meet the editor

Dr. Salim Newaz Kazi is a professor of mechanical engineering at the University of Malaya. He obtained his BSc, MSc, and MS in mechanical engineering and his Ph.D. in chemical and materials engineering. After a long career in the petrochemical industry and consultancy work for different engineering companies, he has been working as an academic since 2009, specializing in heat transfer, fluid mechanics, particle characterization, heat exchanger fouling mitigation, nanofluid synthesis and applications, and renewable energy. He is an active member of many professional bodies and an editorial member of many journals. He has supervised many postgraduate theses and published many technical papers.

Contents

Preface

Heat transfer, its mechanism, augmentation, and applications are an aid to various aspects of engineering design. With energy saving and environmental protection high on the scientific community's agenda, the augmentation and application of heat transfer are highly significant. *Heat Transfer - Fundamentals, Enhancement and Applications* is a contribution to these endeavors.

This book has four sections: "Fundamentals of Heat Transfer and Mechanism", "Different Forms of Heat Transfer and Applications in Various Aspects", "Heat Transfer Enhancement and Techniques", and "Fouling and Its Mitigation in Heat Exchangers".

Boiling and condensation heat transfer are manyfold higher than convective heat transfer. Chapter 1, "Boiling and Condensation", provides a brief overview of these two heat transfer phenomena. A detailed explanation is given of different classifications of boiling, including pool, flow, subcooled and saturated boiling, and different boiling regimes (natural convection boiling, nucleate boiling, transition boiling, film boiling). Types of condensation (drop and film-wise), their heat transfer aspects and applications are also presented. The chapter also includes information about advanced heat transfer enhancement techniques available for boiling and condensation.

Energy conversion is a very contemporary concern. Thermal radiation is converted into heat in many activities: drying agricultural products, space heating, desiccant regeneration, timber seasoning and natural ventilation. Indigenous materials can be used for solar air heating. Solar thermal air heaters have low performance because the convection heat transfer coefficient of the air is low, and the heat loss to the environment is considerable. Chapter 2, "Heat Transfer in Double-Pass Solar Air Heater: Mathematical Models and Solution Strategy", presents two mathematical models of convection and radiant heat transfer in double-pass solar air heaters, which are commonly used because they produce higher air temperature than that of a single pass heater. In this case, the average temperature model was solved by dealing with a system of linear algebraic equations and the other model was derived as ordinary differential equations and solved by numerical integration.

Chapter 3, "Transfer of Heat through a Thin Liquid Film", discusses heat transfer in thin liquid film past a stretching surface in the presence of Joule heating, radiation, viscous dissipation, and magnetic effect. The nonlinear coupled partial differential equations were reduced to nonlinear ordinary differential equations, which were then solved using the shooting method. The effects of Prandtl and Eckert numbers over the temperature distribution were incorporated. Increasing unsteadiness parameter values on the thermal conductivity of the fluid are reported.

Chapter 4, "Heat and Mass Transfer of a Decoupling Cooling System: A Desiccant-Coated Heat Exchanger and a Dew-Point Evaporative Cooler", investigates a de-coupling cooling technology where latent cooling load and sensible cooling load were handled separately by a desiccant-coated heat exchanger-based dehumidifier and a dew-point evaporative cooler. The undesired moisture in the outdoor air was discarded through adsorption and the dehumidified air was then cooled by the dew-point evaporative cooler to the desired temperature. Performance was investigated numerically by analyzing the heat and mass transfer.

Chapter 5, "Dropwise Condensation and Heat Transfer", discusses the fundamentals of condensation, heat transfer and relevant equations for dropwise condensation. Dropwise condensation is much more effective than film-wise condensation because water droplets formed on the nucleation sites accumulate, enhancing heat transfer as they detach from the surface. As the contact angle of the water droplet increases, the heat flow resistance between the substrate and the vapor decreases, and the heat transfer coefficient and the condensation rate both increase. Augmentation in boiling heat transfer was performed by modifying the texture of the bare surface using the thermo-solution immersion method.

Forced cooling in a very large-scale integrated circuit became a turning point to reconsider how to improve the capacity to cool down highly energy-consuming devices and systems. Researchers have sought to enhance equipment performance in heat and mass transfer engineering, and these techniques are discussed in Chapter 6, "Boiling Heat Transfer on the Micro-Textured Interfaces". Micro-/nano-textured aluminum and copper devices can enhance the boiling heat transfer process to the subcooled water. Forced water cooling over a surface with convex micro-textures was achieved by varying the Reynolds number, and the resulting heat transfer characteristics were investigated. The newly constructed boiling curve on the micro-textured interfaces is also presented here.

The use of multiple laser tracks leads to nonuniform hardness in the surface of components. In Chapter 7, "Multi-Track Overlapping by Laser-Treated and Its Effects on the Microstructural Behavior of Al-Fe Alloy Assessed by FEM", a numerical simulation was applied, using the finite element method, optimized through the multigrid technique, to study the influence of multi-track overlapping on the microstructure processed by laser surface remelting. The experimental and numerical simulation results of the multi-track overlapping were validated.

Chapter 8, "A Review on Condensation Process of Refrigerants in Horizontal Microfin Tubes: A Typical Example", explores the application of micro fins in refrigeration systems. Proper refrigerant with good thermodynamic properties helps optimize the equipment. Micro fin tubes can be used to improve heat transfer performance while reducing the size, weight, and amount of refrigerant in the system and maintaining a suitable environment.

Thermal management of modern electronic equipment is discussed in Chapter 9, "Internal and External Influences on Hydro-Thermal Behavior of Micro-channel Flow". Heat generation has become a recurring problem in high-powered electronic systems, energy storage batteries, fuel cells, etc. Operation of these electronic devices at higher temperatures reduces their performance, lifespan and reliability. Microchannel flow is one of the effective solutions for many thermal engineering problems.

Global energy production and consumption levels from fossil fuels (coal, oil, and natural gas) are unsustainable, with energy consumption expected to increase three-fold in the next 30 years. Fossil-fueled power plants convert heat into mechanical energy and then to electrical energy using a heat exchanger. Researchers are investigating various ways of saving energy by increasing the heat transfer coefficient, and hence the performance, of a heat exchanger with smaller dimensions. Chapter 10, "The Combined Method to Improve Heat Transfer Coefficient on Heat Exchanger", focuses on the enhancement of thermal properties of the cooling fluids by introducing nanofluids to enhance the overall performance of the heat exchanger. The combination of nanofluids and ultrasonic vibrations was found to improve overall heat transfer performance in heat exchangers.

Chapter 11, "Enhancing Surface Heat Transfer Characteristics Using Laser Texturing", reports the use of a pulsed laser system to manufacture parallel streamwise riblets on the plates of a heat exchanger. The alteration of surface morphology for enhancement of heat transfer or drag reduction on a surface has been the subject of research for some time. Although skin drag reduction positively lowers energy consumption and enhances heat transfer, there has been little notable research in this area. Surface roughness is thought to keep heat transfer efficiency high in two flow regimes, laminar and turbulent.

Heat exchangers are used extensively in everyday life. The outlet temperature on the secondary side of the heat exchanger can be controlled by the flow on the primary side under given inlet temperature conditions on the primary and secondary sides, using an electronic controller. Chapter 12, "Digital Twin of Heat Exchanger", has investigated alternative models in the trade-off between the accuracy of the calculation and speed.

Fouling material deposits on the heat exchanger surface are usually gradually solidified by the action of heat, and since they have lower thermal conductivity than the substrate, heat exchanger performance is gradually reduced, eventually causing it to be shut down for maintenance. Chapters 13 and 14, "Fouling and Mechanism", and "Fouling in Industrial Heat Exchangers: Formation, Detection and Mitigation", discuss fouling, its formation mechanisms and mitigation approaches. Design concepts, and the operation, cleaning, and maintenance of heat exchangers in industrial practices are also covered.

Although this work was an addition to my routine academic workload, it was thoroughly enjoyable. I would like to thank my students for their patience in receiving delayed support from me. Lastly, I would like to thank my wife, Nilufa Parveen, and son, Kazi Mehrab Newaz, for their sacrifices in our daily life for the duration of this work.

Salim Newaz Kazi
Faculty of Engineering,
Department of Mechanical Engineering,
University of Malaya,
Kuala Lumpur, Malaysia

Section 1

Fundamentals of Heat Transfer and Mechanism

Chapter 1

Boiling and Condensation

Bijoy Kumar Purohit, Zakir Hussain and PVR Sai Prasad

Abstract

This chapter contains a brief overview of both boiling and condensation heat transfer phenomena. Boiling and condensation are the two convective heat transfer phenomena that involve phase change from liquid to vapour and vapour to liquid, respectively. The chapter starts with the basis of heat transfer with an emphasis on the boiling and condensation phenomenon. Next, the overview of the boiling phenomenon and its different classifications like pool, flow, and subcooled and saturated boiling are discussed in detail. Different boiling regimes (natural convection boiling, nucleate boiling, transition boiling and film boiling) with the observed heat transfer rate in the case of pool boiling are mentioned in detail using the boiling curve. The heat transfer aspect and basics of condensation with types (drop and film-wise condensation) and application are also presented. The derivation for the calculation of the rate of heat transfer during film condensation with the correlations for heat transfer coefficient on vertical, horizontal and inclined plates is explained. Some numerical for the calculation of the rate of heat transfer and heat transfer coefficient for condensation phenomena has been also been mentioned. Apart from a basic overview, this chapter also includes information about the advanced heat transfer enhancement techniques available for boiling and condensation.

Keywords: Heat transfer, boiling, condensation, film-wise condensation, boiling regimes, convection, Heat transfer coefficient

1. Introduction

Heat is a type of energy that is in transit between a hot body (source at a higher temperature) and a cool body (receiver at a lower temperature). The driving force for heat energy transport between two points is the temperature difference between them. Calorie and joule are the most frequent units for expressing heat energy [1].

Heat transfer is the branch of science concerned with determining the rates of heat energy transfers. Conduction, convection and radiation are three modes, by which the transfer of heat occurs from a hot source to a cold recipient. In the conduction mode of heat transfer, the heat energy is generally transferred within the substance or to another substance in physical contact and is caused by lattice vibration and free electron movement. Convection is the transmission of heat due to the macroscopic motion of molecules within the medium. In general, conduction heat transfer is observed within solid mediums and convection occurs within fluids (gases or liquids) mediums by the mixing of hot and cold portions of the fluid. In the radiation mode, heat is transferred through electromagnetic waves produced by a hot body. Radiation heat can be transferred over the medium within vacuums and space [2].

Sensible heat is the heat that must be transferred to raise or lower the temperature of a system when no phase change is observed within the medium. The latent heat of phase change is the thermal energy associated with a unit amount of matter at a fixed temperature and pressure when it experiences a phase transition (from solid to liquid and vapour to liquid or vice versa) [2].

2. Boiling and condensation phenomenon

Boiling and condensation both are under the convection heat transmission process in which the system undergoes a phase transition and are opposite to each other. These processes include the involvement of both sensible and latent heat.

Boiling is the process of transferring a medium from a liquid to a vapour state by applying heat. When a liquid medium is applied to heat, then the medium will start to boil at a certain temperature (boiling point temperature). At this particular temperature, the liquid phase changes to the vapour state, known as the boiling phenomenon.

In reverse, condensation is the process of transferring a medium from a vapour to a liquid by removing the heat from the medium. When a medium initially in the vapour state is cooled, then its phase will change from vapour to liquid state, known as the condensation phenomenon. Vapours are generated by boiling, and liquid droplets are formed by condensation [1, 3].

3. Heat transfer to boiling liquids

Boiling phenomenon is generally observed in unit operations such as evaporation, distillation and steam generation and is the opposite of the condensation phenomenon. When the liquid medium is exposed to a surface, at a temperature above the saturation temperature of the liquid, the phase of the medium changes from liquid to vapour.

Suppose a liquid medium is kept within a solid vessel, to which heat is supplied to boil the liquid. Let the temperature of the solid surface be 'Ts' and the liquid medium have the saturation temperature of 'Tsat'. Initially, let the solid surface temperature be below the saturation temperature of the liquid. The boiling will start, when the temperature of the supplied liquid increases from 'Ts' to the saturation temperature 'Tsat'. Further on increasing the temperature of the supplied liquid, the boiling rate will also increase.

According to Newton's law of convection, heat transferred from a solid surface to the liquid (through convection mode) is

$$Q = hA(T_S - T_{Sat}) = hA\Delta T_{excess}$$

Here, $(T_S - T_{Sat}) = \Delta T_{excess}$ = excess temperature = temperature of the supplied liquid – saturation temperature of liquid = the extra heat supplied in excess above the saturation temperature of the fluid, during boiling of the liquid medium [3, 4].

3.1 Classification of the boiling phenomenon

According to the bulk fluid motion and the bulk fluid temperature, the boiling phenomenon can be classified into two basic categories.

i. Based on the bulk fluid motion within the liquid medium

Pool boiling: Boiling phenomenon within a liquid medium, which is at a stationary or non-flow condition, is called pool boiling. In pool boiling, heat is generally supplied through a submerged solid surface (by placing a heating coil inside the liquid) or boiling water within a solid container from external heat. Bubbles generate during the heating process travel in the liquid medium, due to buoyance, and the heat gets transferred through the natural convection process.

Flow boiling: Boiling phenomenon within a liquid medium, which is at flowing condition, is called flow boiling. In the case of flow boiling, heat is generally transferred to a flowing liquid medium through the forced convection process. Boiling in a liquid medium when it is flowing over a hot surface or within a heated pipe is an example of flow boiling [3, 5].

ii. Based on the bulk liquid temperature within the liquid medium

Subcooled or local boiling: The boiling phenomenon is said to be subcooled or local boiling if the temperature of the bulk liquid medium above the heating surface is less than the saturation temperature of the liquid.

Saturated boiling: The boiling phenomenon is said to be saturated if the temperature of the liquid medium above the heating surface is about the saturation temperature of the liquid.

At the early stages of pool boiling, the liquid adjacent to the hot solid surface vaporises and the bubbles are formed by absorbing heat from the hot solid surface. The bubbles contain more heat energy and travel within the liquid medium due to the convection phenomenon.

But during the initial stage of boiling, the bulk fluid (a certain height above a hot solid surface) will be at a temperature much less than the liquids saturation temperature. These bubbles when they move away from the hot surface and come in contact with cold liquid, they condense and collapse by transferring the absorbed heat (from the hot surface) into the liquid medium. This phenomenon happens when the bulk liquid is at a temperature much lower than the saturation temperature (subcooled or local boiling).

Further, when the temperature of the entire liquid medium reaches about the saturation temperature, the bubbles will not condense and will start rising to the top (saturated boiling) (**Figures 1** and **2**) [3, 6].

3.2 Boiling regimes and the boiling curve in case of pool boiling

Boiling process in the pool of a liquid medium will start, when the supplied temperature exceeds the saturation temperature (ΔTexcess >0). Depending on this excess temperature ΔTexcess supplied to the liquid medium, different types of boiling regimes are observed in a pool of liquid [1, 3, 4]. Those regimes are

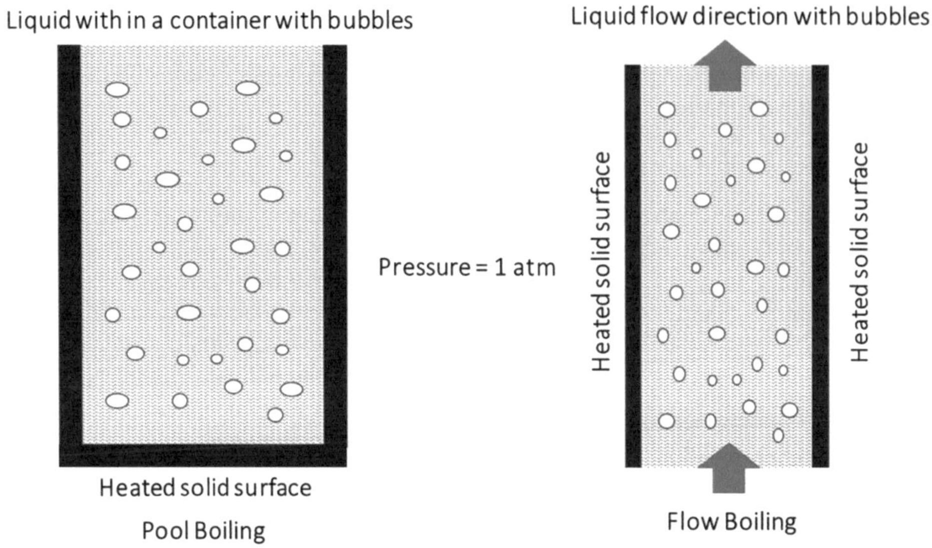

Figure 1.
Pool and flow boiling phenomena [3, 6].

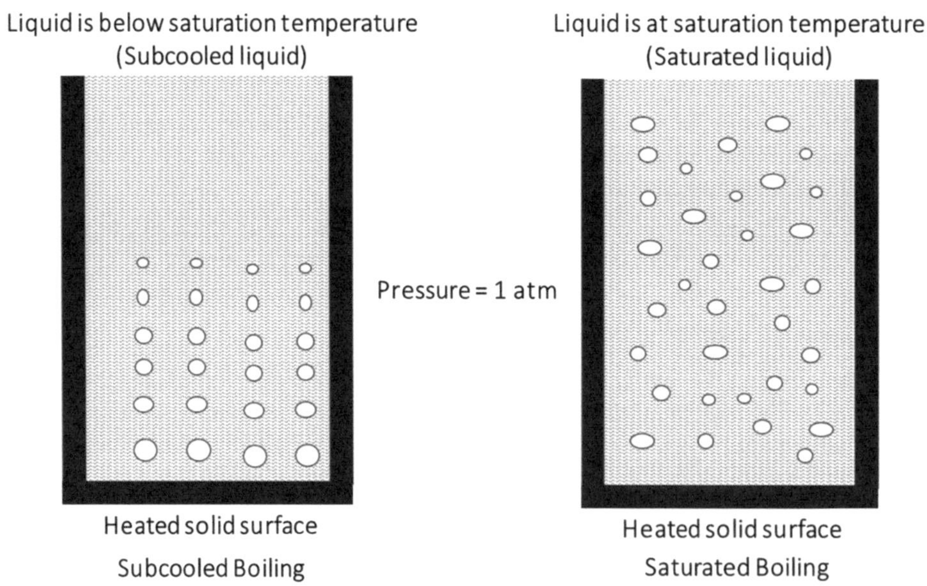

Figure 2.
Subcooled and saturated boiling phenomena [3, 6].

- Natural convection boiling
- Nucleate boiling
- Transition boiling
- Film boiling

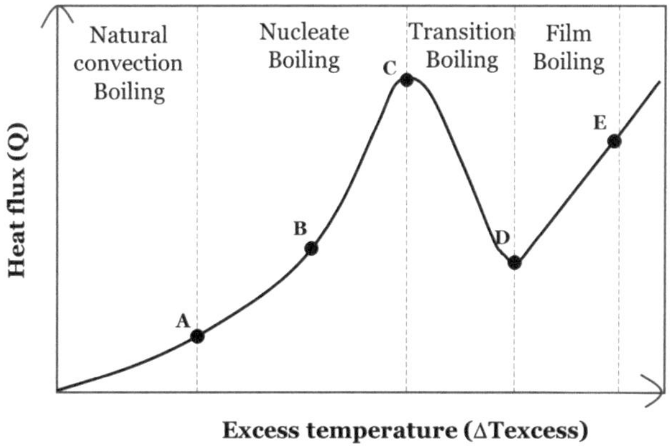

Figure 3.
Boiling curve with the boiling regimes for a pool of liquid [1–5].

To demonstrate these boiling regimes for a pool of liquid, a plot between boiling heat flux (rate of heat transfer per unit area) versus the excess temperature supplied is shown and known as the boiling curve (**Figure 3**).

3.2.1 Natural convection boiling (to point a on the boiling curve)

Boiling or saturation temperature of a pure liquid substance depends on the applied pressure. But in practice, the bubbles are forming on the heating surface only after being heated to a few more degrees above its saturation temperature (up to 6°C for water).

The transfer of heat within the fluid (from the heating surface to the bulk fluid) in this step is by natural convection, and hence the heat flux curve increases slowly. During this condition, the liquid will be at a slightly superheated state and the superheated liquids will evaporate when it rises to the free surface.

3.2.2 Nucleate boiling (between points A and C on the boiling curve)

Upon further increasing the excess temperature (ΔTexcess), the bubbles will start forming at the temperature with respect to point A of the boiling curve. With further increase in excess temperature, the rate of formation of bubbles and hence the heat flux will increase till the point C in the boiling curve.

The nucleate boiling regime (from A to C) can be further divided into two separate regions (from A to B and from B to C). In the region from A to B, with excess heat supplied beyond point A, isolated bubbles will start forming on the heated surface. As soon the bubbles start to move, these bubbles will dissolute in the liquid (subcooled boiling). The formation and dissolution of the bubble will be repeated till the temperature of the liquid reaching to saturated temperature. During nucleate boiling, the movement of the bubbles is responsible for the increase in heat transfer coefficient and heat flux.

In the region from B to C, with excess heat supplied beyond point B, the bubbles form at a great rate and a continuous column of vapour in the liquid will be observed. These bubbles move to the free surface (saturated boiling) where the vapour got released from the bubbles. The heat flux observed in this region will be larger due to the combined effect of liquid entrainment and evaporation.

High heat transfer rates are observed in the case of the nucleate boiling regime compared to other regimes; hence, it is the most desirable boiling regime in practice. For water, it can be achieved with ΔTexcess within about 30°C. The correlation for the boiling heat flux for this region was proposed by Rohsenow, which is [6–9].

$$q = \mu_l h_{fg}\left[\frac{g(\rho_l - \rho_v)}{\sigma}\right]^{1/2}\left[\frac{C_{pl}\Delta T_{excess}}{C_{sf}h_{fg}(Pr_l)^n}\right]^3$$

3.2.3 Transition boiling (between points C and D on the boiling curve)

With Δ Texcess value near reaching the point C, the rate of evaporation of bubbles at the heater surface will be at a very high rate throughout the entire solid surface. These bubbles may cover the heater surface; hence, the contact between solid surface and liquid will be difficult. This formed vapour film acts as an insulation due to the low thermal conductivity of the vapour relative to that of the liquid, and hence the heat flux decreases beyond point C.

The heat flux till point C reaches a maximum value and is called the *critical (or maximum) heat flux*, q_{max}. The correlation for the boiling critical heat flux for this region for boiling was proposed by Lienhard et al. and is expressed as [6–8]:

$$q_{max} = 0.149\rho_l h_{fg}\left[\frac{\sigma g(\rho_l - \rho_v)}{\rho_v^2}\right]^{1/4}$$

3.2.4 Film boiling (beyond point D on the boiling curve)

During the transition boiling, the heater surface will be completely covered by a continuous stable layer of vapour film (with increasing ΔTexcess). As the vapour film separates the liquid from the heater surface and will be responsible for transferring less heat flux, the heat flux will reach a minimum value (point D), called the *Leidenfrost point*. The correlation for the boiling minimum heat flux at the *Leidenfrost point* for a horizontal plate was proposed by Zuber, which is [7].

$$q_{max} = 0.09\rho_v h_{fg}\left[\frac{\sigma g(\rho_l - \rho_v)}{(\rho_l + \rho_v)^2}\right]^{1/4}$$

As the ΔTexcess further increases, the heat transfer will start due to the radiation from hot solid surface to liquid through the vapour medium. The heat flux will again rise slowly and as the heating takes place over a film of vapour, it is known as film boiling. The correlation for the boiling heat transfer coefficient for this region for boiling above a horizontal tube was proposed by Bromley, which is given as follows [6–8]. However, for a vertical plate, the constant 0.62 and D will be replaced by 0.7 and L, respectively:

$$h = 0.62\left[\frac{k_v^3\rho_v(\rho_l - \rho_v)g\left(h_{fg} + 0.4C_{pv}\Delta T_{excess}\right)}{D\mu_v\Delta T_{excess}}\right]^{1/4}$$

Burnout point: As the heat flux is decreasing beyond point C, the boiling process will not be advised to continue further. Beyond point C, the power that needs to be provided to the heater surface will be more (as the heat flux decreases). However, in this condition of excess power supply, the temperature of the nichrome wire (heater) immersed in the liquid will abruptly rise to the melting point of the wire, resulting in burnout. Burnout can be avoided by using platinum wire, which has a much higher melting point.

4. Heat transfer in condensation

Condensation is a convection process of changing a vapour medium to a liquid state and generally occurs when a saturated vapour comes into contact with a cold solid surface at a temperature less than the saturation temperature of the vapour. The latent heat of vaporisation must be removed during the condensation.

Condensers are widely used in the chemical industry. The process of condensation occurs by two distinct mechanisms/modes and at different rates of heat transfer. Those are film-wise and drop-wise condensation [2, 3].

4.1 Difference between drop-wise and film-wise condensation

4.1.1 Drop-wise condensation

When a saturated vapour comes into contact with a cold solid surface (a surface at a lower temperature than the saturated temperature of vapour), it condenses to liquid form. If condensate does not wet the surface, then the droplets of liquid are formed on the surface.

The size of droplets expands with time and eventually drops down the surface in a random pattern (due to the effect of gravity), leaving the metal surface bare on which further condensation develops.

4.1.2 Film-wise condensation

When a saturated vapour comes into contact with a cold surface (a surface at a lower temperature than the saturated temperature of vapour), it condenses to liquid form. If condensate wets the surface, then it forms a continuous film of condensate.

These condensates completely cover the solid surface, and then heat must be transported through the condensed liquid layer. Then the vapours have to condense into the liquid film rather than direct contact with the surface. Under the action of gravity, the condensate eventually flows down the surface. The condensation caused by this technique is termed as film-wise condensation [2, 3].

The film covering the solid surface serves as a heat transmission barrier in film-wise condensation, but in drop-wise condensation, a considerable section of the surfaces is exposed directly to the vapour. Hence, the heat transfer coefficients (and thus the heat transfer rates) in drop-wise condensation are generally four to eight times greater than in film-wise condensation. The presence of dirt on the surfaces (where condensate drops develop), which appear to favour drop-wise condensation, is known as nucleation sites. Because most surfaces become wet after being exposed to the condensing vapours, film-wise condensation is very common (**Figure 4**) [2, 3, 6].

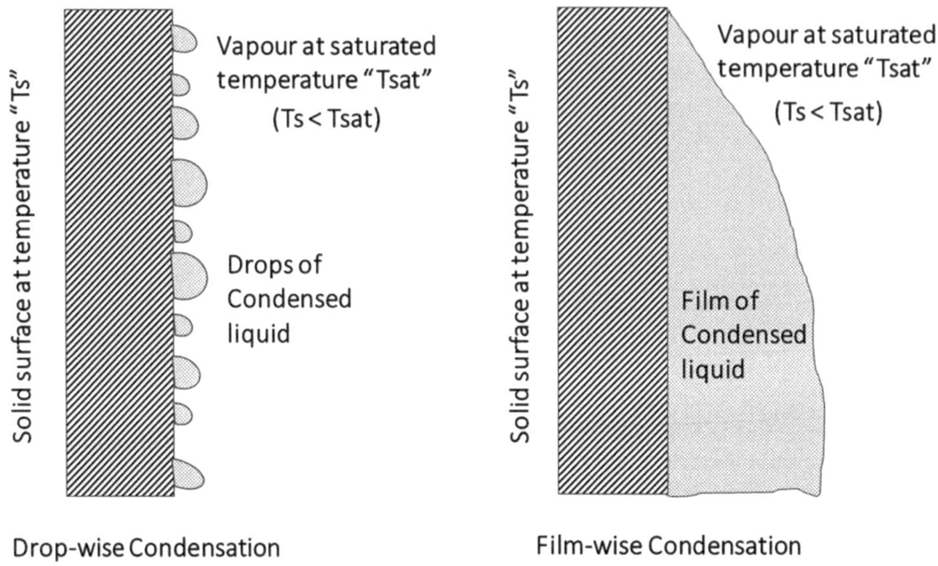

Figure 4.
Drop-wise and film-wise condensation phenomena [2, 3, 6].

Drop-wise condensation	Film-wise condensation
In drop-wise condensation, the condensate liquids partially wet the complete solid surface by forming droplets of condensate on the surface.	In film-wise condensation, the condensed liquid wets the solid surface by forming a continuous film of condensate on the surface.
These droplets then fall down the surface under the action of gravity, leaving the bare solid surface to condense further.	Condensate flows down the surface under the action of gravity by forming a continuous film, and further heat transfer takes place through this layer.
As the bare solid surface is further available to condense, the heat transfer coefficients and thus heat transfer rate are very high compared to film-wise condensation. The heat transfer coefficient value is difficult to predict.	Due to the presence of a continuous liquid film of condensate between the vapour and solid surface, the heat transfer coefficients and thus heat transfer rate are very low compared to drop-wise condensation. The heat transfer coefficient value can be predicted easily.
Drop-wise condensation is difficult to achieve and generally occurs on oily or greasy surfaces.	Film-wise condensation is easily obtainable and generally occurs on smooth, clean and uncontaminated surfaces.
Drop-wise condensation condition is difficult to maintain and is unstable. Drop-wise condensation is commonly not used industrially.	Film-wise condensation conditions can be easily maintained and stable. Film-wise condensation is commonly used industrially.

4.2 Heat transfer for film-wise condensation on vertical plate

Let us consider a vertical plate maintained at a constant surface temperature 'Ts' with a height 'L' and width 'b'. Let us consider a single vapour medium, at the saturation temperature 'Tsat', exposed to this surface [3, 4]. The surface temperature of the solid surface is below the saturation temperature (Ts < Tsat). When this saturated vapour comes in contact with the cold surface, then the vapour will condense on it.

In the case of film-wise condensation, it will form a continuous film of condensate on the surface of the vertical plate. The condensate liquid film layer ultimately flows down and will obtain a state as shown in the figure under the influence of viscosity and gravity.

Let the downward direction is taken as the positive x-direction with the origin placed at the top of the plate where condensation initiates, as shown in **Figure 5**. The film thickness of condensate 'δ' and thus the mass flow rate of the condensate increase with respect to the length of plate 'x'.

Heat must be transferred from the vapour to the plate through the film, which provides heat transfer resistance. The greater the thermal resistance of the film, the slower the rate of heat transfer will be. Nusselt first derived the analytical relationship between the heat transfer coefficient throughout the length of the plate in film condensation on a vertical plate in 1916, using the following assumptions:

- The surface temperature of solid surface ('Ts') and the vapour is at saturation temperature ('Tsat') and is more than solid surface temperature (Ts < Tsat).
- The temperature of condensate liquid varies linearly across the liquid film.
- The flow within the condensate liquid layer is laminar and the acceleration of the condensate liquid layer is negligible.
- Heat transfer within the condensate liquid film is through conduction mode only (no convection phenomenon in the film).
- All the properties of the condensate liquid are constant throughout the film.

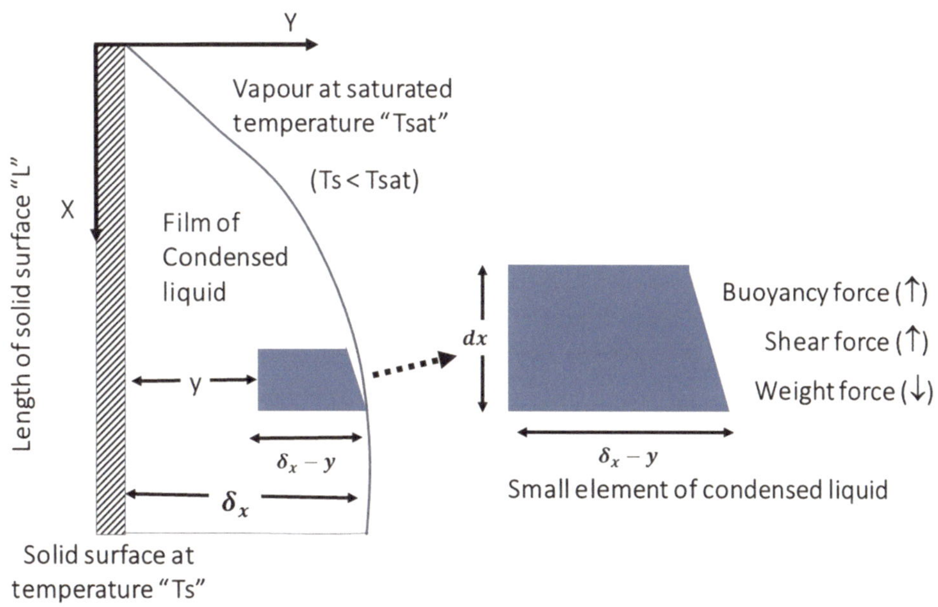

Figure 5.
Heat transfer phenomena for film condensation on vertical plate [3, 4].

- The viscous shear on the liquid–vapour interface is negligible (velocity of the saturated vapour is maintained very low to avoid drag on the condensate film).

The rate of heat transfer from the vapour phase to the solid surface with respect to vertical direction x can be expressed as:

$$Q = h_x A(T_{sat} - T_s) = k_l A \frac{(T_{sat} - T_s)}{\delta_x}$$

$$Then, Heat\ flux = q = \frac{Q}{A} = h_x(T_{sat} - T_s) = k_l \frac{(T_{sat} - T_s)}{\delta_x}$$

$$h_x = \frac{k_l}{\delta_x} \tag{1}$$

The heat transfer coefficient value for the heat transfer from the vapour to the plate is changing along the length of the plate due to the thermal resistance offered by the varying thickness of condensate liquid film. If the thickness of the film is more, then more will be the thermal resistance for the flow of heat from vapour to solid and thus lower the rate of heat transfer.

Let us consider a small selected volume element of condensate in the vertical x-direction. Since the acceleration of the small section fluid is assumed zero, then according to Newton's second law of motion, it can be written as:

$$\sum F_x = ma_x = 0$$

$$\sum F_x(Downward\ direction, \downarrow) = \sum F_x(Upward\ direction, \uparrow)$$

$$Weight\ force\ (\downarrow) = \text{Viscous shear force}\ (\uparrow) + \text{Buoyancy force}\ (\uparrow)$$

The forces that act on this small elemental volume will be the weight of the liquid element (acting downward), viscous shear or fluid friction force (acting upward) and buoyancy force (acting upward).

$$Weight\ (\downarrow)\ or\ gravity\ force\ on\ small\ liquid\ element = \rho_l g(\delta_x - y)dx$$

$$\text{Viscous shear force}\ (\uparrow)\ on\ small\ liquid\ element = \mu_l \frac{du}{dy} dx$$

$$\text{Buoyancy force}\ (\uparrow)\ on\ small\ liquid\ element = \rho_V g(\delta_x - y)dx$$

Thus,

$$\rho_l g(\delta_x - y)dx = \mu_l \frac{du}{dy} dx + \rho_V g(\delta_x - y)dx$$

$$\mu_l \frac{du}{dy} dx = \rho_l g(\delta_x - y)dx - \rho_V g(\delta_x - y)dx$$

$$\frac{du}{dy} = \frac{(\rho_l - \rho_V)g(\delta_x - y)}{\mu_l}$$

$$du = \frac{(\rho_l - \rho_V)g}{\mu_l}(\delta_x - y)dy$$

Integrating the aforementioned equation from y = 0 to y, we will have the relationship between the velocity along the length of the vertical pipe:

[At y = 0; u = 0 (no-slip boundary condition) and at y = y; u = u (y) (not zero)]

$$\int_0^{u(y)} du = \frac{(\rho_l - \rho_V)g}{\mu_l} \int_0^y (\delta_x - y)dy$$

$$u = \frac{(\rho_l - \rho_V)g}{\mu_l} \left(y\delta_x - \frac{y^2}{2} \right)$$

Then, the mass flow rate of the condensate with the boundary layer thickness is 'δ', at any location 'x' over the solid surface will be

$$\dot{m} = \int \rho_l u dA = \int_0^\delta \rho_l \frac{(\rho_l - \rho_V)g}{\mu_l} \left(y\delta_x - \frac{y^2}{2} \right) bdy$$

$$\dot{m} = \frac{\rho_l(\rho_l - \rho_V)gb}{\mu_l} \int_0^\delta \left(y\delta_x - \frac{y^2}{2} \right) dy$$

$$\dot{m} = \frac{\rho_l(\rho_l - \rho_V)gb}{\mu_l} \left(\frac{\delta_x^2}{2}\delta_x - \frac{\delta_x^3}{2 \times 3} \right) = \frac{\rho_l(\rho_l - \rho_V)gb\delta_x^3}{3\mu_l} = \frac{\rho_l(\rho_l - \rho_V)gb\delta_x^3}{3\mu_l}$$

Then

$$\frac{d\dot{m}}{dx} = \frac{d}{dx}\left(\frac{\rho_l(\rho_l - \rho_V)gb\delta_x^3}{3\mu_l} \right) = \frac{\rho_l(\rho_l - \rho_V)gb}{3\mu_l} \frac{d}{dx}\delta_x^3 = \frac{\rho_l(\rho_l - \rho_V)gb\delta_x^2}{\mu_l} \frac{d\delta_x}{dx} \tag{2}$$

Again, the rate of heat transfer from the vapour to the solid surface through the liquid film layer will be equal to the amount of heat released when vapour is condensed and is expressed as:

$$dQ = d\dot{m}\ \mathrm{x}\lambda = k_l A \frac{(T_{sat} - T_s)}{\delta_x} = h_x A (T_{sat} - T_s)$$

$$dQ = d\dot{m}\ \mathrm{x}\lambda = k_l (b\ \mathrm{x}dx) \frac{(T_{sat} - T_s)}{\delta_x} = h_x (b\ \mathrm{x}dx)(T_{sat} - T_s)$$

$$\frac{d\dot{m}}{dx} = k_l b \frac{(T_{sat} - T_s)}{\delta_x \lambda} \tag{3}$$

Equating the aforementioned two Eqs. (2) and (3), we have

$$\frac{d\dot{m}}{dx} = k_l b \frac{(T_{sat} - T_s)}{\delta_x \lambda} = \frac{\rho_l(\rho_l - \rho_V)gb\delta_x^2}{\mu_l} \frac{d\delta_x}{dx}$$

$$k_l b \frac{(T_{sat} - T_s)}{\delta_x \lambda} = \frac{\rho_l(\rho_l - \rho_V)gb\delta_x^2}{\mu_l} \frac{d\delta_x}{dx}$$

$$\frac{k_l b \mu_l (T_{sat} - T_s)}{\rho_l(\rho_l - \rho_V)gb\lambda} dx = \delta_x^3 d\delta_x$$

$$\delta_x^3\, d\delta_x = \frac{k_l \mu_l (T_{sat} - T_s)}{\rho_l(\rho_l - \rho_V)g\lambda} dx$$

The liquid film thickness at any location x can be determined by integrating the aforementioned equation from x = 0 (δ_x = 0 at the top of the plate) to x = x (δ_x = x):

$$\int_0^{\delta_x} \delta_x^3 d\delta_x = \frac{k_l \mu_l (T_{sat} - T_s)}{\rho_l(\rho_l - \rho_V) g\lambda} \int_0^x dx$$

$$\frac{\delta_x^4}{4} = \frac{k_l \mu_l (T_{sat} - T_s)}{\rho_l(\rho_l - \rho_V) g\lambda} x$$

$$\delta_x = \left(\frac{4k_l \mu_l (T_{sat} - T_s)\mathrm{x}}{\rho_l(\rho_l - \rho_V) g\lambda}\right)^{1/4}$$

Then the heat transfer rate from the vapour to the solid plate surface at any location x along the length of the plate can be expressed as (from Eq. (1)):

$$h_x = \frac{k_l}{\delta_x}$$

$$h_x = \frac{k_l}{\left(\frac{4k_l \mu_l (T_{sat} - T_s)\mathrm{x}}{\rho_l(\rho_l - \rho_V) g\lambda}\right)^{1/4}} = \left(\frac{\rho_l(\rho_l - \rho_V) g\lambda k_l^4}{4k_l \mu_l (T_{sat} - T_s)\mathrm{x}}\right)^{1/4}$$

$$h_x = \left(\frac{\rho_l(\rho_l - \rho_V) g\lambda k_l^3}{4\mu_l (T_{sat} - T_s)\mathrm{x}}\right)^{1/4}$$

Upon integrating the aforementioned equation for the local heat transfer coefficient over the entire length of the plate (L), the value average heat transfer coefficient value is determined:

$$h_{avg} = h_{Vertical} = \frac{1}{L}\int_0^L h_x dx = 0.943\left(\frac{\rho_l(\rho_l - \rho_V) g\lambda k_l^3}{\mu_l (T_{sat} - T_s)\mathrm{L}}\right)^{1/4}$$

Note: In general, the density of the vapour medium will be negligible compared to the density of the liquid:

$$\rho_l \gg \rho_V \; or \; \rho_l \approx (\rho_l - \rho_V)$$

$$h_{Vertical} = 0.943\left(\frac{\rho_l^2 g\lambda k_l^3}{\mu_l (T_{sat} - T_s)\mathrm{L}}\right)^{1/4}$$

Note 1: In case, the vertical plate is inclined at an angle Θ. The heat transfer coefficient is

$$h_{Inclined} = h_{Vertical}(cos\theta)^{1/4}$$

Note 2: In the case of the horizontal tube or sphere with diameter D, the heat transfer coefficient is

$$h_{Horizontal} = 0.729\left(\frac{\rho_l(\rho_l - \rho_V) g\lambda k_l^3}{\mu_l (T_{sat} - T_s)\mathrm{D}}\right)^{1/4} = 0.729\left(\frac{\rho_l^2 g\lambda k_l^3}{\mu_l (T_{sat} - T_s)\mathrm{D}}\right)^{1/4}$$

Note 3: The aforementioned equations are applicable for a single-tube system, in case N number of tubes are arranged in the system or a stack of tubes are present. h_1 is the heat transfer coefficient for the top tube. Then the heat transfer coefficient for N tubes (h_N) is

$$h_N = h_1(\mathrm{N})^{1/4}$$

5. Research and techniques of enhancement of boiling and condensation heat transfer

Enhancement of boiling heat transfer: Different advanced techniques that can be used to improve heat transfer in pool boiling are classified as active or passive techniques. Active approaches regulate the fluid movement by different techniques such as mixing the fluid using mechanical agitation, pumping the fluid, vibrating the surface of the container, rotating the container continuously, and adding an external electrostatic or magnetic field [6, 7, 10]. On the other hand, passive heat transfer enhancement techniques focus on changing fluid characteristics and/or heat transfer surfaces, such as increasing the number of active nucleation sites and the rate of bubble formation at each site [11, 12]. A rough, dirt-covered surface produces more nucleation sites than a smooth surface. The rate of nucleation can also be promoted by applying a thin porous layer to the surface or constructing mechanical voids on the surface to allow for continuous vapour production.

Different surface modification approaches for improving channel flow boiling heat transfer were reviewed by Liang et al. [12] and Kim et al. [13]. It covers macroscale (the use of cylindrical pins, macro ribs and twisted tape inserts), microscale (the use of micro-fins, micro-pin-fins, artificial cavities porous coating) and nanoscale (the use of nanotubes or nanowires to coat a heating surface) [13, 14] techniques to enhance the rate of heat transfers. Nanostructure approaches are reported to be less effective than macroscale and microscale improvement techniques.

Shah et al. [15] investigated the flow pattern, nucleate boiling, bubble growth, void fraction, liquid layer thickness, critical heat flux, pressure drop and heat transfer models for boiling fluid in microchannels. Adnan et al. [16] investigated the usage of nanofluids (h-BN/DCM and SiO_2/DCM) to improve heat transmission in pool boiling. These nanoparticles were found to greatly improve the thermal properties of the base fluid, with a 27.59% improvement in the rate of heat transfer coefficient for saturation boiling. Heat transfer enhancement employing ZnO-water, TiO_2-water, and Al_2O_3-water nanofluids has also been reported [17–19]. Amiri et al. [20] reported that multi-walled carbon nanotubes treated with cysteine, silver nanoparticles and Gum Arabic exhibited significant enhancement in the pool boiling heat transfer coefficients and critical heat fluxes when added with different concentrations to the aqueous media. Chen et al. [21] reported a recent and detailed review on the boiling heat transfer enhancement using different nanofluid solutions [22].

Enhancement of condensation heat transfer: Many recent reviewed research articles [23–25] are reported for advanced condensation phenomenon techniques. For condensation, the two major conditions that are nucleation on the surface and departure of liquid droplets are greatly influenced by the hydrophilicity and hydrophobicity properties of solid surfaces. Similar to boiling, numerous surface modification techniques are used to improve the heat transfer rate for condensation. Among them, constructing low free energy surfaces and building micro-nano structure surfaces are two possibilities to enhance the nucleation of liquid droplets [26].

Metallisation, ion implantation and organic polymer coating are available reported methods for lowering surface free energy. Some precious metals, such as gold, silver, palladium, rhodium, and chromium, can be plated on solid surfaces to produce metallisation. Ion implantation procedures involve the ionisation of gases such as nitrogen, argon, helium, and hydrogen using a high-voltage electric field before bombarding them into a metal surface. A thin covering of organic polymer such as fluorocarbon polymer, silica gel polymer, hexamethyldisiloxane polymer, fluorinated propylene polymer, polyhexafluoropropylene polymer and poly(p-xylene) polymers can also be used to induce drop-wise condensation on a solid metal surface [27–30].

In micro-nano structure surfaces techniques, different micro-nano structures like nanowires, nanocons, nanosheets, nanoblocks of carbon nanotube, nanographene particles, ZnO, Ni and polystyrene are fabricated on the solid surfaces that provide the nucleation sites and promoting drop-wise condensation [22, 26, 30, 31].

6. Conclusion

This chapter provides a comprehensive overview of the boiling and condensation phenomenon. These are the two opposite phenomena related to convective heat transfer which is the heat transfer involved during changing phase from liquid to vapour and from vapour to liquid, respectively. Boiling occurs when the temperature of liquids raises above its saturation temperature. Boiling can be classified as pool and flow boiling and as subcooled and saturated boiling. Depending on the value of the excess temperature supplied (above saturation temperature) to the liquid medium, different types of boiling regimes are observed in a pool of liquid. Those regimes include natural convection boiling, nucleate boiling, transition boiling, and film boiling.

Again, the condensation process deals with changing a vapour to a liquid state with two distinct mechanisms, that is, film-wise condensation and drop-wise condensation. The heat transfer coefficients (and thus the heat transfer rates) in drop-wise condensation are greater than in film-wise condensation. Drop-wise condensation is difficult to achieve and generally occurs on oily or greasy surfaces. Film-wise condensation is easily obtainable and generally occurs on smooth, clean uncontaminated surfaces.

7. Numerical related to condensations

Example 1: Consider a vertical tube within which hot gas is flowing at 80°C. The tube has a diameter of 40 mm and 1 m in length. This tube is now utilised to condense steam at atmospheric pressure. Determine the mass of condensate or the rate of condensation per hour that will generate in this system. Given the properties of condensate:

$k = 0.67$ W/m.K; $\rho = 972$ kg/m^3; $\lambda = 2310$ kJ/kg; $\mu = 3.55 \times 10^{-04}$ (N.s)/m^2.

Answer:

We know that the heat transfer coefficient over a vertical surface will be

$$h_{Vertical} = 0.943\left(\frac{\rho_l^2 g \lambda k_l^3}{\mu_l (T_{sat} - T_s)\mathrm{L}}\right)^{1/4}$$

L = height of the vertical plate = 1 m; D = diameter of the vertical plate = 40 mm = 0.04 m; g = gravitational acceleration = 9.81 m/s^2; ρ_l = densities of the condensed liquid = 972 kg/m^3; μ_l = viscosity of the condensed liquid = 3.55 x 10^{-04} (N.s)/m^2 = 355 x 10^{-06} kg/m·s; λ = latent heat of vaporisation = 2310 kJ/kg = 2310 x 10^3 J/kg; k_l =thermal conductivity of the condensed liquid = 0.67 W/m.K; Ts = surface temperature of the plate = 60°C = 273.15 + 60 = 333.15 K; Tsat = saturation temperature (at atmospheric pressure) of the condensing fluid = 100° C = 273.15 + 100 = 373.15 K.

Then

$$h_{Vertical} = 0.943\left(\frac{972^2\ x9.81\ x2310\ x10^3\ x0.67^3}{355\ x10^{-6}x(373.15 - 333.15)\ x1}\right)^{1/4} = 4352\frac{W}{m^2.K}$$

We know that the rate of heat transfer from the vapour to the solid surface through the liquid film layer will be equal to the amount of heat released when vapour is condensed and is expressed as:

$$dQ = d\dot{m}\ x\lambda = k_l A\frac{(T_{sat} - T_s)}{\delta_x} = h_x A(T_{sat} - T_s)$$

$$Q = \dot{m}\ x\lambda = h_x A(T_{sat} - T_s)$$

Area of heat transfer = π x D x L = π x 0.04 x 1 = 0.12564 m^2

$$Mass\ of\ condensate = \dot{m} = \frac{h_x A(T_{sat} - T_s)}{\lambda}$$

$$= \frac{4352\ x0.12564\ x(373.15 - 333.15)}{2310\ x10^3} = 9.46\ x10^{-03}\frac{kg}{s}$$

$$= 34.08\frac{kg}{hr}$$

Example 2: Consider a horizontal tube within which hot gas is flowing at 80°C. The tube has a diameter of 40 mm and 1 m in length. This tube is now utilised to condense steam at atmospheric pressure. Determine the mass of condensate that will generate in this system. Given the properties of condensate:

k = 0.67 W/m.K; ρ = 972 kg/m^3; λ = 2310 kJ/kg; μ = 3.55 x 10^{-04} (N.s)/m^2.

Answer:

We know that the heat transfer coefficient over a horizontal surface will be

$$h_{Horizontal} = 0.729\left(\frac{\rho_l^2 g\lambda k_l^3}{\mu_l(T_{sat} - T_s)D}\right)^{1/4}$$

L = height of the horizontal plate = 1 m; D = diameter of the vertical plate = 40 mm = 0.04 m; g = gravitational acceleration = 9.81 m/s^2; ρ_l = densities of the condensed liquid = 972 kg/m^3; μ_l = viscosity of the condensed liquid = 3.55 x 10^{-04} (N.s)/m^2 = 355 x 10^{-06} kg/m·s; λ = latent heat of vaporisation = 2310 kJ/kg = 2310 x 10^3 J/kg; k_l =thermal conductivity of the condensed liquid = 0.67 W/m.K; Ts = surface temperature of the plate = 60°C = 273.15 + 60 = 333.15 K; Tsat = saturation temperature (at atmospheric pressure) of the condensing fluid = 100°C = 273.15 + 100 = 373.15 K.

Then

$$h_{Horizontal} = 0.729\left(\frac{972^2 x9.81\ x2310\ x10^3\ x0.67^3}{355\ x10^{-6}\ x(373.15 - 333.15)\ x0.04}\right)^{1/4} = 7522\frac{W}{m^2.K}$$

We know that the rate of heat transfer from the vapour to the solid surface through the liquid film layer will be equal to the amount of heat released when vapour is condensed and is expressed as:

$$dQ = d\dot{m}x\lambda = k_l A\frac{(T_{sat} - T_s)}{\delta_x} = h_x A(T_{sat} - T_s)$$
$$Q = \dot{m}x\lambda = h_x A(T_{sat} - T_s)$$

Area of heat transfer = π x D x L = π x 0.04 x 1 = 0.12564 m^2

$$Mass\ of\ condensate = \dot{m} = \frac{h_x A(T_{sat} - T_s)}{\lambda} =$$
$$= \frac{75222\ x0.12566\ x(373.15 - 333.15)}{2310\ x10^3} = 0.016\frac{kg}{s} = 58.91\frac{kg}{hr}$$

Example 3: Consider a vertical tube with a temperature at 96°C and is exposed to steam at saturation temperature at atmospheric pressure. The tube has a diameter of 20 mm and 30 cm in length. Obtain the outside film heat transfer coefficient and the rate of heat transfer. Given the properties of condensate:

k = 0.57 kcal/hr-m-°C; ρ = 950 kg/m^3; λ = 540 kcal/kg; μ = 1.02 kg/m.hr.

Answer:

We know that the heat transfer coefficient over a vertical surface will be

$$h_{Vertical} = 0.943\left(\frac{\rho_l^2 g\lambda k_l^3}{\mu_l(T_{sat} - T_s)L}\right)^{1/4}$$

L = height of the vertical plate = 30 cm = 0.3 m; D = diameter of the vertical plate = 20 mm = 0.02 m; g = gravitational acceleration = 9.81 m/s^2; ρ_l = densities of the condensed liquid = 950 kg/m^3; μ_l = viscosity of the condensed liquid = 1.02 kg/m · hr. = 2.83 x 10^{-4} kg/m.s; λ = latent heat of vaporisation = 540 kcal/kg = 540 x 10^3 x 4.184 J/kg = 2,259,360 J/kg; k_l =thermal conductivity of the condensed liquid = 0.57 kcal/hr-m-°C = (0.57 x 10^3 x 4.184/3600) J/s-m-°C = 0.66 W/m. °C; Ts = surface temperature of the plate = 96°C; Tsat = saturation temperature (atmospheric pressure) of condensing fluid =100°C.

Then

$$h_{Vertical} = 0.943\left(\frac{950^2 x9.81\ x2259360\ x0.66^3}{2.83\ x10^{-4}\ x(100 - 60)\ x0.3}\right)^{1/4} = 6049.28\frac{W}{m^2.^\circ C}$$

Area of heat transfer = π x D x L = π x 0.002 x 0.3 = 0.018846 m^2.

We know that

$$Rate\ of\ heat\ transfer$$
$$= Q = h_x A(T_{sat} - T_s)$$
$$= 6049.28\ x0.018846\ x(100 - 96) = 456.018\ W$$

Example 4: A tube 40 mm in diameter and 1 m in length is used to condense steam at 100°C. The tube surface is at 60°C. Determine for which arrangement of the tube, the rate of heat transfer and the mass of condensate will be maximum. (a) Vertical, (b) horizontal, (c) inclined at an angle of 45°C and (d) ten number of horizontal tubes in the vertical direction. Properties of condensate are.

k = 0.67 W/m.K; ρ = 972 kg/m^3; λ = 2310 kJ/kg; μ = 3.55 x 10^{-04} (N.s)/m^2.

Answer:

Given data, L = height of the vertical plate = 1 m; D = diameter of the vertical plate = 40 mm = 0.04 m; g = gravitational acceleration = 9.81 m/s^2; ρ_l = densities of the condensed liquid = 972 kg/m^3; μ_l = viscosity of the condensed liquid = 3.55 x 10^{-04} (N. s)/m^2 = 355 x 10^{-06} kg/m·s; λ = latent heat of vaporisation = 2310 kJ/kg = 2310 x 10^3 J/ kg; k_l =thermal conductivity of the condensed liquid = 0.67 W/m.K; Ts = surface temperature of the plate = 60°C = 273.15 + 60 = 333.15 K; Tsat = saturation temperature (atmospheric pressure) of condensing fluid =100°C = 273.15 + 100 = 373.15 K.

Area of heat transfer = A = π x D x L = π x 0.04 x 1 = 0.12564 m^2

a. We know that the heat transfer coefficient over a vertical surface will be

$$h_{Vertical} = 0.943\left(\frac{\rho_l^2 g \lambda k_l^3}{\mu_l (T_{sat} - T_s) \mathrm{L}}\right)^{1/4}$$

$$h_{Vertical} = 0.943\left(\frac{972^2 \mathrm{x} 9.81 \ \mathrm{x} 2310 \ \mathrm{x} 10^3 \ \mathrm{x} 0.67^3}{355 \ \mathrm{x} 10^{-6} \mathrm{x} (373.15 - 333.15) \ \mathrm{x} 1}\right)^{1/4} = 4352 \frac{W}{m^2.K}$$

We know that

$$Rate\ of\ heat\ transfer = Q = h_x A (T_{sat} - T_s)$$
$$= 4352 \ \mathrm{x} 0.12564 \ \mathrm{x} (373.15 - 333.15) = 21871.42\ W$$

$$Mass\ of\ condensate = \dot{m} = \frac{h_x A (T_{sat} - T_s)}{\lambda} = \frac{21871.42}{2310 \ \mathrm{x} 10^3} = 9.46 \ \mathrm{x} 10^{-3} \frac{kg}{s} = 34.08 \frac{kg}{hr}$$

b. Heat transfer coefficient over a horizontal surface will be

$$h_{Horizontal} = 0.729\left(\frac{\rho_l^2 g \lambda k_l^3}{\mu_l (T_{sat} - T_s) \mathrm{D}}\right)^{1/4}$$

$$h_{Horizontal} = 0.729\left(\frac{972^2 \mathrm{x} 9.81 \ \mathrm{x} 2310 \ \mathrm{x} 10^3 \ \mathrm{x} \ 0.67^3}{355 \ \mathrm{x} \ 10^{-6} \ \mathrm{x} \ (373.15 - 333.15) \ \mathrm{x} \ 1}\right)^{1/4} = 3364.37 \frac{W}{m^2.K}$$

We know that

$$Rate\ of\ heat\ transfer = Q = h_x A (T_{sat} - T_s)$$
$$= 3364.37 \ \mathrm{x} \ 0.12564 \ \mathrm{x} (373.15 - 333.15) = 16907.98\ W$$

$$Mass\ of\ condensate = \dot{m} = \frac{h_x A (T_{sat} - T_s)}{\lambda} = \frac{16907.98}{2310 \ \mathrm{x} \ 10^3} = 7.31 \ \mathrm{x} \ 10^{-3} \frac{kg}{s} = 26.35 \frac{kg}{hr}$$

c. Heat transfer coefficient for the inclined angle of 45°C will be

$$h_{Inclined} = h_{Vertical}(cos\theta)^{1/4}$$

$$h_{Inclined} = 4352 \text{ x } (cos\ 45)^{1/4} = 3990\frac{W}{m^2.K}$$

We know that

$$Rate\ of\ heat\ transfer$$
$$= Q = h_x A(T_{sat} - T_s)$$
$$= 3990 \text{ x } 0.12564 \text{ x}(373.15 - 333.15) = 20052.15\ W$$

$$Mass\ of\ condensate = \dot{m} = \frac{h_x A(T_{sat} - T_s)}{\lambda} = \frac{20052.15}{2310 \text{ x } 10^3} = 8.68 \text{ x } 10^{-3}\frac{kg}{s} = 31.25\frac{kg}{hr}$$

d. Heat transfer coefficient for 10 number horizontal tubes in the vertical direction will be

$$h_N = h_1(\text{N})^{1/4}$$

$$h_N = 3364.37 \text{ x } (10)^{1/4} = 5982\frac{W}{m^2.K}$$

We know that

$$Rate\ of\ heat\ transfer$$
$$= Q = h_x A(T_{sat} - T_s)$$
$$= 5982 \text{ x } 0.12564 \text{ x}(373.15 - 333.15) = 30063.14\ W$$

$$Mass\ of\ condensate = \dot{m} = \frac{h_x A(T_{sat} - T_s)}{\lambda} = \frac{30063.14}{2310 \text{ x } 10^3} = 0.014\frac{kg}{s} = 46.85\frac{kg}{hr}$$

Mass of condensate and the heat transfer rate will be maximum for the arrangement with 10 horizontal tubes in the vertical direction.

Example 5: Consider a stainless steel pan with water at atmospheric pressure. Externally heat is supplied, to boil the liquid, through the bottom of pan (diameter is 40 cm) and a temperature of 106°C is maintained at the inner surface of the bottom of the pan. Considering steady state condition, determine the rate of heat transfer to the water and the rate of evaporation of water.

Data: The saturation temperature of Tsat = 100°C and the properties of water at this condition are.

$\sigma = 0.0589$ N/m	**$h_{fg} = 2257 \text{ x } 10^3$ J/kg**	$Pr_l = $ **1.75**
$\rho_l = 957.9$ kg/m^3	$\rho_v = 0.6$ kg/m^3	$\mu_l = 0.282 \text{ x } 10^{-3}$ kg · m/s
$C_{pl} = 4217$ J/kg · °C	$C_{sf} = 0.013$	n = 1

Answer:

The rate of heat transfer will be equal to the product of heat flux and heat transfer area.

Here, $\Delta T_{excess} = (T_S - T_{Sat})$ = excess temperature = temperature of the supplied liquid – saturation temperature of liquid = 106–100 = 6°C.

Excess temperature resembles the stage of nucleate boiling of liquid. Hence, the heat flux at this nucleate boiling condition will be (Rohsenow, [9])

$$q = \mu_l h_{fg} \left[\frac{g(\rho_l - \rho_v)}{\sigma}\right]^{1/2} \left[\frac{C_{pl}\Delta T_{excess}}{C_{sf} h_{fg} (Pr_l)^n}\right]^3$$

$$q = 0.282\text{x}10^{-3}\text{x}2257\text{x } 10^3 \left[\frac{9.81 \text{ x } (957.9 - 0.6)}{(0.0589)}\right]^{1/2} \left[\frac{4217 \text{ x } 6}{(0.013)\text{x } \left(2257 \text{ x } 10^3\right)\text{x}(1.75)^1}\right]^3$$

$$q = 30409.35 W/_{m^2}$$

Heat transfer area or the surface area of the bottom of the pan:

$$A = \frac{\pi}{4}D^2 = \frac{\pi}{4} \times (0.4)^2 = 0.125m^2$$

The rate of heat transfer will be = Q = q x A = $30409.35 W/_{m^2} \times 0.12m^2 = 3821\ W$.

Again, the rate of evaporation of water $(\dot{m})$ can be determined from the formula:

$$Q = \dot{m} * \lambda$$

$$\dot{m} = \frac{Q}{\lambda} = \frac{3821}{2257 \text{ x } 10^3} = 1.692 \text{ x } 10^{-3}\frac{kg}{s} \approx 1.7\, gram\ of\ water\ evaporates\ per\ sec$$

Nomenclature

Q	Amount of heat transferred from a solid surface to the fluid, W
a_x	Acceleration of small selected volume element of condensate, m/s^2
h_N	Average heat transfer coefficient for N tubes, W/m^2.K
h_1	Average heat transfer coefficient for top horizontal tube, W/m^2.K
h_{avg} or h	Average heat transfer coefficient value, W/m^2.K
$h_{Inclined}$	Average heat transfer coefficient value for inclined surface, W/m^2.K
$h_{Horizontal}$	Average heat transfer coefficient value for horizontal surface, W/m^2.K
$h_{Vertical}$	Average heat transfer coefficient value for vertical surface, W/m^2.K
q_{max}	Critical (or maximum) heat flux, W/m^2
D	Diameter of the horizontal tube or sphere, m
x	Distance along the length of the solid surface, m
y	Distance along the width of the condensate, m
ρ_l	Densities of the condensed liquid, kg/m^3
ρ_V	Densities of the condensed vapour, kg/m^3
h_{fg} or λ	Enthalpy or latent heat of vaporisation, J/kg
C_{sf}	Experimental constant depends on the fluid-surface combination
n	Experimental constant depends on the fluid
ΔT_{excess}	Excess temperature supplied to fluid, °C, K

δ	Film thickness of condensate, m
δx	Film thickness of condensate along the L direction, m
F_x	Force on small selected volume element of condensate, kg/m.s^2
g	Gravitational acceleration, m/s^2
q	Heat flux, W/m^2
h_x	Heat transfer coefficient along the L direction, W/m^2·°C, W/m^2·K
L or H	Height of the solid surface, m
Θ	Inclination angle of a solid surface, ^{0}
$\dot{m}$	Mass flow rate of condensate along the length of condensate, kg/s
m	Mass of small selected volume element of condensate, kg
N	Number of horizontal tubes arranged in stacks
Pr_l	Prandtl number of liquid
C_{pl}	Specific heat of the liquid, J/kg. °C
C_{pv}	Specific heat of the vapour, J/kg. °C
A	Surface area of solid surface, m^2
σ	Surface tension in the vapour-liquid interface, N/m
Tsat	Saturation temperature of the condensing fluid, °C or K
Ts	Surface temperature of the solid surface, °C or K
k_l	Thermal conductivity of the condensed liquid, W/m·°C or W/m·K
k_v	Thermal conductivity of the vapour, W/m·°C or W/m·K
u	Velocity rate of condensate along the length of condensate, m/s
μ_l	Viscosity of the condensed liquid, kg/m·s
μ_v	Viscosity of the vapour, kg/m·s
b	Width of the solid surface, m

Author details

Bijoy Kumar Purohit*, Zakir Hussain and PVR Sai Prasad
Department of Chemical Technology, Loyola Academy Degree and PG College, Secunderabad, Telangana, India

*Address all correspondence to: pch16001@rgipt.ac.in

References

[1] Kern DQ, Kern DQ. Process Heat Transfer. Vol. 5. New York: McGraw-Hill; 1950

[2] McCabe WL, Smith JC, Harriott P. Unit Operations of Chemical Engineering. Vol. 5. New York: McGraw-hill; 1993

[3] Holman JP. Heat Transfer. 10th ed. New York: Mc-GrawHill Higher Education; 2010

[4] Bergman TL, Bergman TL, Incropera FP, Dewitt DP, Lavine AS. Fundamentals of Heat and Mass Transfer. United States: John Wiley & Sons; 2011

[5] Welty J, Rorrer GL, Foster DG. Fundamentals of Momentum, Heat, and Mass Transfer. John Wiley & Sons; 2020

[6] Cengel Y, Heat TM. A Practical Approach. New York, NY, USA: McGraw-Hill; 2003

[7] Dutta BK. Heat Transfer: Principles and Applications. India: PHI Learning Pvt. Ltd.; 2000

[8] Coulson JM. Coulson & Richardson Chemical Engineering, Volume 1. Disp, 10, 32. 2000

[9] Rohsenow WM, Hartnett JP, Ganic EN. Handbook of Heat Transfer Fundamentals. 1985

[10] Tillery SW, Heffington S, Smith MK, Glezer A. Boiling heat transfer enhancement by submerged vibration induced jets. In: The Ninth Intersociety Conference on Thermal and Thermomechanical Phenomena In Electronic Systems (IEEE Cat. No. 04CH37543). Vol. 2. New York: IEEE; 2004. pp. 17-22

[11] Bergles AE, Nirmalan V, Junkhan GH, Webb RL. Bibliography on Augmentation of Convective Heat and Mass Transfer-II (No. ISU-ERI-AMES-84221). Ames (USA): Iowa State Univ. of Science and Technology; 1983 Heat Transfer Lab

[12] Liang G, Mudawar I. Review of channel flow boiling enhancement by surface modification, and instability suppression schemes. International Journal of Heat and Mass Transfer. 2020; **146**:118864

[13] Kim DE, Yu DI, Jerng DW, Kim MH, Ahn HS. Review of boiling heat transfer enhancement on micro/nanostructured surfaces. Experimental Thermal and Fluid Science. 2015;**66**:173-196

[14] Barber J, Brutin D, Tadrist L. A review on boiling heat transfer enhancement with nanofluids. Nanoscale Research Letters. 2011;**6**(1): 1-16

[15] Saha SK, Celata GP, Kandlikar SG. Thermofluid dynamics of boiling in microchannels. In: Advances in Heat Transfer. Vol. 43. Amsterdam: Elsevier; 2011. pp. 77-226

[16] Çiftçi E, Sözen A. Heat transfer enhancement in pool boiling and condensation using h-BN/DCM and SiO2/DCM nanofluids: Experimental and numerical comparison. International Journal of Numerical Methods for Heat & Fluid Flow. 2020;**31**(1):26-52

[17] Prajapati OS, Rohatgi N. Flow boiling heat transfer enhancement by using ZnO-water nanofluids. Science and Technology of Nuclear Installations. 2014;**2014**

[18] He Y, Jin Y, Chen H, Ding Y, Cang D, Lu H. Heat transfer and flow behaviour of

aqueous suspensions of TiO2 nanoparticles (nanofluids) flowing upward through a vertical pipe. International Journal of Heat and Mass Transfer. 2007;**50**(11–12):2272-2281

[19] Prajapati OS, Rajvanshi AK. Al2O3-water nanofluids in convective heat transfer. In: Applied Mechanics and Materials. Vol. 110. Switzerland: Trans Tech Publications Ltd.; 2012. pp. 3667-3672

[20] Amiri A, Shanbedi M, Amiri H, Heris SZ, Kazi SN, Chew BT, et al. Pool boiling heat transfer of CNT/water nanofluids. Applied Thermal Engineering. 2014;**71**(1):450-459

[21] Chen J, Ahmad S, Cai J, Liu H, Lau KT, Zhao J. Latest progress on nanotechnology aided boiling heat transfer enhancement: A review. Energy. 2021;**215**:119114

[22] Kandlikar SG. Handbook of Phase Change: Boiling and Condensation. England: Routledge, Taylor & Francis; 2019

[23] Kharangate CR, Mudawar I. Review of computational studies on boiling and condensation. International Journal of Heat and Mass Transfer. 2017;**108**: 1164-1196

[24] Wang SP, Chato JC. Review of Recent Research on Boiling and Condensation Heat Transfer with Mixtures. Urbana, Illinois: Air Conditioning and Refrigeration Center TR-23, College of Engineering. University of Illinois at Urbana-Champaign; 1992

[25] Moreira TA, Furlan G, e Oliveira GHDS, Ribatski G. Flow boiling and convective condensation of hydrocarbons: A state-of-the-art literature review. Applied Thermal Engineering. 2021;**182**:116129

[26] Hu X, Yi Q, Kong X, Wang J. A review of research on dropwise condensation heat transfer. Applied Sciences. 2021;**11**(4):1553

[27] Erb RA. Wettability of metals under continuous condensing conditions. The Journal of Physical Chemistry. 1965; **69**(4):1306-1309

[28] Qi Z, Dongchang Z, Jifang L. Surface materials with dropwise condensation made by ion implantation technology. International Journal of Heat and Mass Transfer. 1991;**34**(11):2833-2835

[29] Ma X, Xu D, Lin J. Dropwise condensation on superthin polymer surface. Journal of Chemical Industry and Engineering-China. 1993;**44**:165

[30] Tanasawa I. Advances in condensation heat transfer. In: Advances in Heat Transfer. Vol. 21. Amsterdam: Elsevier; 1991. pp. 55-139

[31] Akbari A, Alavi Fazel SA, Maghsoodi S, et al. Pool boiling heat transfer characteristics of graphene-based aqueous nanofluids. Journal of Thermal Analysis and Calorimetry. 2019; **135**:697-711

Chapter 2

Heat Transfer in Double-Pass Solar Air Heater: Mathematical Models and Solution Strategy

Nguyen Van Hap and Nguyen Minh Phu

Abstract

Multipass air collectors are commonly used because they produce higher air temperature than that of a single-pass one due to reduced top heat loss. In this chapter, two mathematical models of convection and radiant heat transfer in a double-pass solar air heater were presented. They included an average temperature model and a model of temperature variation along the airflow direction. The method for solving these two mathematical models was reported. The average temperature model was solved by dealing with a system of linear algebraic equations, whereas the other model was derived as ordinary differential equations and solved by a numerical integration. The calculation programs were developed in EES software. The computation time of temperature variation model was about 0.9 s, but that of the average temperature model was negligible. Outcomes from two solutions were almost identical. The largest error of the outlet air temperature was 2.1%. The models are applicable to multipass collectors with or without recycling airflow.

Keywords: temperature gradient, numerical integration, multiple-pass heat exchanger, convection heat transfer, radiation heat transfer

1. Introduction

Energy conversion is a current concern for developing countries because the industrial development is premium strategic target and the traditional fuel is increasingly scarce. Solar thermal energy conversion is deployed massively in countries near the equator due to the availability and stability of solar intensity. Converting thermal radiation into hot air is widely applied in drying agriculture, space heating, desiccant regeneration, timber seasoning, and natural ventilation. A solar air heater (SAH) has a simple structure and is easily made from local materials [1]. However, it has the disadvantage that the convection heat transfer coefficient (HTC) of the air is low, and the heat loss is considerable. Therefore, solutions to enhance heat transfer are proposed and applied. It may be the addition of inserts into the SAH duct to remove the viscous sublayer close to the absorption plate. They can be vortex generators [2–4], baffles [5–7], fins [8], or ribs [9]. Increasing the number of air passes in the SAH duct is also a measure to reduce heat loss due to the high temperature of the absorber plate. SAH having two, three, or four passes with or without airflow recycling has demonstrated a high thermohydraulic performance. The solution method for the mathematical model of a multiple-pass SAH has always been of interest to researchers since the SAH has several glasses, absorber plate, back plate, and multiple airflows. There are two commonly used analytical models: local solution and mean solution. The local solution establishes ordinary differential equations (ODEs) along with the air temperature boundary conditions. The mean solution approximates the temperature gradients to the temperature differences in each air pass to form a system of linear algebraic equations. It is obvious that the mean solution is more straightforward than the local solution and able to calculate by hand. However, the local solution can predict the temperature of fluid flow and heat exchanger surfaces along the collector length. From these temperature profiles, temperature cross or temperature meet phenomena can be detected and local corrections can be made. **Table 1** presents literature review on solutions studied on multiple-pass SAH during the last 15 years. Both the solutions are almost equally used by researchers. Typically, there are four research groups on the analytical model of multiple-pass SAH. The research group of Ho et al. employed the local solution. Meanwhile, the research teams of Velmurugan et al. and Matheswaran et al. applied the mean solution. Our research group (Phu et al.) used both the solutions.

No.	Researchers (year)	Local solution	Mean solution
1	Ho et al. (2005) [10]	✓	
2	Ramani et al. (2010) [11]	✓	
3	Yeh and Ho (2013) [12]	✓	
4	Hernandez et al. (2013) [13]	✓	
5	Ho et al. (2013) [14]	✓	
6	Karim et al. (2014) [15]		✓
7	Velmurugan and Kalaivanan (2015) [16]		✓
8	Velmurugan and Kalaivanan (2015) [17]		✓
9	Velmurugan and Kalaivanan (2015) [18]		✓
10	Velmurugan and Kalaivanan (2016) [19]		✓
11	Singh et al. (2018) [20]	✓	
12	Ho et al. (2018) [21]	✓	
13	Matheswaran et al. (2018) [22]		✓
14	Matheswaran et al. (2019) [23]		✓
15	Luan and Phu (2021) [24]		✓
16	Phu et al. (2021) [25]	✓	
17	Ahmadkhani et al. (2021) [26]	✓	
18	Ho et al. (2021) [27]	✓	
19	Phu and Tu (2021) [28]	✓	
20	Phu et al. (2021) [29]	✓	

Table 1.
Analytical studies on the multiple-pass SAH.

From the above extensive literature review, there has not been a study on comparing the heat transfer model for a multiple-pass SAH using both the solutions. Hence, in this chapter, both the solutions are applied to a typical double-pass SAH to realize the mathematical model, the solution method, and the result. From the analysis in this chapter, it can be developed for a variety of air collectors and further research on hydraulics and energy performance.

2. Model description

Figure 1 depicts the physical model of a double-pass solar air heater. It consists of a glass cover, an absorber plate, and a back plate. Solar radiation penetrates the glass to the absorber plate and heats the absorber plate. Airflow travels over the plate surface and receives heat. In addition, the air can receive more heat from the glass and back plate because these surfaces absorb thermal radiation from the absorber plate. The air moves from the top to the bottom channel forming two passes. The following subsection presents the five-temperature calculation models of the double-pass SAH, including glass (T_g), air in first pass (T_{f1}), absorption plate (T_p), air in second pass (T_{f2}), and the back plate (T_b).

2.1 Local solution

In this solution, the temperature of the collector components is a function of x-direction, i.e., $T(x)$. The energy balance for the glass cover is shown in Eq. (1). Solar energy arriving at the collector (I) and absorbed by the glass is balanced with the heat exchanged by convection and radiation above the glass surface, convection to the air in the first pass, and radiation to the absorber plate [24]:

$$I\alpha_g + h_w(T_a - T_g(x)) + h_{ra}(T_{sky} - T_g(x)) + h_{r_{g,p}}(T_p(x) - T_g(x)) + h_{f1_g}(T_{f1}(x) - T_g(x)) = 0. \tag{1}$$

The interpretation of the symbols can be found in **Table 2** and **Figure 1**. The heat transfer coefficients (h) can be viewed in Eqs. (16)–(20). The variation of the air temperature in the first pass in the direction of flow is described by Eq. (2). The

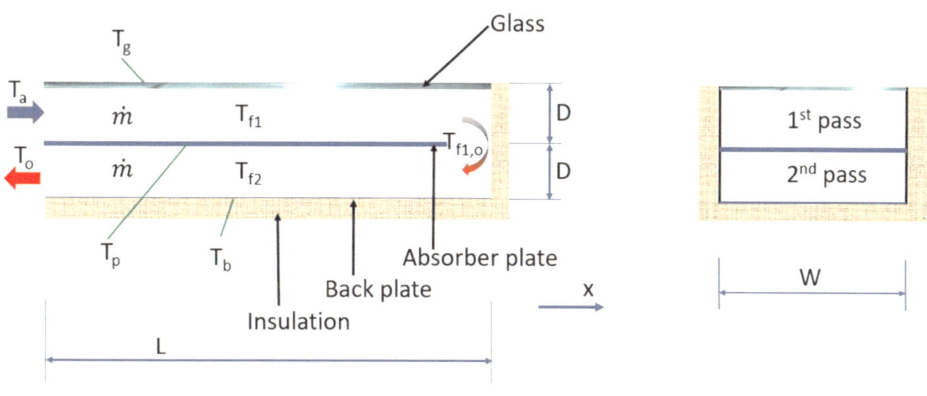

Figure 1.
Double-pass solar air heater.

Parameter	Symbol	Value
Collector length	L	2 m
Collector width	W	0.46 m
Air channel depth	D	25 mm
Solar radiation	I	1000 W/m^2
Inlet air temperature	T_a	27°C
Absorptivity of glass cover	α_g	0.06
Absorptivity of absorber plate	α_p	0.95
Emissivity of glass cover	ε_g	0.94
Emissivity of absorber plate	ε_p	0.94
Emissivity of back plate	ε_b	0.94
Transmissivity of glass cover	τ_g	0.84
Air specific heat	c_p	1005 J/kg K
Air thermal conductivity	k	0.02566 W/m K
Thermal conductivity of insulation layer	k_i	0.025 W/m K
Air viscosity	μ	0.00001858 kg/m s
Air density	ρ	1.176 kg/m^3
Thickness of insulation layer	t_i	50 mm
Wind velocity	V_{wind}	1.5 m/s
Stefan constant	Σ	5.67×10^{-8} W/m^2-K^4
Air mass flow rate	$\dot{m}$	0.01 kg/s

Table 2.
Input parameters.

change in air temperature is due to the convective heat exchange with the glass cover and the absorber plate [25]:

$$\frac{dT_{f1}(x)}{dx} = \frac{Wh_{f1,g}\left(T_g(x) - T_{f1}(x)\right) + Wh_{f1,p}\left(T_p(x) - T_{f1}(x)\right)}{\dot{m}c_p}. \tag{2}$$

Like the glass cover, the heat balance for the absorber plate is expressed as follows:

$$\begin{aligned} &I\tau_g\alpha_p + h_{f1,p}\left(T_{f1}(x) - T_p(x)\right) + h_{f2,p}\left(T_{f2}(x) - T_p(x)\right) \\ &\quad + h_{r,g,p}\left(T_g(x) - T_p(x)\right) + pxh_{r,b,p}\left(T_b(x) - T_p(x)\right) = 0. \end{aligned} \tag{3}$$

The variation of air temperature in the second pass is

$$\frac{dT_{f2}(x)}{dx} = \frac{Wh_{f2,b}\left(T_b(x) - T_{f2}(x)\right) + Wh_{f2,p}\left(T_p(x) - T_{f2}(x)\right)}{\dot{m}c_p}. \tag{4}$$

The energy balance of the back plate is

$$h_{r,b,p}\left(T_b(x) - T_p(x)\right) + h_{f2,b}\left(T_b(x) - T_{f2}(x)\right) + h_b\left(T_b(x) - T_a\right) = 0. \tag{5}$$

In this model, the air temperature equations are ordinary differential equations (ODEs); boundary conditions for ODEs (2) and (4) are [25]

$$T_{f1(x=0)} = T_a \tag{6}$$

$$T_{f1(x=L)} = T_{f2(x=L)}. \tag{7}$$

2.2 Mean solution

In this solution, the collector temperature components (T_g, T_{f1}, T_p, T_{f2}, and T_b) are the mean temperatures. Like the five governing equations in the local solution, the five corresponding governing equations in the mean solution are written as follows:

Glass cover:

$$I\alpha_g + h_w(T_a - T_g) + h_{ra}(T_{sky} - T_g) + h_{r,g,p}(T_p - T_g) + h_{f1,g}(T_{f1} - T_g) = 0. \tag{8}$$

The airflow in the first pass, spatial derivative of temperature is discretized as (17):

$$\frac{T_{f1,o} - T_a}{L} = \frac{Wh_{f1,g}(T_g - T_{f1}) + Wh_{f1,p}(T_p - T_{f1})}{\dot{m}c_p}. \tag{9}$$

Absorber plate:

$$I\tau_g\alpha_p + h_{f1,p}(T_{f1} - T_p) + h_{f2,p}(T_{f2} - T_p) + h_{r,g,p}(T_g - T_p) + h_{r,b,p}(T_b - T_p) = 0. \tag{10}$$

The air in the second pass:

$$\frac{T_o - T_{f1,o}}{L} = \frac{Wh_{f2,b}(T_b - T_{f2}) + Wh_{f2,p}(T_p - T_{f2})}{\dot{m}c_p} \tag{11}$$

Back plate:

$$h_{r,b,p}(T_b - T_p) + h_{f2,b}(T_b - T_{f2}) + h_b(T_b - T_a) = 0. \tag{12}$$

Two new variables ($T_{f1,o}$ and T_o) can be resolved by the arithmetic mean of terminal temperatures [16]:

$$T_{f1} = 0.5(T_a + T_{f1,o}) \tag{13}$$

$$T_{f2} = 0.5(T_o + T_{f1,o}). \tag{14}$$

2.3 Common equations

The common equations of the two models above are presented in this section. The sky temperature (T_{sky}) is calculated as follows [22]:

$$T_{sky} = 0.0552 T_a^{1.5}. \tag{15}$$

The convective heat transfer coefficient of the wind on the glass cover is written as follows [17]:

$$h_w = 5.7 + 3.8V_{wind}. \tag{16}$$

The radiant heat transfer coefficient of the glass cover and the sky is

$$h_{ra} = \sigma\varepsilon_g\left(T_g{}^2 + T_{sky}{}^2\right)\left(T_g + T_{sky}\right). \tag{17}$$

The radiant heat transfer coefficient of the absorber plate and the back plate is

$$h_{r,b,p} = \sigma\left(T_b{}^2 + T_p{}^2\right)\frac{T_b + T_p}{1/\varepsilon_b + 1/\varepsilon_p - 1}. \tag{18}$$

The radiant heat transfer coefficient of the absorber plate and the glass cover is

$$h_{r,g,p} = \sigma\left(T_g{}^2 + T_p{}^2\right)\frac{T_g + T_p}{1/\varepsilon_g + 1/\varepsilon_p - 1}. \tag{19}$$

It should be noted that in the local solution, the temperature of the collector components varies in the x-direction. Hence, the radiant heat transfer coefficients also change with the x-direction. The forced convection heat transfer coefficients of the air in the passes and surfaces ($h_{f1,g}$, $h_{f1,p}$, $h_{f2,p}$, and $h_{f2,b}$) is calculated from the following equation [22]:

$$h = 0.018\,Re^{\,0.8}Pr^{0.4}k/D_e \tag{20}$$

where Re is the Reynolds number, Pr is the Prandtl number, and D_e is the hydraulic diameter. These quantities can be estimated as

$$Re = \rho D_e V/\mu \tag{21}$$

$$Pr = \mu c_p/k \tag{22}$$

$$D_e = \frac{4WD}{2(W + D)}. \tag{23}$$

The air velocity in a pass (V) and air mass flow rate ($\dot{m}$) are correlated by

$$\dot{m} = WD\rho V. \tag{24}$$

The heat transfer coefficient from the back plate to the surroundings is evaluated as follows:

$$h_b = k_i/t_i. \tag{25}$$

The heat gain of air received when passing through the collector is

$$Q = \dot{m}c_p(T_o - T_a). \tag{26}$$

The thermal efficiency of the collector is

$$\eta_{th} = \frac{Q}{LWI}. \tag{27}$$

2.4 Solution strategy and validation

Table 2 presents the fixed parameters to be entered into the mathematical models to calculate the five temperatures in the double-pass solar air heater. For the local solution, the mathematical model consists of ODEs, so a certain understanding of numerical methods and computer skills is required. In our study, we use the integral function in EES software (F-chart software). The program will calculate iteratively until the difference between the temperatures $T_{f1(x\,=\,L)}$ and $T_{f2(x\,=\,L)}$ is very small. For the mean solution, Eqs. (8)–(12) can be rearranged into a system of linear algebraic

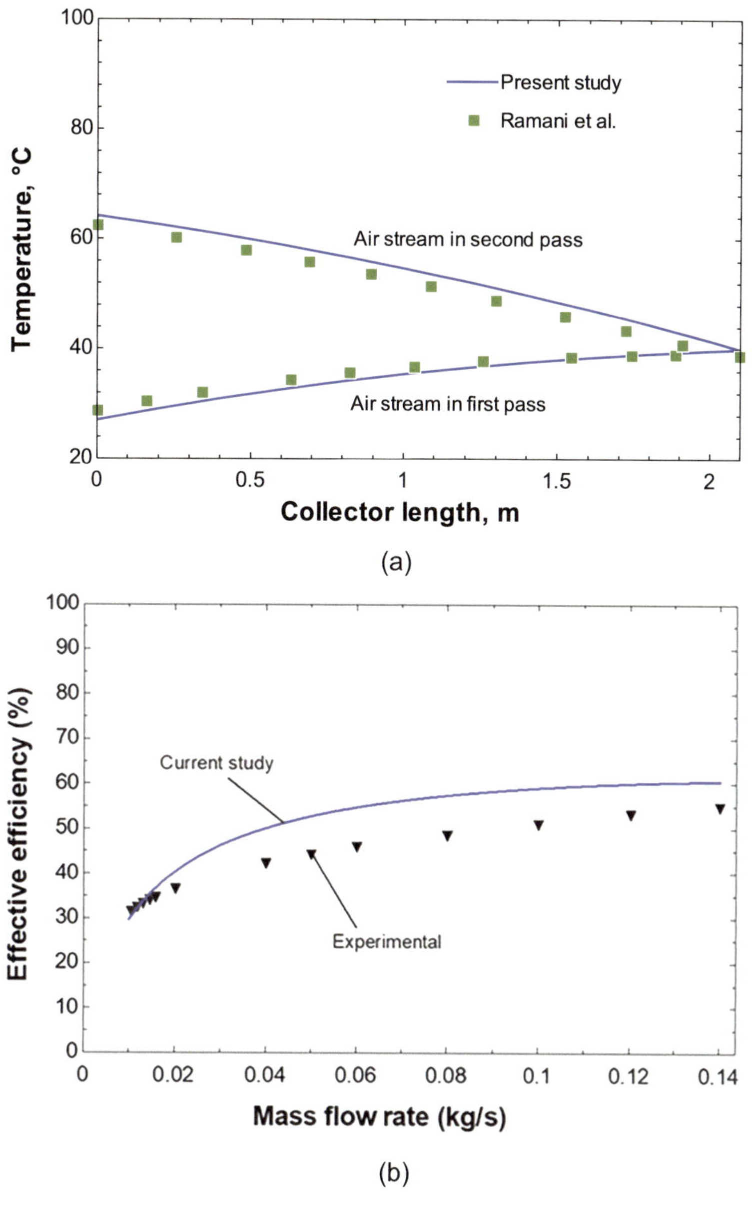

Figure 2.
Validation of local air temperature in local solution (a) [25] and collector efficiency in mean solution (b) [24].

equations that can be solved by hand or by applying an iterative method. We also write the code in EES software because of its convenience in solving a system of equations. The computation time of the local solution is about 0.9 s. Meanwhile, the results can be obtained immediately for the mean solution. The validation of both the solutions can be seen in our previous studies [24, 25]. **Figure 2a** presents a comparison of the air temperature distribution along the collector length between the local solution and that of Ramani et al. [11]. **Figure 2b** confirms the calculation results of the collector using mean solution with the published data by Ramadan et al. [30]. It is clear that there is good agreement between the present study and the published results.

3. Results and discussion

Figure 3 presents the fluid temperature through the two approaches. For the local solution (curves), it is possible to clearly observe that the temperature profiles rise along the flow direction. The air temperature leaving the first pass coincides with the air temperature entering the second pass, i.e., the temperatures at x = 2 m. This proves that the iterative solution has high accuracy. For the mean solution (horizontal lines), the air temperature in a pass is an average value. It is obviously deduced that the mean temperature is in the middle of the temperature distribution of the local solution, which confirms the accuracy of both the approaches.

Figure 4 is to compare the temperatures of the heat transfer surfaces obtained by the two solutions. The glass and back plate temperatures of the two methods are quite similar. However, the absorber plate temperature of the mean solution is insignificantly higher than that of the local solution. Observing the absorber plate temperature profile in the local solution, the maximum temperature is at x = 0.7 m. This is because the two air currents across the surface of the absorber plate are in opposite directions. This complexity leads to small deviations between the two approaches in predicting the absorber plate temperature. The maximum deviation of the absorber plate

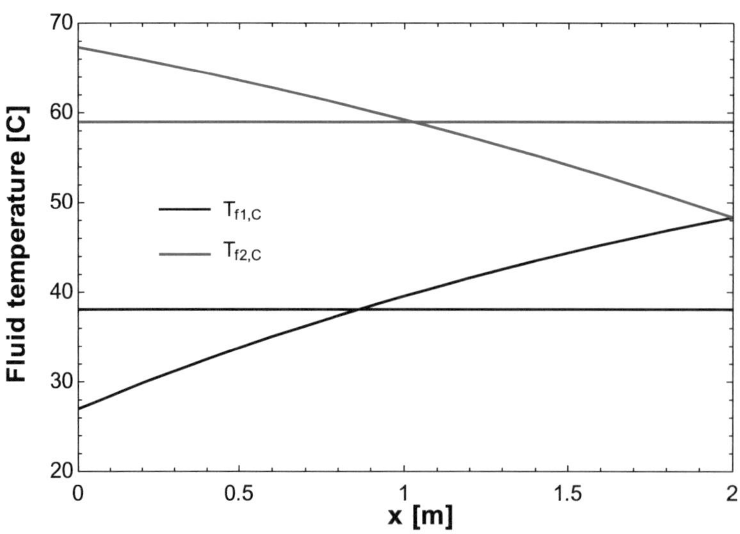

Figure 3.
Air temperature in passes for mean solution (horizontal lines) and local solution (curves).

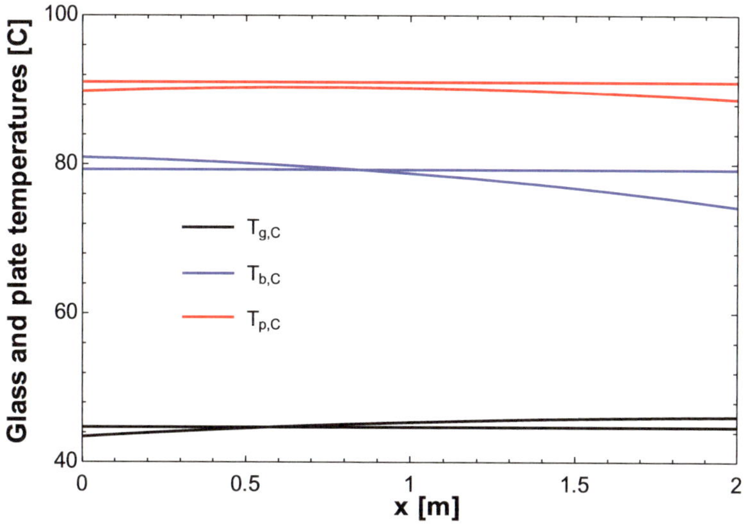

Figure 4.
Temperature of plates and glass cover for mean solution (horizontal lines) and local solution (curves).

temperature is 2.6%, which occurs at x = 2 m. The glass cover and back plate temperatures increase with the airflow direction, as expected. It is suggested that the back plate insulation should be as well as possible, i.e., increasing t_i and decreasing k_i. Thereby, the heat loss to the environment can be reduced, and the back plate temperature approaches that of the absorber plate.

The comparison of radiant heat transfer coefficient (HTC) is shown in **Figure 5**. In terms of magnitude, the radiant heat transfer coefficient between the absorber plate and the back plate ($h_{r,b,p}$) is the largest, and the radiant heat transfer coefficient between the glass and the sky (h_{ra}) is the smallest. These are because the temperature of the absorber plate is the highest and the temperature of the glass is the lowest. The

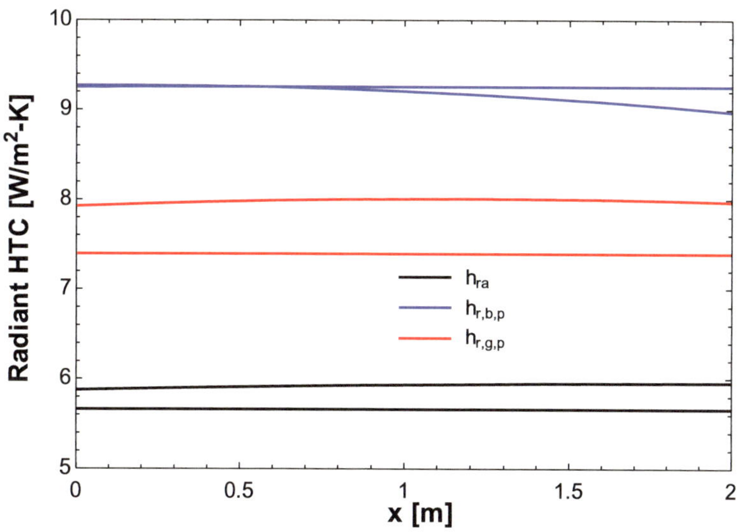

Figure 5.
Radiant heat transfer coefficients for mean solution (horizontal lines) and local solution (curves).

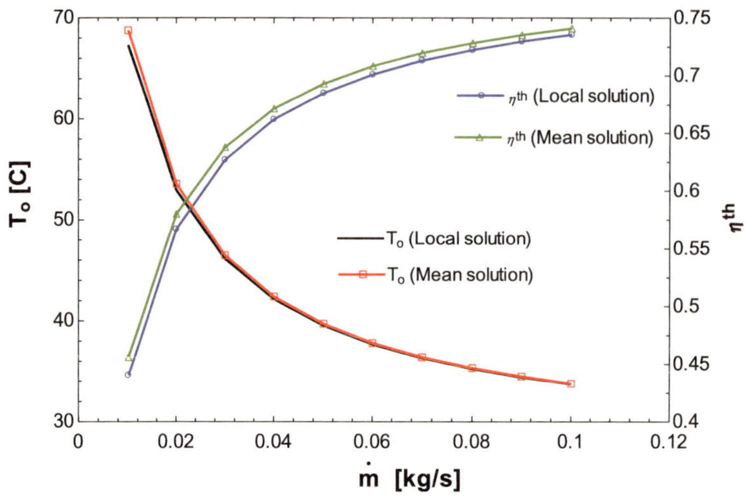

Figure 6.
Outlet air temperature and thermal efficiency for local and mean solutions.

error in the prediction of the radiant heat transfer coefficient between the absorber plate and the glass cover is the largest. This can be explained by the highest difference between the glass and absorber plate temperatures compared to other temperature differences (**Figure 4**).

Expanding the investigation, the influence of air mass flow rate in the range 0.01–0.1 kg/s on the outlet temperature and the thermal efficiency can be seen in **Figure 6**. As the flow increases, the temperature is reduced because the heater is well cooled. The increased flow rate increases the convection heat transfer rate, thus increasing the thermal efficiency. It can be concluded that the error in predicting the outcome parameters of the two methods is negligible. The air temperature of the mean solution is higher than that of the local solution, resulting in a higher thermal efficiency of the mean solution. The highest deviation of 2.1% is observed at the lowest airflow rate. The deviation reduced sharply with the flow rate. However, this difference is not of concern for a practical application.

4. Conclusion

Two analytical models for calculating heat transfer in a double-pass solar air heater were presented in this chapter. These included local solution and mean solution, which were commonly used to predict the performance of multiple-pass solar heaters. Some comparative results of the two models were drawn as follows:

- The deviation in the two solutions was not significant. Designers can choose one of them depending on calculation requirements and computer skills.
- The mean solution was simple in calculation.
- The local solution predicted the temperature distribution of plates, fluids, and glass covers.

- The mean solution predicted a slightly higher temperature than that of the local solution.

Both the models can be easily customized to predict heat transfer in a solar air heater with various modifications, such as multiple-pass collector, finned or ribbed or baffled absorber plate, and airflow recirculation as well.

Acknowledgements

The first author acknowledges the support of time and facilities from Ho Chi Minh City University of Technology (HCMUT), VNU-HCM for this study.

Conflict of interest

The authors declare no conflict of interest.

Author details

Nguyen Van Hap[1,2] and Nguyen Minh Phu[3*]

1 Faculty of Mechanical Engineering, Ho Chi Minh City University of Technology, Ho Chi Minh City, Viet Nam

2 Viet Nam National University, Ho Chi Minh City, Viet Nam

3 Faculty of Heat and Refrigeration Engineering, Industrial University of Ho Chi Minh City, Ho Chi Minh City, Viet Nam

*Address all correspondence to: nguyenminhphu@iuh.edu.vn

References

[1] Phu NM, Luan NT. A review of energy and exergy analyses of a roughened solar air heater. Journal of Advanced Research in Fluid Mechanics and Thermal Sciences. 2021;**77**(2): 160-175

[2] Menni Y, Ameur H. Improvement of the performance of solar channels by using vortex generators and hydrogen fluid. Journal of Thermal Analysis and Calorimetry. 2020;**147**:1-22

[3] Phu NM, Bao TT, Hung HN, Tu NT, Van Hap N. Analytical predictions of exergoeconomic performance of a solar air heater with surface roughness of metal waste. Journal of Thermal Analysis and Calorimetry. 2021;**144**(5):1727-1740

[4] Ngo TT, Zhou T, Nguyen HV, Nguyen PM, Lee GS. New design of a hot mixing chamber for lowering its surface temperature by adopting a perforated inner cylinder. Journal of Mechanical Science and Technology. 2021;**35**(12): 5723-5731

[5] Thao PB, Truyen DC, Phu NM. CFD analysis and Taguchi-based optimization of the Thermohydraulic performance of a solar air heater duct baffled on a Back plate. Applied Sciences. 2021;**11**(10): 4645

[6] Phu NM, Thao PB, Van Hap N. Effective efficiency assessment of a solar air heater having baffles spaced with different successive ratios. Case Studies in Thermal Engineering. 2021;**28**: 101486

[7] Menasria F, Zedairia M, Moummi A. Numerical study of thermohydraulic performance of solar air heater duct equipped with novel continuous rectangular baffles with high aspect ratio. Energy. 2017;**133**:593-608

[8] Abo-Elfadl S, Yousef MS, Hassan H. Assessment of double-pass pin finned solar air heater at different air mass ratios via energy, exergy, economic, and environmental (4E) approaches. Environmental Science and Pollution Research. 2021;**28**(11):13776-13789

[9] Deo NS, Chander S, Saini J. Performance analysis of solar air heater duct roughened with multigap V-down ribs combined with staggered ribs. Renewable Energy. 2016;**91**:484-500

[10] Ho C, Yeh C, Hsieh S. Improvement in device performance of multi-pass flat-plate solar air heaters with external recycle. Renewable Energy. 2005; **30**(10):1601-1621

[11] Ramani B, Gupta A, Kumar R. Performance of a double pass solar air collector. Solar energy. 2010;**84**(11): 1929-1937

[12] Yeh H-M, Ho C-D. Collector efficiency in downward-type internal-recycle solar air heaters with attached fins. Energies. 2013;**6**(10): 5130-5144

[13] Hernandez AL, Quiñonez JE. Analytical models of thermal performance of solar air heaters of double-parallel flow and double-pass counter flow. Renewable Energy. 2013; **55**:380-391

[14] Ho C-D, Lin C-S, Chuang Y-C, Chao C-C. Performance improvement of wire mesh packed double-pass solar air heaters with external recycle. Renewable Energy. 2013;**57**:479-489

[15] Karim M, Perez E, Amin ZM. Mathematical modelling of counter flow v-grove solar air collector. Renewable Energy. 2014;**67**:192-201

[16] Velmurugan P, Kalaivanan R. Energy and exergy analysis of multi-pass flat plate solar air heater—An analytical approach. International Journal of Green Energy. 2015;**12**(8):810-820

[17] Velmurugan P, Kalaivanan R. Energy and exergy analysis of solar air heaters with varied geometries. Arabian Journal for Science and Engineering. 2015;**40**(4): 1173-1186

[18] Velmurugan P, Kalaivanan R. Thermal performance studies on multi-pass flat-plate solar air heater with longitudinal fins: An analytical approach. Arabian Journal for Science and Engineering. 2015;**40**(4):1141-1150

[19] Velmurugan P, Kalaivanan R. Energy and exergy analysis in double-pass solar air heater. Sādhanā. 2016; **41**(3):369-376

[20] Singh S, Dhiman P. Analytical and experimental investigations of packed bed solar air heaters under the collective effect of recycle ratio and fractional mass flow rate. Journal of Energy Storage. 2018;**16**:167-186

[21] Ho C-D, Chang H, Hsiao C-F, Huang C-C. Device performance improvement of recycling double-pass cross-corrugated solar air collectors. Energies. 2018;**11**(2):338

[22] Matheswaran M, Arjunan T, Somasundaram D. Analytical investigation of solar air heater with jet impingement using energy and exergy analysis. Solar Energy. 2018;**161**:25-37

[23] Matheswaran M, Arjunan T, Somasundaram D. Energetic, exergetic and enviro-economic analysis of parallel pass jet plate solar air heater with artificial roughness. Journal of Thermal Analysis and Calorimetry. 2019;**136**(1): 5-19

[24] Nguyen Thanh L, Nguyen MP. First and second law evaluation of multipass flat-plate solar air collector and optimization using preference selection index method. Mathematical Problems in Engineering. 2021;**2021**:5563882

[25] Phu NM, Tu NT, Hap NV. Thermohydraulic performance and entropy generation of a triple-pass solar air heater with three inlets. Energies. 2021;**14**(19):6399

[26] Ahmadkhani A, Sadeghi G, Safarzadeh H. An in depth evaluation of matrix, external upstream and downstream recycles on a double pass flat plate solar air heater efficacy. Thermal Science and Engineering Progress. 2021;**21**:100789

[27] Ho C-D, Chang H, Hsiao C-F, Lin Y-C. Optimizing thermal efficiencies of double-pass cross-corrugated solar air heaters on various configurations with external recycling. Energies. 2021; **14**(13):4019

[28] Phu NM, Tu NT. One-dimensional modeling of triple-pass concentric tube heat exchanger in the parabolic trough solar air. The Collector. 2021. DOI: 10.5772/intechopen.100008

[29] Phu NM, Van Hap N, Nhan PT, Hien HP. A one-dimensional analysis and optimum air flow rate of a triple-pass solar air heater. In: 2021 24th International Conference on Mechatronics Technology (ICMT). Singapore: IEEE; 2021

[30] Ramadan MRI, El-Sebaii AA, Aboul-Enein S, El-Bialy E. Thermal performance of a packed bed double-pass solar air heater. Energy. 2007;**32**(8): 1524-1535

Chapter 3

Transfer of Heat through a Thin Liquid Film

Tahir Naseem

Abstract

The present work discusses heat transfer in thin liquid films past a stretching surface in the presence of Joule heating, radiation, viscous dissipation, and magnetic effects. Using a suitable similarity transformation, the nonlinear coupled partial differential equations are reduced to nonlinear ordinary differential equations. The coupled nonlinear ordinary differential equations are then solved by using the shooting method. The physical quantities, such as the unsteadiness parameter, the Prandtl number, the magnetic field parameter, the radiation parameter, and the viscous dissipation parameter on temperature distributions, are depicted graphically. It is observed that the effects of Prandtl and Eckert numbers are same over the temperature distribution. Furthermore, it is also evident that the increasing values of the unsteadiness parameter enhance the thermal conductivity of the fluid.

Keywords: thin film, viscous dissipation, joule heating, radiation effect, Rosseland number

1. Introduction

Due to industrial uses, researchers have developed a strong interest in boundary-layer flow and heat transmission in thin liquid films on stretching surfaces. Understanding heat transfer inside a thin liquid film is critical in engineering for operations, such as wire and fiber coating, food processing, cooling metallic plates, reactor fluidization, drawing polymer sheets, and aerodynamic extrusion of plastic sheets. This knowledge is critical during the extrusion process to ensure the extrudate's surface quality remains consistent. By minimizing friction, transparency, and strength, a smooth surface enhances the aesthetic and performance of the product [1]. The flow and heat transfer of laminar thin films across an unstable stretched sheet with a changing transverse magnetic field is investigated. The graph shows how various parameters affect the temperature distribution. Listed below are some of the key findings from our graphic research. Sakiadas' research began with an investigation of MHD flow's boundary-layer behavior on a stretching surface [2]. Wang [3] made significant contributions to the study of thin-film flow across a stretched sheet. Additionally, Andersson et al. [4, 5] improved Wang's work by including a power-law fluid with changeable physical characteristics. Wang and Pop [6] later devised a homotopy analytic approach for analyzing the heat transfer features of a power-law fluid's liquid film flow.

Khan et al. [7] showed that when the thin film parameter is increased, the temperature profiles get lower and the concentration profiles get higher. They did this while also looking at the thermal conductivity and viscosity of fluids moving through a stretching sheet. Pop et al. [8] used computational methods to explore the effect of injection and suction on the flow of a continuous laminar gravity-driven film down a vertical wall, emphasizing the importance of mass transfer and the Prandtl number. According to Khan et al. [9], a second-grade fluid flowing through a porous medium through a stretched sheet exhibits thin-film flow and heat transfer.

Sun et al. [10] investigated the thin-film flow over a rotating disc subjected to heat source and nonlinear radiation and found that increase in the rotation parameter result in a decrease in the azimuthal component of velocity, whereas increase in the temperature function of the radiation parameter. Ma et al. [11] studied how liquids flow and evaporate in a rocket combustion chamber with high temperatures and high shear forces. He investigated the flow and evaporation of a liquid film in a rocket combustion chamber with high temperatures and high shear forces. Khan et al. [12] explored Darcy–Forchheimer laminar thin-film flow with MHD and the investigation of heat transfer on an unstable horizontal stretched surface. Thermal radiation and viscous dissipation effects on thin-film flow are also examined. Using boundary-layer flow, you can look at a source of heat that emits thermal radiation. This could be very useful in industrial engineering operations like electric power generation and solar energy modernization and in astrophysical flow. Shahzad et al. [13] explored thin-film flow for radially expanding surfaces using magnetic effects and viscous dissipation for the first time. They analyzed that as the magnetic and unsteadiness parameters are raised, the film thickness decreases.

To the best of the author's knowledge, the provided facts on thermal analysis have not been investigated previously for the radially stretching surface by taking in account the effects of Joule heating and radiation effects. Section 2 detailed the governing coupled PDEs, which were later changed into coupled nonlinear ODEs by suitable transformations; Section 3 explained the approach used to solve the problem; Section 4 summarised the findings and discussed them, while Section 5 expanded on the final remarks.

2. Physical problem and its mathematical form

If an incompressible fluid flows through a flat plate with a uniform temperature T_w that is different from the ambient temperature T_∞, then we have magnetohydrodynamic boundary-layer flow. **Figure 1** depicts fluid flow in a simplified manner.

$$\frac{\partial u}{\partial r}+\frac{u}{r}+\frac{\partial w}{\partial z}=0, \tag{1}$$

$$\frac{\partial u}{\partial t}+u\frac{\partial u}{\partial r}+w\frac{\partial u}{\partial z}=\upsilon\frac{\partial^2 u}{\partial z^2}+B(t)^2 u, \tag{2}$$

$$\begin{gathered}\frac{\partial T}{\partial t}+u\frac{\partial T}{\partial r}++w\frac{\partial T}{\partial z}=\\ \alpha\frac{\partial^2 T}{\partial z^2}+\frac{\nu}{C_p}\left(\frac{\partial u}{\partial z}\right)^2+\frac{1}{\sigma\rho C_P}B(t)^2u^2-\frac{1}{\sigma\rho C_p}\frac{\partial}{\partial z}q_r.\end{gathered} \tag{3}$$

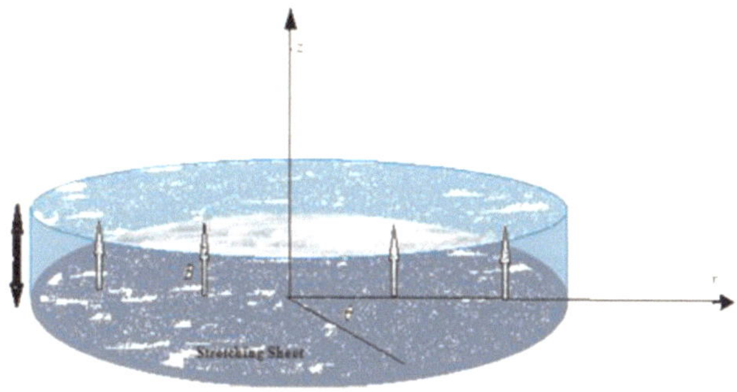

Figure 1.
Schematic diagram of fluid flow

We obtain $q_r = -\frac{4\sigma^*}{3k^*}\frac{\partial T^4}{\partial y}$, using the Rosseland approximation for radiation, where k^* and σ^* denote the absorption coefficient and Stefan Boltzmann constant, respectively. By expanding T^4 in a Taylor's series in the vicinity of T_∞ and neglecting higher-order terms, we get

$$T^4 = 4T_\infty{}^3T - 3T_\infty{}^4.$$

So Eq. (3) can be rewritten as

$$\frac{\partial T}{\partial t} + u\frac{\partial T}{\partial r} + +w\frac{\partial T}{\partial z} = \left(\alpha + \frac{1}{\rho C_p}\frac{16\sigma^*}{3k^*}T_\infty^3\right)\frac{\partial^2 T}{\partial z^2} + \frac{\nu}{C_p}\left(\frac{\partial u}{\partial z}\right)^2 + \frac{\sigma}{\rho C_P}B(t)^2u^2, \quad (4)$$

where $u(x,y)$ and $v(x,y)$ are velocities; T represents the temperature. Also, ν and α, respectively, are the thermal diffusivity and kinematics viscosity of the fluid. The boundary conditions may be expressed as

$$u = U, w = 0, T = T_s, \text{at } z = 0, \frac{\partial u}{\partial z} = \frac{\partial T}{\partial z} = 0, w = \frac{dh}{dt}, \text{at } z = h, \quad (5)$$

An acceptable stream function $\psi = \psi(r,z)$ satisfying continuity Eq. (1) identically, with similarity transformations, so that the flow equations are translated to nonlinear ODEs.

$$\eta = \frac{z}{r}Re^{\frac{1}{2}}, \psi = -r^2URe^{-\frac{1}{2}}f(\eta), \text{and } \theta(\eta) = \frac{T_0 - T(r,z)}{T_s - T_0}, \quad (6)$$

where η is the independent variable, $\mathrm{Re} = \frac{\mathrm{rU}}{\upsilon}$ is the local Reynold number, and $\psi(r,z)$ is Stokes stream function such that $\mathrm{u} = \left[-\frac{1}{\mathrm{r}}\left(\frac{\partial\psi}{\partial z}\right)\right]$ and $\mathrm{w} = \left[\frac{1}{\mathrm{r}}\left(\frac{\partial\psi}{\partial r}\right)\right]$. Eqs. (2)–(4) becomes after using the aforementioned modification.

$$f'''(\eta) + 2f(\eta)f''(\eta) - S\left(f'(\eta) + \frac{\eta}{2}f''(\eta)\right) - (f'(\eta))^2 - Mf'(\eta) = 0, \quad (7)$$

$$(1 + Rd)\theta''(\eta) = Pr\left(\frac{S}{2}(3\theta(\eta) + \eta\theta'(\eta)) + 2f'(\eta)\theta(\eta) - 2\theta'(\eta)f(\eta) + 2Ec(f''(\eta))^2 + EcMf'^2\right), \quad (8)$$

$$f'(0) = 1, f(0) = 0, \theta(0) = 1, f'(\beta) = 0, \theta'(\beta) = 0, f(\beta) = \frac{S\beta}{2}, \tag{9}$$

where $Pr = \frac{\nu}{\alpha}$ and $Ec = \frac{U^2}{C_{p(T_w - T_\infty)}}$ are dimensionless Prandtl and Eckert numbers. $S = \alpha/b$ is the measure of unsteadiness parameter and β (unknown constant) is the dimensionless thickness of the film and is given as $\beta = (b/\upsilon(1 - \alpha t))(1/2)\, h$. The rate at which film thickness varies can be obtained as $dh/dt = (-\alpha\beta/2)(\upsilon/b(1 - \alpha t))(1/2)$. The important physical quantity of interest in this problem is Nusselt number $Nu_r = (-r/Tref)\left(\frac{\partial T}{\partial Z}\right)_{Z=0}$. In non-dimensional form, it can be expressed as

$$2(1 - \alpha t)^{1/2} Re^{-3/2} Nur = -(1 + Rd)\theta'(0).$$

The rate of heat transfer is represented by the quantity of physical interest $-(1 + Rd)\theta'(0)$. As a result, our goal is to figure out how the controlling factors $Pr, Ec, M,$ and n affect these values.

3. Solution methodology

The system of nonlinear PDEs is given in Eqs. (2) and (4) subject to given boundary conditions given by Eq. (5) are transformed to ODEs via similarity transformation and solved via the shooting method. The following are the third-order differential equations for the second-order nonlinear ODE and the second-order nonlinear ordinary differential for first order.

$$f = y(1), f' = y(2), f'' = y(3), f''' = yy1, \tag{10}$$

$$yy1 = S\left(y(2) + \frac{\eta}{2}y(3)\right) + y(2)^2 - 2y(1)y(3) + My(2) \tag{11}$$

$$\theta = y(4), \theta' = y(5), \theta'' = yy2, \tag{12}$$

$$yy2 = \frac{Pr}{(1 + Rd)}\left(\frac{S}{2}(3y(4) + \eta y(5)) + 2y(4)y(2) - 2y(1)y(5) + 2Ecy(3)^2 + EcMy(2)^2\right) \tag{13}$$

The iterative procedure will end with the appropriate level of precision.

4. Result and discussion

This work fundamentally alters our understanding of thin films. Unsteadiness, magnetic parameters, Eckert number, Prandtl number, and radiation parameters are the regulating parameters. Numerical simulation is utilized in conjunction with analytical solutions of nonlinear differential equations. Additionally, dimensionless parameters were investigated in axisymmetric MHD thin liquid film only for heat transfer past the stretching surface.

4.1 Temperature profile influenced by unsteadiness parameters

According to **Figure 2**, a higher unsteadiness parameter (S) results in a decrease in temperature, which in turn enhances heat transmission in thin liquid films.

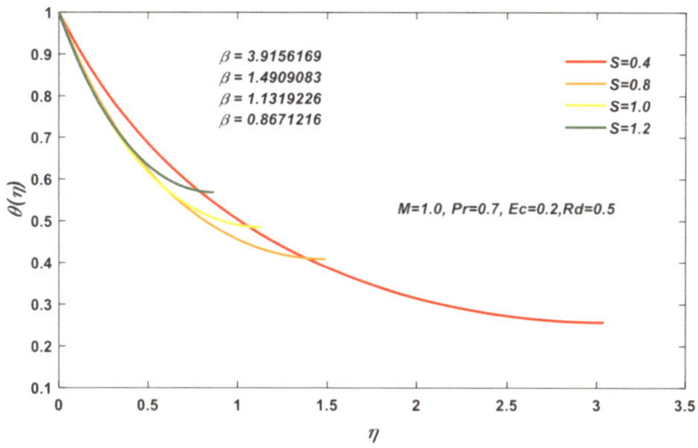

Figure 2.
Illustration of S on the $\theta(\eta)$.

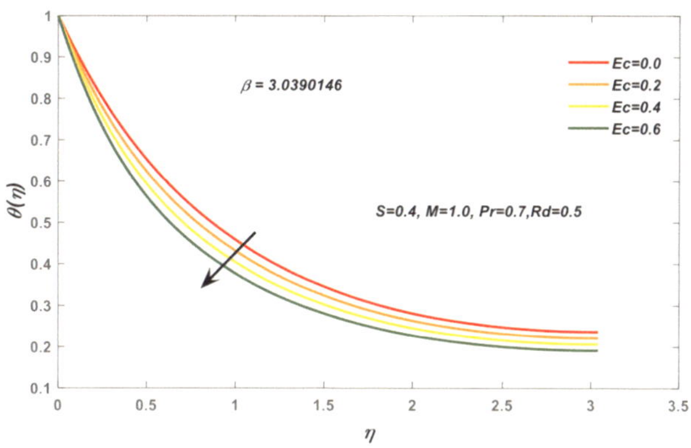

Figure 3.
Illustration of Ec on $\theta(\eta)$.

4.2 Temperature profile influenced by Eckert number

The temperature distribution for various Eckert numbers is depicted in **Figure 3**. This physical characteristic characterizes the relationship between kinetic and internal energy.

4.3 Temperature profile influenced by Prandtl number

The Prandtl number presents the relationship between momentum and thermal diffusivity. When the Prandtl number is decreased, the thickness of the boundary-layer drops, further restricting heat transfer. The Prandtl number is plotted against the temperature distribution in **Figure 4**. The thermal boundary layer was decreased as a result of the enhanced dissipation.

4.4 Temperature profile influenced by radiation parameter

The influence of the radiation parameter *Rd* on the temperature profile for the unsteadiness parameter *S* is seen in **Figure 5**. It is clearly demonstrated that radiation

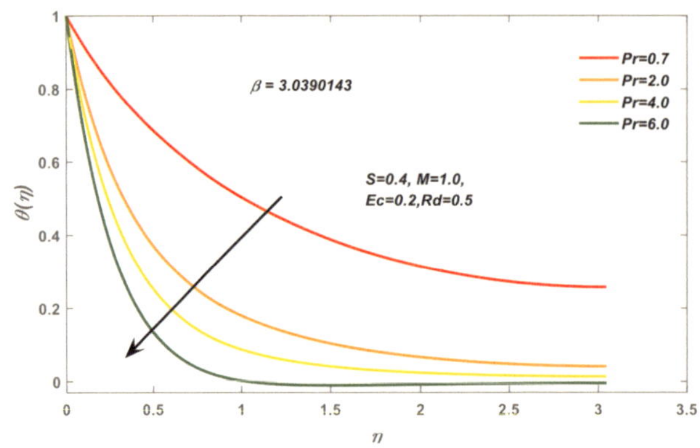

Figure 4.
Illustration of Pr on $\theta(\eta)$.

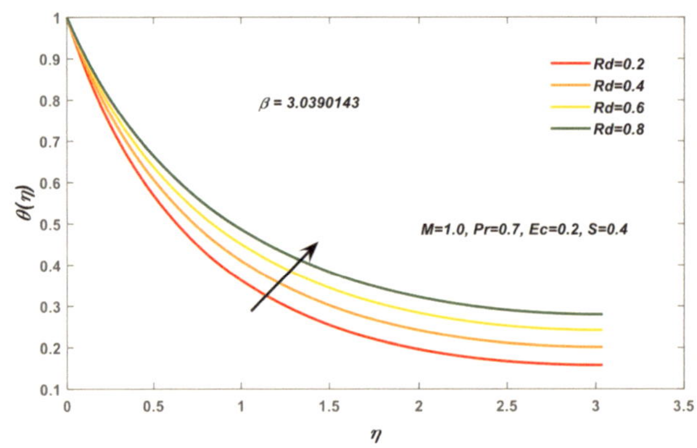

Figure 5.
Illustration of Rd on $\theta(\eta)$.

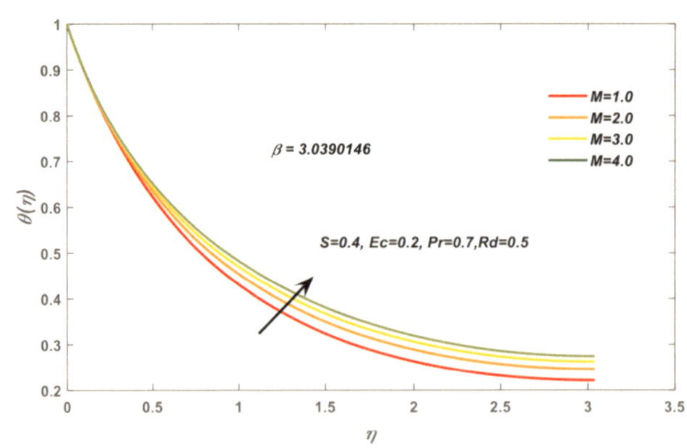

Figure 6.
Illustration of M on $\theta(\eta)$.

M	*S*	Pr	Ec	$-\theta'(0)$	
				[13]	**Present**
1	0.8	0.7	0.2	1.4105271	1.528545
	1.0			1.4200617	1.5319487
	1.2			1.3936895	1.4958518
1	0.4	0.7	0.2	1.3214547	1.4443904
		4		3.81002363	4.3083451
		6		4.7918785	5.4647752
1	0.4	0.7	0.0	1.2314224	1.2314194
1			0.2	1.3214547	1.3214547
1			0.4	1.4114869	1.6573614

Table 1
Effect of pertinent parameters on heat transfer rate.

raises the temperature in the fluid film's boundary- layer region, lowering the cooling rate for thin-film flow.

4.5 Temperature profile influenced by magnetic parameter

For a fixed unsteadiness parameter S, the temperature distribution is shown in **Figure 6** in relation to the magnetic field value M. As M rises, so does the thickness of the thermal boundary layer. As a result of the Lorentz force, the thermal boundary layer becomes thicker because frictional drag increases. In turn, this causes an increase in the thermal boundary-layer thickness. The purpose of **Table 1** is to examine the influence of several important factors on the dimensionless heat transfer rate at the surface. The Prandtl number and the unsteadiness parameter have an effect on the rate of heat transfer at the surface.

5. Conclusion

The flow and heat transfer through a stretching sheet with a variable transverse magnetic field is studied in this work. The shooting strategy is used to solve the governing equations numerically. The temperature distribution is shown in the graph below as a function of several parameters. In summary, the following are some of the most significant discoveries from our graphical investigation:

i. Enhance the thermal boundary layer, decrease fluid particle movement, and increase the temperature field when a magnetic field is applied to an electrically conducting fluid.

ii. The temperature decreases for the increasing values of Prandtl and Eckert numbers, i.e., both have the same effect.

iii. The influence of radiation parameter Rd. rises in thermal boundary- layer thickness, which enhances heat transmission.

Nomenclature

(r, θ, z)	polar coordinate
$\alpha = \left(\frac{k}{\rho c_p}\right)$	thermal diffusivity
(u, v, w)	velocity component
$\beta(t)$	applied magnetic field
T	fluid's Temperature
T_{ref}	reference temperature
c_p	specific heat
P	pressure
$\psi(r, z)$	Stokes stream function
PDF	Partial differential equations
σ	electrical conductivity
ρ	density of fluid
Re	Local Reynolds number
Cf	Local skin-friction coefficient
Nu	Nusselt number
Pr	Prandtl number
Ec	Eckert-number
β	thickness of the film
υ	kinematic viscosity
ODE	Ordinary differential equations

Author details

Tahir Naseem
Department of Mathematics, Government Postgraduate College Haripur, Pakistan

*Address all correspondence to: tahir.gch@gmail.com

References

[1] Aziz RC, Hashim I, Alomari AK. Thin film flow and heat transfer on an unsteady stretching sheet with internal heating. Meccanica. 2011;**46**(2):349-357

[2] Sakiadis BC. Boundary-layer behavior on continuous solid surfaces: II. The boundary layer on a continuous flat surface. AICHE Journal. 1961;7(2): 221-225

[3] Wang CY. Liquid film on an unsteady stretching surface. Quarterly of Applied Mathematics. 1990;**48**(4):601-610

[4] Andersson HI, Aarseth JB, Braud N, Dandapat BS. Flow of a power-law fluid film on an unsteady stretching surface. Journal of Non-Newtonian Fluid Mechanics. 1996;**62**(1):1-8

[5] Andersson HI, Aarseth JB, Dandapat BS. Heat transfer in a liquid film on an unsteady stretching surface. International Journal of Heat and Mass Transfer. 2000;**43**(1):69-74

[6] Wang C, Pop I. Analysis of the flow of a power-law fluid film on an unsteady stretching surface by means of homotopy analysis method. Journal of Non-Newtonian Fluid Mechanics. 2006; **138**(2–3):161-172

[7] Khan NS, Gul T, Islam S, Khan W. Thermophoresis and thermal radiation with heat and mass transfer in a magnetohydrodynamic thin-film second-grade fluid of variable properties past a stretching sheet. The European Physical Journal Plus. 2017;**132**(1):1-20

[8] Pop I, Watanabe T, Konishi H. Gravity-driven laminar film flow along a vertical wall with surface mass transfer. International Communications in Heat and Mass Transfer. 1996;**23**(5): 687-695

[9] Khan NS, Islam S, Gul T, Khan I, Khan W, Ali L. Thin film flow of a second grade fluid in a porous medium past a stretching sheet with heat transfer. Alexandria Engineering Journal. 2018; **57**(2):1019-1031

[10] Sun TC, Uddin I, Raja MAZ, Shoaib M, Ullah I, Jamshed W, et al. Numerical investigation of thin-film flow over a rotating disk subject to the heat source and nonlinear radiation: Lobatto IIIA approach. Waves in Random and Complex Media. 2022:1-15

[11] Ma X, Wang Y, Tian W. A novel model of liquid film flow and evaporation for thermal protection to a chamber with high temperature and high shear force. International Journal of Thermal Sciences. 2022;**172**:107300

[12] Khan Z, Jawad M, Bonyah E, Khan N, Jan R. Magnetohydrodynamic thin film flow through a porous stretching sheet with the impact of thermal radiation and viscous dissipation. Mathematical Problems in Engineering. 2022

[13] Shahzad A, Gulistan U, Ali R, Iqbal A, Benim AC, Kamran M, et al. Numerical study of axisymmetric flow and heat transfer in a liquid film over an unsteady radially stretching surface. Mathematical Problems in Engineering. 2020

Chapter 4

Heat and Mass Transfer of a Decoupling Cooling System: A Desiccant-Coated Heat Exchanger and a Dew-Point Evaporative Cooler

Seung Jin Oh, Yeongmin Kim, Yong-yoo Yang and Yoon Jung Ko

Abstract

This study presents an innovative de-coupling cooling technology where latent and sensible cooling loads are handled separately by a desiccant-coated heat exchanger (DCHE)-based dehumidifier and a dew-point evaporative cooler (DEC). The DCHE first removes the moisture of humid outdoor air by adsorption process. Subsequently, the DEC sensibly cools down the dehumidified air, maintaining the humidity ratio. Their performances are investigated numerically by analyzing the heat and mass transfer. The cyclic average outlet values of DCHE are applied to the inlet conditions for DEC simulation. Comparison between the simulation results and the experimental data was carried out and showed good agreement and a similar trend with a maximum discrepancy of 8.6% for DCHE and 3.2% for DEC, respectively. Key results revealed that moisture removal capacity (MRC) and latent cooling capacity (Q_L) for DCHE are largely affected by varying air dry-bulb and air wet-bulb temperatures, while the almost constant COP_{th} was observed regardless of the variation of temperatures. For the DEC, the higher dew-point effectiveness and wet-bulb effectiveness were observed at the higher dry-bulb temperature and higher humidity ratio, while the higher sensible cooling capacity was observed at the higher dry-bulb temperature and lower humidity ratio.

Keywords: decoupling cooling system, desiccant-coated heat exchanger, dew-point evaporative cooler, moisture removal capacity, dew-point effectiveness

1. Introduction

The growing population and rising energy demand in many countries are imposing significant challenges to energy and environmental sustainability. The global energy and environmental scenarios are closely interlinked. That is, the problems of supply and use of energy are related to global warming and climate change [1]. The electrical energy consumption for air conditioning accounts for 20–40% of the total electricity used in buildings around the world today [2, 3]. Global energy demand for air conditioning is expected to triple by 2050, requiring new electricity capacity that is the equivalent to the combined electricity capacity of the United States, the EU, and Japan today. Accordingly, the global stock of air conditioners in buildings will grow to 5.6 billion by 2050, up from 1.6 billion today, which means new air conditioners are sold every second for the next 30 years [4].

In conventional air-conditioning systems, the compressor efficiency has shown a significant improvement chronologically from 1.2 kW/RT in the 1990s to 0.85 ± 0.05 kW/RT [5]. Since 2000, however, this improvement trend has leveled asymptotically, implying that efficiency improvement for compressor stages and heat exchangers has been saturated. Despite massive investment in the R&D of cooling technologies by chillers' manufacturers, there are physical and material limits to which efficiency improvement of the major components in the cooling cycle can be attained. If one were to continue to improve the cooling efficiency of conventional chillers (with HFC refrigerants), the improvement in kW/RT may improve only marginally by less than 5% and no quantum efficiency improvement can be expected. Therefore, we believe that there is a need for an out-of-box solution for cooling where the consumption of energy and water can be reduced significantly, by as much as 50% so as to attain sustainable cooling for the future, addressing the space cooling needs of the present and future generations.

Recently, many studies have been focused on the development of a desiccant-coated heat exchanger (DCHE) to improve dehumidification performance by employing various desiccant materials and heat exchangers. Sunhor et al. [6] investigated experimentally the dehumidification behavior of AlPO-zeolite-coated heat exchanger by varying operating conditions. Chai et al. [7] presented a desiccant heat pump by combining a heat pump and a desiccant-coated heat exchanger. They employed a commercial silica gel to fabricate the desiccant-coated heat exchanger. Liang et al. [8] developed a desiccant-coated heat exchanger by using a microchannel heat exchanger and explored the performance under various operating conditions.

Many investigations also have been carried out mathematically and experimentally to study the effectiveness and COP of various dew-point evaporative coolers (DECs). Xu et al. [9] proposed a novel complex heat and mass exchanger (HMX) to improve the performance of a dew-point evaporative cooler. Duan et al. [10] explored the operational performance of a regenerative evaporative cooler under various conditions by analyzing the wet-bulb effectiveness and COP. Balyani et al. [11] conducted an economic, environmental, and thermodynamic analysis for the best cooling technology for various climates. Duan et al. [10] investigated experimentally the energy-saving potential of a prototype of a counter-flow regenerative cooler when it applied to China's various regions. Jafarian et al. [12] developed GMDH-type neural networks for modeling and optimization of a flat plate dew-point counter flow IEC. Their results indicated that the COP was improved by 36.6% in hot and dry climate. Sohani et al. [13] conducted a multiobjective optimization to find the best design for two different dew-point evaporative coolers (i.e., counter-flow and cross-flow configurations).

In this study, firstly, we present an innovative decoupling cooling technology where latent cooling load and sensible cooling load are handled separately by a desiccant-coated heat exchanger (DCHE)-based dehumidifier and a dew-point evaporative cooler (DEC). Secondly, the performance is investigated numerically by analyzing the heat and mass transfer for both DCHE and the DEC.

DOI: http://dx.doi.org/10.5772/intechopen.105876

2. De-coupling cooling system

In order to achieve a large reduction in the kW/RT of cooling, we propose a de-coupling cooling system that thermodynamically hybridizes a desiccant coated heat exchanger (DCHE)-based dehumidifier with a dew-point evaporative cooler (DEC). The decoupling cooling system handles a sensible cooling load and a latent cooling load separately. **Figure 1** shows schematically the working principle of the decoupling system and its thermodynamic process on a psychrometric chart. The DCHE first removes the undesired moisture of humid outdoor air by the adsorption process. During this process, the desiccant adsorbs water vapor up to its equilibrium state, which is determined by the water vapor partial pressure and the temperature of the desiccant. At this point, the air temperature rises slightly due to adsorption heat generation. Cooling water (~25°C) is supplied to remove the adsorption heat. Subsequently, the DEC cools down the dehumidified air up to the desired temperature, maintaining moisture level. The DEC consists of two different channels, namely a dry air channel and a wet-air channel. The dehumidified air coming out from the DCHE enters the dry air channel and releases the heat to the adjacent wet channel where a certain portion (20–35%) of the dehumidified air flows. The inner surface of the wet channel is always wetted by water film, and the evaporation takes place on the surface of the water film by absorbing the heat from itself, which causes the reduction of the temperature. It is noteworthy that the proposed system can handle all climate conditions successfully as well as it employs only waste heat for the regeneration of adsorbent, and the IEC uses only water as a refrigerant.

2.1 Desiccant-coated heat exchanger-based dehumidification

The desiccant-coated heat exchanger (DCHE) has been tested successfully for dehumidification application using only the low-temperature waste heat, as proven in the many case studies [2, 5, 15–17]. It could remove effectively moisture from humid air at a relative humidity ratio of 95% or a humidity ratio of above 20 g/kg using commercial adsorbents such as the silica gel, Zeolite, and MOF.

The DCHE was first introduced in order to improve the efficiency of a conventional solid desiccant dehumidifier, where solid desiccant was packed within a finned

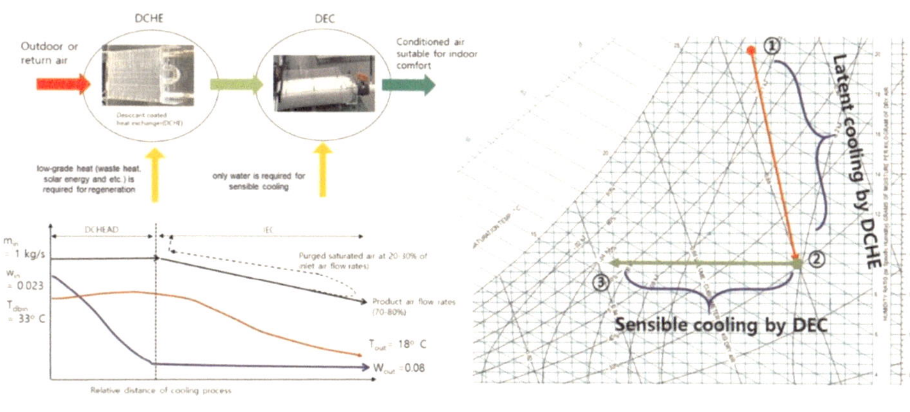

Figure 1.
The working principle of the decoupling cooling system and its thermodynamic process on a psychrometric chart [2, 14].

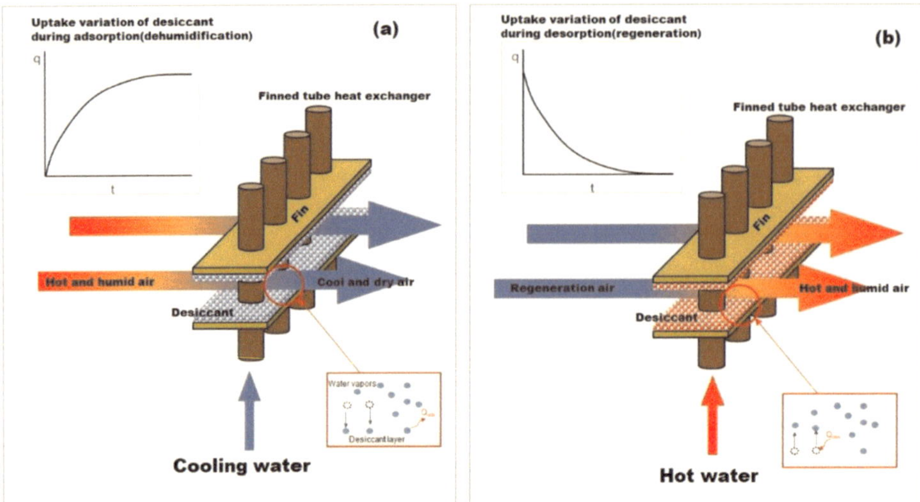

Figure 2.
The concept of the desiccant-coated heat exchanger (DCHE) for the dehumidification: (a) adsorption/dehumidification process and (b) desorption/regeneration process.

tube heat exchanger using a wire mesh. It was found that the packing method is inefficient due to poor heat transfer between solid desiccant and cooling water flowing through the heat exchanger. To coat the fins of the heat exchanger with the desiccant powder (ex. silica gel), hydroxyethylcellulose (HEC) is used as a binder since it exhibits the best binding force between the silica gel powder and the metal fins as well as it does not affect the adsorbate uptake capacity. During the dip-coating process, the heat exchanger is mounted onto a rotation machine to ensure a uniform coating layer. After that, the heat exchanger is placed into an oven for curing at 120°C for 12 h. The optimal coating thickness of the desiccant layer is found to be 0.1 mm in many studies [2, 14, 17–20]. As depicted in **Figure 2**, the DCHE operates alternatively between two different modes, namely dehumidification (adsorption) mode and regeneration (desorption) mode. During the dehumidification mode, cooling water is supplied through the tubes in order to remove the adsorption heat, keeping the desiccant layer at a lower temperature and thus ensuring the highest uptake. On the other hand, hot water is supplied to release the water vapor that was adsorbed in the previous process during the regeneration mode. In this mode, each process continues until the uptake reaches the equilibrium uptake given the temperature and vapor pressure.

2.2 Dew-point evaporative cooler (DEC)

As one of evaporative cooling methods, the dew-point evaporative cooler (DEC) is an energy-efficient and environmentally friendly sensible cooling device that employs only water as a refrigerant [21–24]. The key feature of the DEC is that it employs the process air as the working air by purging the process air at a specified rate (20–30%) to reduce the supply air temperature below the wet-bulb temperature. The working principles for three different evaporative coolers are represented in **Figure 3**.

The DEC differs from the conventional adiabatic cooling or commonly known as the cooling towers, swarm coolers, etc., where the air stream experience changes in both the temperature and absolute humidity. In the DEC, however, the product air

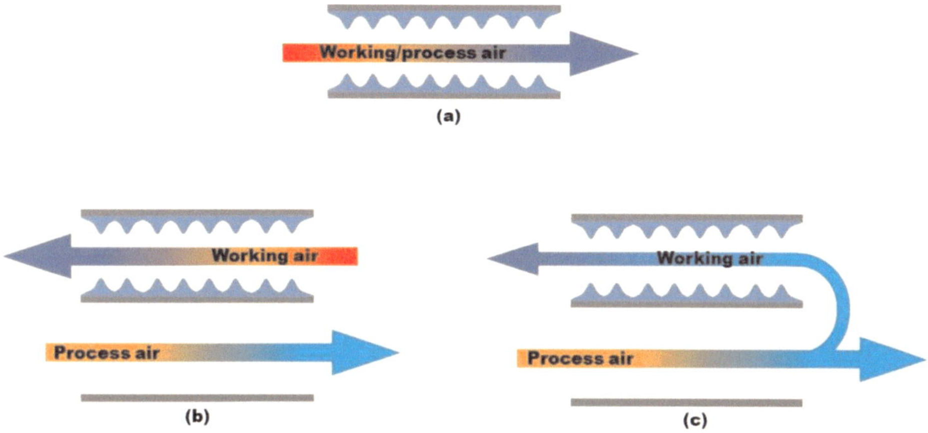

Figure 3.
The working principles of evaporative cooling devices: (a) direct evaporative cooler, (b) indirect evaporative cooler, and (c) dew-point evaporative cooler.

(conditioned air for space cooling) is flowing in a dry channel while a small fraction (20–30%) of the product air is purged into a wet channel, where the purged air picks up the water vapor from a hydrophilic membrane that physically separates the evaporative moist air flowing in a counter-flow direction from the product air. The evaporative cooling in the wet channel cools the air flowing in the adjacent dry channel by heat transfer, and the processes are repeated in succession to achieve an overall cooling of primary air. At the end of a section, the saturated secondary air is purged while a part of the conditioned primary air (but at the original inlet humidity) is directed into the wet channel to repeat the evaporative cooling. Thermodynamically, it incurs two major energy losses to achieve cooling of the primary air stream: Firstly, energy is transferred to the air by a fan for maintaining the necessary air flows in all channels. Secondly, it continuously purges a fraction of the conditioned air of the non-wetted channel so as to reinitiate the evaporative cooling in successive wetted channels. The key feature here is that it continuously purges a fraction of the conditioned air so that the wet-bulb temperature of the conditioned stream can be lowered. The novelty of the decoupling cycle is the harnessing of evaporative potential of inlet air as well as waste heat, but the overall electricity incurred is a quantum lower than that of conventional chillers.

3. Numerical modeling of a decoupling cooling system

3.1 Latent cooling by DCHE

In order to investigate the heat and mass transfer for the DEC, a mathematical model was established. The differential control volume includes half the height of the dry and wet channels, the separating plate, and the water film. To simplify the heat and mass transfer processes, the following assumptions were made:

1. The air streams in both the dry and wet channels are at a steady state.

2. Air flow is laminar and fully developed in both channels.

3. The heat and mass transfer coefficients are constant.

4. Heat losses to the surroundings are negligible.

5. The water is continuously supplied to maintain the water film.

Mass transfer between the adsorbent (desiccant) and adsorbate (water vapor) can be expressed by LDF (linear driving force) model as follows:

$$\frac{dq}{dt} = k(q^* - q) \tag{1}$$

where dq/dt is the rate of the uptake of water vapor, q^* is the equilibrium uptake. k is a reaction coefficient and can be obtained by

$$k = \frac{15D_{so}\,exp\left(-\frac{E_a}{RT_d}\right)}{R_p^2} \tag{2}$$

where D_{so} a diffusion coefficient and E_a is an activation energy, R is an universal gas constant, T_d is the desiccant temperature, R_p is an average radius of desiccant particles, respectively. These values are listed in **Table 1**. The equilibrium uptake, q^*, can be approximated with Tóth model.

$$\frac{q^*}{q_0} = \frac{K_T P exp\left(\frac{Q_{st}}{RT}\right)}{\left\{1 + \left[K_T\,exp\left(\frac{Q_{st}}{RT}\right)\right]^t\right\}^{1/t}} \tag{3}$$

where q_0 is the maximum uptake of the desiccant and can be obtained by the measured isotherm. Q_{st} is the isosteric heat of the desiccant, t is heterogeneity factor, KT is Tóth constant, P is the vapor partial pressure.

The energy conservation equation of DCHE during adsorption/desorption process can be expressed by the following equation:

$$\left[M_d c_{p,d} + M_{hx} c_{p,hx} + M_d c_{p,v} q(t)\right] \frac{dT_d}{dt} = M_d \frac{dq}{dt} Q_{st} - \dot{m}_a c_{p,a}(T_{a,o} - T_{a,i}) - \dot{m}_w c_{p,w}(T_{w,o} - T_{w,i}) \tag{4}$$

Operation parameters	Baseline	Values
Process air inlet dry-bulb temperature [°C]	35	27, 30, 40, 46
Process air inlet wet-bulb temperature [°C]	24	19, 22, 24, 26
Process/regeneration air mass flow rate [kg/h]	13	
Cooling water inlet temperature [°C]	30	25, 30, 35
Hot water inlet temperature [°C]	80	50, 60, 70, 80
Cycle time [sec.]	150	

Table 1.
Simulation conditions for DCHE.

The first term on the left-hand side represents the rate of the change of energy content of the desiccant. The first term on the right-hand side represents the rate of adsorption heat generation. The second and the third terms are the rates of energy of air and water streams, respectively. Mass conservation equation of adsorbed phase of water vapor in the desiccant can be expressed by the following equation:

$$M_d \frac{dq}{dt} = \dot{m}_{v,in} - \dot{m}_{v,out} \quad (5)$$

The first term on the left-hand side represents the rate of the change of moisture content in the desiccant. $\dot{m}_{v,in}$ and $\dot{m}_{v,out}$ are the mass flow rate of the moisture in the air at the inlet and outlet of DCHE, respectively. The humidity ratio of the air at the outlet can be calculated by Eq. (6).

$$\omega_o = \frac{\dot{m}_{v,o}}{\dot{m}_{da}} \quad (6)$$

where $\dot{m}_{da}$ is the mass flow rate of dry air. It should be noted that the mass flow rate of dry air remains constant during the dehumidification and regeneration processes.

The temperature of the process air at the outlet of DCHE is obtained by the following equation:

$$T_{a,o} = (T_d - T_{a,i})\, exp\left(-\frac{A_{c,a}h_{c,a}}{\dot{m}_a c_{p,a}}\right) \quad (7)$$

$$T_{w,o} = (T_d - T_{w,i})\, exp\left(-\frac{A_{c,w}h_{c,w}}{\dot{m}_w c_{p,w}}\right) \quad (8)$$

In order to develop the temperature and velocity profile along the channel, Nusselt number is approximated by the following equation [25].

$$Nu = Nu_{FD} + \frac{0.0841}{0.002907 + Gz^{-0.6504}} \quad (9)$$

$$Gz = \frac{RePrP_e}{4L_w} \quad (10)$$

$$Re = \frac{\rho u D_h}{\mu} \quad (11)$$

$$Pr = \frac{\mu c_p}{k} \quad (12)$$

The convective heat transfer coefficient for air side is calculated by Dittus-Boelter correlation:

$$h_a = \frac{N_{u,a} k_a}{D_{h,a}} \quad (13)$$

The same correlation can be applied to predict the convective heat transfer coefficient for water side of the DCHE

$$h_w = \frac{N_{u,w} k_w}{D_{h,w}} \tag{14}$$

The cooling/heat water flow inside the tube at a mass flow rate of 4 LPM and thus turbulent flow is formed with Reynolds number greater than 2300. Therefore, the Nusslet number is calculated from an empirical relation,

$$Nu_w = 0.023\, Re^{0.8} Pr^n \tag{15}$$

The exponent n equals to 0.4 for heating the water bulk flow (regeneration) and 0.33 for cooling conditions (dehumidification).

3.2 Sensible cooling by DEC

The mathematical model was for a differential control volume including half the height of the dry and wet channels, the separating plate, and the water film. To simplify the heat and mass transfer processes, the following assumptions were made:

1. The air streams in both the dry and wet channels are at steady state.
2. The air streams are fully developed laminar flow in both channels.
3. The heat and mass transfer coefficients are constant.
4. Heat losses to the surroundings are negligible.
5. The water is continuously supplied to maintain the water film.

In the dry channel of DEC, only a sensible cooling effect takes place by forced convective heat transfer without changing the humidity ratio. The energy conservation equation for air flowing in the dry channel is expressed by

$$k_a \frac{d^2 T_{da}}{dx^2} - u_{da} \rho_{da} c_{p,m} \frac{\partial T_{da}}{dx} = \frac{h_{da}}{\bar{H}} \left(T_{da} - T_p \right) \tag{16}$$

where $c_{p,m}$ is the specific heat of air-water vapor mixture. $\bar{H}$ represents half the height of the dry channel. The first term on the left-hand side of the above equation represents the longitudinal heat transfer by conduction, and the second term denotes heat transfer due to advection. The term on the right-hand side of the equation indicates convective heat transfer between the air and the separator.

Unlike the dry channel, both sensible and latent cooling effects take place in the wet channel. That is, both heat and mass transfer mechanism are involved to create cooling effect between the air and water film layer, and it is expressed as follows:

$$k_a \frac{d^2 T_{wa}}{dx^2} + u_{da} \rho_{da} c_{p,m} \frac{dT_{wa}}{dx} = \frac{h_{wa}}{\bar{H}} \left(T_{wa} - T_f \right) + \frac{h_m \rho_{da}}{\bar{H}} \left[\omega_{sat} \left(T_f \right) - \omega_v c_{p,v} \left(T_{wa} - T_f \right) \right. \tag{17}$$

The first term on the left-hand side represents conduction heat transfer in the longitudinal direction, and the second term denotes advection heat transfer. The first term on the right-hand side refers to convective heat transfer, and the second term is heat transfer due to mass transfer between the air and the water film surface.

The mass transfer takes place only in the wet channel by the driving force of vapor partial pressure difference, and the mass conservation equation is written as

$$\rho_{da} D_v \frac{d^2\omega_v}{dx^2} = -\rho_{da} u_{wa} \frac{d\omega_v}{dx} + h_m \bar{H} \rho_{da} \left(\omega_{v,sat}(T_f) - \omega_v\right) \tag{18}$$

On the other hand, the energy balances for the water film and the separating plate are given as

$$k_f \frac{d^2T_f}{dx^2} = -\frac{h_w(T_{wa} - T_f)}{\delta_f} - k_f \frac{T_p - T_f}{\delta_f^2} + \frac{h_{fg} h_m \rho_{da}\left[\omega_{sat}(T_f) - \omega\right]}{\delta_f} \tag{19}$$

$$k_p \frac{d^2T_p}{dx^2} = -\frac{h_d(T_{da} - T_p)}{\delta_p} + \frac{k_p(T_p - T_f)}{\delta_p^2} \tag{20}$$

Given the geometry of the channels and the operating conditions, the air streams are laminar flow with the maximum Reynolds number of 1048. The convective heat transfer coefficients for the dry and wet channels are expressed by Eq. (21).

$$h_{wa} = \frac{N_{u,wa} k_{wa}}{D_{h,wa}} \tag{21}$$

The Nusselt number can be calculated by the correlation by Eq. (8) for fully developed laminar flow in the rectangular dry channel [26].

$$Nu = 8.235\left(1 - 2.042AR + 3.085AR^2 - 2.477AR^3 + 1.058AR^4 - 0.186AR^5\right) \tag{22}$$

where AR is the aspect ratio of the channel (i.e., the ratio of the minimum to the maximum dimensions). The Nusselt number in the wet channel, in which both heat and mass transfer processes are involved, can be calculated by the following Eq. (25).

$$Nu = 0.10\left(\frac{l_e}{l}\right)^{0.12} Re^{0.8} Pr^{1/3} \tag{23}$$

where l_e is the characteristic length, l_e = V/As and, V is the volume occupied by the wick material or water film. As is the wetted surface area of the channel, l is the total thickness of channel plate including the water film.

The convective mass transfer coefficient for the wet channel is approximated using heat and mass transfer analogy and can be expressed by Lewis number (**Table 2**).

$$\frac{h}{h_m} = \rho_a c_p Le^{\frac{2}{3}} \tag{24}$$

4. Numerical method and performance indices

4.1 Numerical method

The heat and mass transfer analysis for the decoupled cooling system has been carried out using the developed mathematical models. Variable-order

	Parameter	Values
DCHE	Height of heat exchanger (mm)	100
	Length of heat exchanger (mm)	25
	Width of heat exchanger (mm)	100
	Pitch of fins (mm)	1.8
	Thickness of the coating layer (mm)	0.1
DEC	Channel length (m)	1.2
	Channel width (m)	0.3
	Channel height (mm)	4
	Thickness of the separating plate(mm)	0.1
	Thickness of the water film(mm)	0.4

Table 2.
Design parameters of DCHE and DEC for numerical study.

method (ode15s) was employed for the transient simulation of DCHE and fourth-order method (bvp4c) for the steady-state simulation of DEC in MATLAB platform.

At first, simulation has been carried out for DCHE and examined the output states of the process air, namely the dry-bulb temperature and humidity ratio. The cyclic average values are then calculated and applied to the inlet conditions for DEC simulation. Simulation conditions were determined on the basis of IISO 5151:2017, which defines outdoor temperatures for three different climates, namely moderate climate, cool climates, and hot climates [27]. As shown in **Table 1**, moderate climate conditions (T_{db} = 35°C T_{wb} = 24°C) were considered as the baseline, and the cool climates are adopted for the lower limit condition and hot climates for the upper limit conditions.

4.2 Performance indices

The performance of DCHE was examined by analyzing two different metrics, namely moisture removal capacity (MRC) and latent cooling capacity (Q_L) and coefficient of performance (COP_{th}). MRC is defined as the total amount of water vapor that is absorbed by the desiccant during the dehumidification process, and Q_L is defined as the cooling power to remove the moisture. They can be calculated by the following equation:

$$MRC = \int_{t_0}^{t} \dot{dv}\, dt \tag{25}$$

$$dv = \dot{m}_{da}(\omega_{in} - \omega_{out}) \tag{26}$$

$$Q_l = H_v \times \dot{dv} \tag{27}$$

COP_{th} is defined as the ratio of MRC to the total amount heat that is consumed by the desiccant during the regeneration process and can be calculated by the following equation:

$$COP_{th} = \frac{Q_l}{Q_{reg}} \tag{28}$$

$$Q_{reg} = \dot{m}_{hw}\left(T_{hw,in} - T_{hw,out}\right) \tag{29}$$

For the evaluation of DEC, the temperature of product air ($T_{da,out}$), sensible cooling capacity (Q_S), wet-bulb effectiveness (ε_{wb}), and dew-point effectiveness (ε_{dp}) are analyzed judiciously, which can be calculated by the following formulas:

$$Q_S = (1-\gamma)\dot{m}_{da}\left(T_{da,in} - T_{da,out}\right) \tag{30}$$

$$\varepsilon_{wb} = \frac{T_{da,in} - T_{da,out}}{T_{da,in} - T_{wb}(T_{da,in})} \tag{31}$$

$$\varepsilon_{db} = \frac{T_{da,in} - T_{da,out}}{T_{da,in} - T_{db}(T_{da,in})} \tag{32}$$

5. Results and discussions

5.1 Validation

The developed numerical models for DCHE and DEC have been validated with our experimental results and the published experimental results from references [28]. To evaluate the accuracy of mathematical models, relative error percentage and root-mean-square error (RMSE) were used for the consistency between simulated results and experimental data, and they are expressed as [29]

$$\mathrm{E_r} = \left(\frac{X_{sim,} - X_{exp,i}}{X_{exp,i}}\right) \times 100 \tag{33}$$

$$\mathrm{RMSE} = \sqrt{\frac{\sum\left[\frac{X_{sim,i} - X_{exp,i}}{X_{exp,i}}\right]^2}{n}} \tag{34}$$

Figure 4 shows the comparison of the transient humidity variation with time during one cycle between the simulation results and the experimental results under the specific inlet air conditions of 32.6°C temperature with 14.0 g_v/kg_{da} humidity ratio. As shown in **Figure 4a**, the outlet humidity ratio obtained from the numerical analysis is quite similar to the experimental result. It is also observed from **Figure 4b** that the highest relative error is 8.6% with an RMSE of 0.0261.

Figure 5 shows the comparison of the results obtained from the numerical analysis with the experimental results obtained from reference [23]. It also shows a good agreement between the simulation and experimental data under different inlet temperatures. The maximum deviation of 3.2% is observed with an RMSE of 0.0159.

5.2 Performance of DCHE

The performance of the DCHE was numerically analyzed, and simulation results are judiciously compared through a series of runs under various operating conditions. The inlet air temperature is 35°C for both dehumidification and regeneration

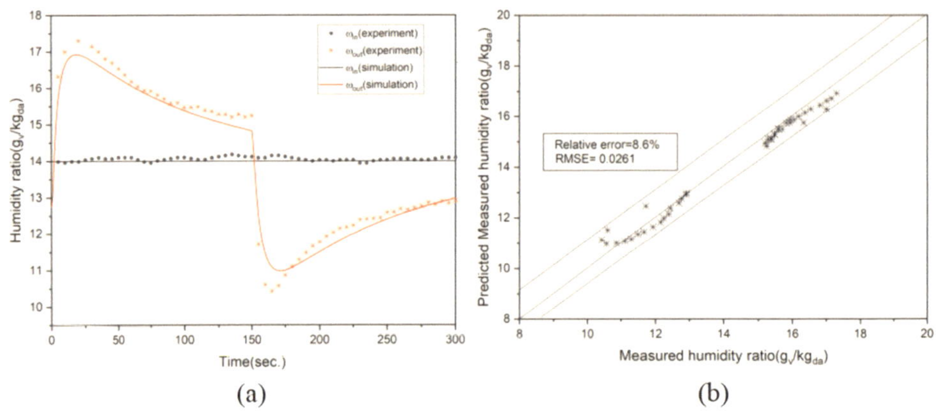

Figure 4.
Validation of the developed the mathematical model for DCHE: (a) transient outlet humidity ratio, (b) fitting of outlet humidity ratios in numerical and experimental results.

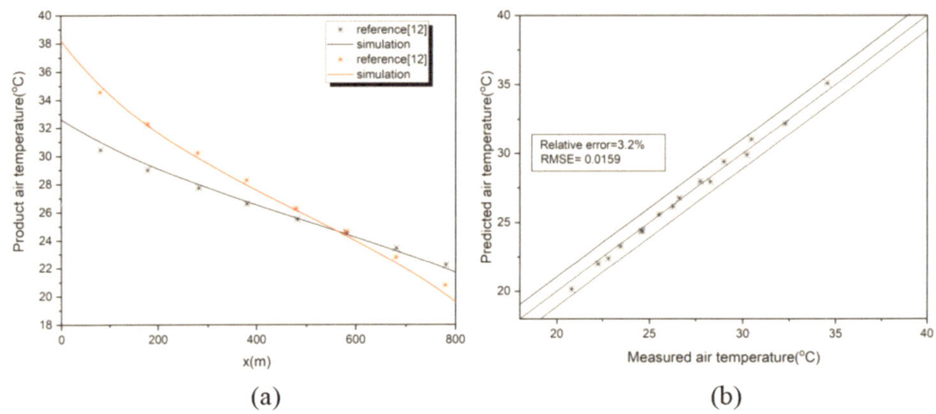

Figure 5.
Validation of the developed the mathematical model for DEC with experimental data from reference [28]: (a) temperature distributions along the channel under specific primary air conditions, (b) fitting of product air temperature in numerical and experimental results.

processes. The temperatures of cooling and hot water are set at 30°C and 50°C, respectively. During the entire process, the inlet humidity ratio is kept to 14 g/kg. **Figure 6** shows the dynamic variation of the performance of DCHE during three cycles under the baseline conditions where. It is easily seen from **Figure 6a** that the equilibrium uptake reaches its saturated state at 0.33 during the dehumidification process and at 0.12 during the regeneration process in a given condition. It is also seen that the actual uptake follows the trajectory of the equilibrium uptake at a different rate. It can also be observed from **Figure 6b** that the outlet humidity ratio reaches its equilibrium state at a faster rate during the regeneration process. However, it is noteworthy that the amount of the vapor that is released and adsorbed by the desiccant is conserved during the cycle so that the system continues to operate at a cyclic steady state. The lowest humidity ratio of 0.0092 kg_v/kg_{da} was observed at 16 s after the dehumidification process starts, and then it increases gradually up to its saturated level. It is also worthy to note that the desorption rate is a function of the adsorbent's surface temperature. Therefore, as the surface temperature rises when the supply of hot water flows through the heat exchanger, the more adsorbates (water vapor) are

DOI: http://dx.doi.org/10.5772/intechopen.105876

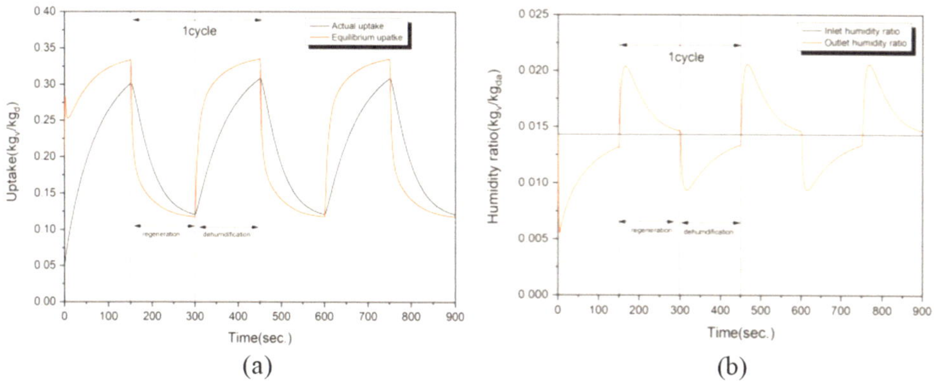

Figure 6.
Dynamic variations of the performance of the DCHE under the baseline conditions (T_{db} = 35°C, T_{wb} = 24°C, $\dot{m}_a$ =13 kg/h, T_{cw} = 30°C, T_{hw} = 50°C, t_{cycle} = 150 s): (a) the variations of instantaneous uptake and the equilibrium uptake; (b) the variation of the air humidity ratio at the outlet of DCHE.

released at faster rate. Hence, the regeneration time becomes shorter at higher surface temperature. This explains the reason for supplying the cooling water to the heat exchanger to enable a cooler adsorbent that is capable of adsorbing more moisture than the desorbing process.

Figure 7 shows the effect of change in the dry-bulb temperature of the process air on the performance of DCHE, namely moisture removal capacity (MRC), COP_{th}, latent cooling capacity (Q_l), and regeneration energy (Q_{reg}). The dry-bulb temperature varied from 27 to 46°C while the other parameters were kept constant as in the baseline conditions. It is observed that MRC, Q_l, and Q_{reg} are markedly affected by process air inlet dry-bulb temperature, whereas COP_{th} remains constantly around 0.125. MRC drops by 23.6% from 1.867 (T_{db} = 35°C) to 1.42 (T_{db} = 46°C). This is mainly because the humidity ratio decreases from 0.0142 to 0.0096, which implies the air carries less water vapor at a higher temperature when wet-bulb temperate is fixed at constant.

Figure 8 shows the effect of change in the wet-bulb temperature of the process air on the performance of DCHE. Unlike the previous results, the performance indices are raised when increasing the wet-bulb temperature except for COP_{th}. The reason for

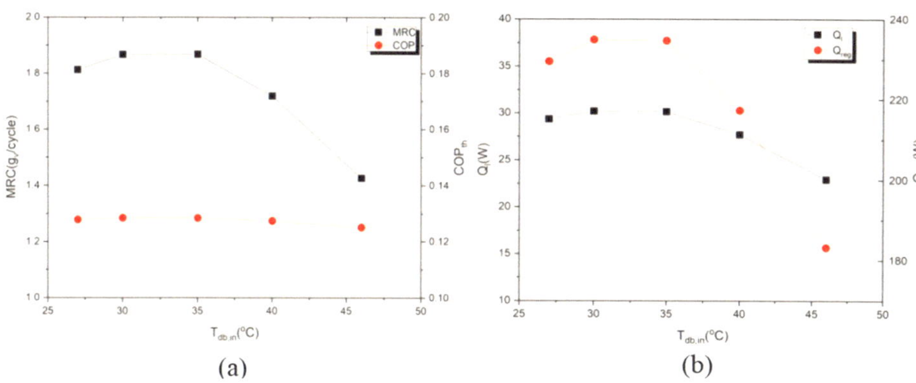

Figure 7.
Effect of change in the process air dry-bulb temperature on the performance of DCHE: (a) moisture removal capacity (MRC) and COP_{th}; (b) latent cooling capacity (Q_l) and regeneration energy (Q_{reg}).

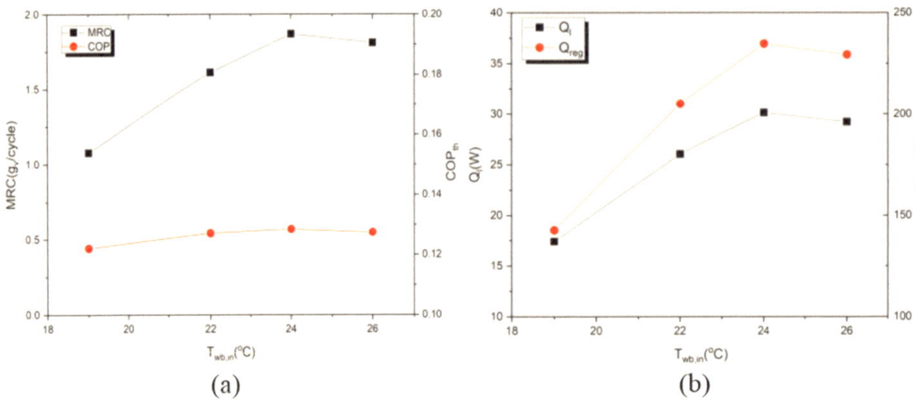

Figure 8.
Effect of change in the process air wet-bulb temperature on the performance of DCHE: (a) moisture removal capacity (MRC) and COP_{th}; (b) latent cooling capacity (Q_l) and regeneration energy (Q_{reg}).

this is that the water vapor partial pressure increases as the dry-bulb temperature rises when the wet-blub temperature is fixed. Hence, the equilibrium uptake of the desiccant increases, which is defined as the ratio of the vapor partial pressure to the saturation pressure at the desiccant temperature. As a result, more water vapors can be absorbed by the desiccant. The highest values for MRC, Q_l and Q_{reg} were 1.87 g/cycle and 30.1 W, respectively when T_{wb} = 24°C. It can be observed that MRC, Q_l, and Q_{reg} decrease slightly at T_{wb} = 26°C.

5.3 Performance of DEC

Figure 9 depicts the temperature and humidity ratio profiles of the product air and the working air for DEC under a specified operating condition. The product air flows along the channel length and the working air flow reversely in a counter-flow manner. It is observed from **Figure 9b** that the product air is purged at the end of the channel and is flowed back to the inlet. The product air temperature drops sensibly from 26°C to 16.18°C while the working air temperature increases from 16.8 to 24°C. It can be found that the temperature drops significantly within a distance of 0.1 m from the end

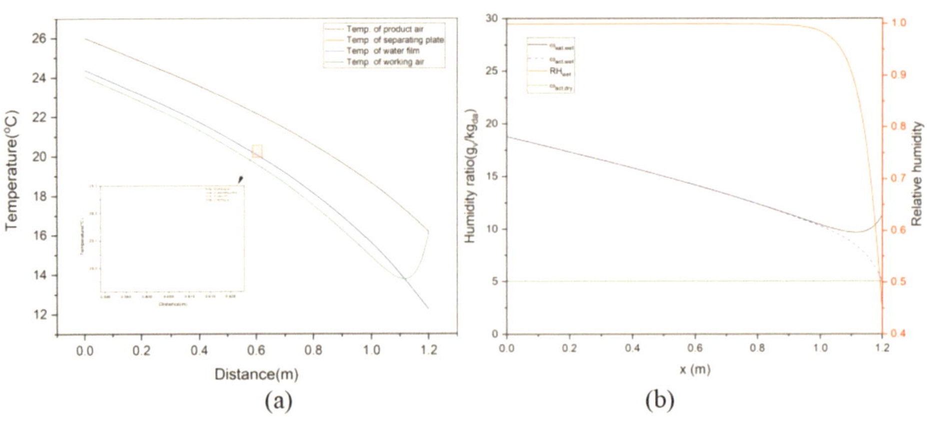

Figure 9.
Temperature profile (a) and humidity ratio profile (b) of each component in DEC.

of the channel immediately after being purged. This is because that instantaneous evaporation takes place from the surface of the water film by the continuously purged air. As depicted n **Figure 9b**, the relative humidity surges rapidly from 0.4 to 0.98 in the immediate vicinity of the exit. In addition, once the actual humidity ratio reaches the saturation level, both humidity ratios increase gradually, which allows for water to evaporate continuously along the channel length.

Figure 10 depicts the effect of change in the air temperature and humidity ratio on the performance of DEC while other parameters are kept constant as in the baseline conditions. The dry-bulb temperature varies from 26 to 38°C, and the humidity ratio varies from 0.005 kg_v/kg_{da} to 0.0187 kg_v/kg_{da} respectively. It can be found that the performance indices are almost linearly dependent on the air dry-bulb temperature and the humidity ratio. For the dew-point effectiveness, it varies between 0.44 and 0.57 when the dry-bulb temperature varies from 26 to 38°C. It can be inferred that the higher the air temperature is, the higher the dew-point effectiveness. It should be noted that the higher dew-point effectiveness and the wet-bulb effectiveness can be expected at a higher humidity ratio, whereas higher cooling capacity can be obtained at a lower humidity ratio. This is because both wet-bulb and dew-point temperatures increase in proportion to the humidity ratio when the dry-bulb temperature is fixed. Therefore, the potential for water evaporation is declined, and thus, the temperature drop in the dry channel gets closer to its maximum (i.e., 1).

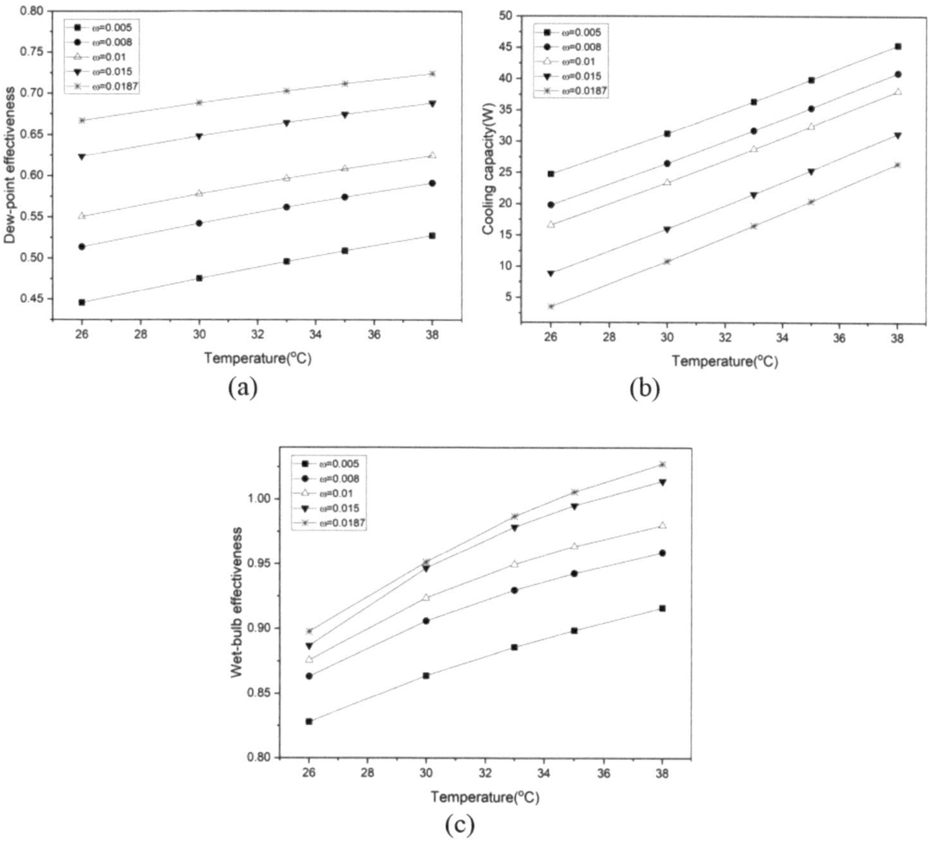

Figure 10.
Effect of change in the air temperature and humidity ratio on the performance of DEC: (a) dew-point effectiveness, (b) wet-bulb effectiveness, and (c) cooling capacity.

6. Conclusions

This study presents an innovative decoupling cooling technology where latent cooling load and sensible cooling load are handled separately by a desiccant-coated heat exchanger (DCHE)-based dehumidifier and a dew-point evaporative cooler (DEC). The DCHE first removes the undesired moisture of humid outdoor air by adsorption process. Subsequently, the DEC sensibly cools down the dehumidified air up to the desired temperature, maintaining the moisture level. Their performances are investigated numerically by analyzing the heat and mass transfer. Simulation has been carried out for DCHE and examined the output states of the process air, namely the dry-bulb temperature and humidity ratio. The following findings can be inferred from this study:

1. The equilibrium uptake reaches its saturated state at 0.33 during the dehumidification process and at 0.12 during the regeneration process at a given condition. The actual uptake follows the trajectory of the equilibrium uptake at a different rate.

2. MRC and Q_l of the DCHE were markedly affected by varying the dry-bulb temperature and wet-bulb temperature, whereas COP_{th} remained constant around 0.125. The MRC drop by 23.6% at higher temperature was observed due to the decreases in the humidity ratio.

3. As the wet-bulb temperature of the process air increases, the improved MRC and Q_l of the DCHE were observed due to the increases in the humidity ratio.

4. For the DEC, dew-point effectiveness, wet-bulb effectiveness, and cooling capacity were linearly dependent on the dry-bulb temperature of the process air as well as the humidity ratio.

5. The higher dew-point effectiveness and the wet-bulb effectiveness can be expected at a higher humidity ratio, whereas higher cooling capacity can be obtained at a lower humidity ratio.

The proposed decoupling system has no moving parts and harmful refrigerant, rendering less maintenance compared with an existing cooling system. Furthermore, it is an energy-efficient means of latent and sensible cooling by adsorption process and water evaporation process with a waste heat source as compared with other conventional air-conditioning processes.

Acknowledgements

This work was supported by the National Research Council of Science & Technology (NST) grant by the Korea government (MSIT) (No. CPS21171-110).

Nomenclature

c_p	specific heat at constant pressure, J/(kg K)
D_v	diffusion coefficient, m^2/s
h	heat transfer coefficient, $W/(m^2\ K)$

h_{fg}	latent heat evaporation, J/kg
h_m	mass transfer coefficient, m/s
$\bar{H}$	half height of channel, m
k	thermal conductivity, W/(m K)
L	channel length, m
$\dot{m}$	mass flow rate, kg/s
P	pressure, Pa
Q	heat transfer rate, W
r	working air ratio
T	temperature, °C
u	velocity, m/s
W	channel width, m
x	x-coordinate, m

Greek symbols

δ	thickness, mm
ρ	density, kg/m^3
ω	humidity ratio, kg/kg dry air

Subscripts

a	air
da	dry air
dp	dew point
f	water film
i	inlet
m	water vapor-air mixture
p	plate
pa	primary air
s	supply
sat	saturated
v	water vapor
w	wet air

Author details

Seung Jin Oh*, Yeongmin Kim, Yong-yoo Yang and Yoon Jung Ko
Sustainable Technology and Wellness R&D Group, Korea Institute of Industrial Technology, Jeju, Jeju Special Self-Governing Province, Korea

*Address all correspondence to: ohs8680@kitech.kr

References

[1] Shahzad MW, Burhan M, Ang L, Ng KC. Energy-water-environment nexus underpinning future desalination sustainability. Desalination. 2017; **413**:52-64. DOI: 10.1016/j.desal.2017.03.009

[2] Oh SJ, Ng KC, Chun W, Chua KJE. Evaluation of a dehumidifier with adsorbent coated heat exchangers for tropical climate operations. Energy. 2017;**137**:441-448. DOI: 10.1016/j.energy.2017.02.169

[3] Oh SJ, Ng KC, Thu K, Chun W, Chua KJE. Forecasting long-term electricity demand for cooling of Singapore's buildings incorporating an innovative air-conditioning technology. Energy and Buildings. 2016; **127**:183-193. DOI: 10.1016/j.enbuild.2016.05.073

[4] International Energy Agency (IEA). The Future of Cooling. 2018. p. 92

[5] Oh SJ, Choon Ng K, Thu K, Kum Ja M, Islam MR, Chun W, et al. Studying the performance of a dehumidifier with adsorbent coated heat exchangers for tropical climate operations. Built Environment. 2017;**23**:127-135

[6] Sunhor S, Adi D, Osaka Y, Tsujiguchi T, Kodama A. Dehumidification behavior of an aluminophosphate zeolite coated crossflow heat exchanger driven with direct hot water heating and evaporative cooling. Applied Thermal Engineering. 2022;**210**:118355

[7] Chai S, Sun X, Zhao Y, Dai Y. Experimental investigation on a fresh air dehumidi fi cation system using heat pump with desiccant coated heat exchanger. Energy. 2019;**171**:306-314. DOI: 10.1016/j.energy.2019.01.023

[8] Liang CP, Ture F, Dai YJ, Wang RZ, Ge TS. Experimental investigation on performance of desiccant coated microchannel heat exchangers under condensation conditions. Energy & Buildings. 2021;**231**

[9] Xu P, Ma X, Zhao X, Fancey K. Experimental investigation of a super performance dew point air cooler. Applied Energy. 2017;**203**:761-777. DOI: 10.1016/j.apenergy.2017.06.095

[10] Duan Z, Zhan C, Zhao X, Dong X. Experimental study of a counter-flow regenerative evaporative cooler. Building and Environment. 2016;**104**:47-58. DOI: 10.1016/j.buildenv.2016.04.029

[11] Hasani Balyani H, Sohani A, Sayyaadi H, Karami R. Acquiring the best cooling strategy based on thermal comfort and 3E analyses for small scale residential buildings at diverse climatic conditions. International Journal of Refrigeration. 2015;**57**:112-137. DOI: 10.1016/j.ijrefrig.2015.04.008

[12] Jafarian H, Sayyaadi H, Torabi F. Modèle numérique d'un refroidisseur évaporatif indirect à contre-courant à point de rosée avec une condition aux limites modifiées et prenant en considération les effets des zones d'entrée. International Journal of Refrigeration. 2017;**84**:36-51. DOI: 10.1016/j.ijrefrig.2017.09.003

[13] Sohani A, Sayyaadi H, Mohammadhosseini N. Comparative study of the conventional types of heat and mass exchangers to achieve the best design of dew point evaporative coolers at diverse climatic conditions. Energy Conversion and Management. 2018;**158**:327-345. DOI: 10.1016/j.enconman.2017.12.042

[14] Hyun M, Oh SJ. Numerical Study on Dehumidification Performance of a Desiccant Coated Heat Exchanger. n.d

[15] Hu LM, Ge TS, Jiang Y, Wang RZ. Performance study on composite desiccant material coated fin-tube heat exchangers. International Journal of Heat and Mass Transfer. 2015;**90**: 109-120. DOI: 10.1016/j.ijheat masstransfer.2015.06.033

[16] Ge TS, Dai YJ, Wang RZ, Li Y. Feasible study of a self-cooled solid desiccant cooling system based on desiccant coated heat exchanger. Applied Thermal Engineering. 2013;**58**: 281-290. DOI: 10.1016/j.applthermal eng.2013.04.059

[17] Tu YD, Wang RZ, Hua LJ, Ge TS, Cao BY. Desiccant-coated water-sorbing heat exchanger: Weakly-coupled heat and mass transfer. International Journal of Heat and Mass Transfer. 2017;**113**: 22-31. DOI: 10.1016/j.ijheatmasstransfer. 2017.05.047

[18] Li Z, Michiyuki S, Takeshi F. Experimental study on heat and mass transfer characteristics for a desiccant-coated fin-tube heat exchanger. International Journal of Heat and Mass Transfer. 2015;**89**:641-651. DOI: 10.1016/ j.ijheatmasstransfer.2015.05.095

[19] Zheng X, Wang RZ, Ge TS, Hu LM. Performance study of SAPO-34 and FAPO-34 desiccants for desiccant coated heat exchanger systems. Energy. 2015; **93**:88-94. DOI: 10.1016/j.energy.2015. 09.024

[20] Panigrahi B, Chen YS, Luo WJ, Wang HW. Dehumidification effect of polymeric superabsorbent SAP-LiCl composite desiccant-coated heat exchanger with different cyclic switching time. Sustain. 2020;**12**:1-16. DOI: 10.3390/su12229673

[21] Lin J, Thu K, Bui TD, Wang RZ, Ng KC, Chua KJ. Study on dew point evaporative cooling system with counter-flow configuration. Energy Conversion and Management. 2016;**109**: 153-165. DOI: 10.1016/j.enconman. 2015.11.059

[22] Riangvilaikul B, Kumar S. An experimental study of a novel dew point evaporative cooling system. Energy and Buildings. 2010;**42**:637-644. DOI: 10.1016/j.enbuild.2009.10.034

[23] Lin J, Bui DT, Wang R, Chua KJ. On the fundamental heat and mass transfer analysis of the counter-flow dew point evaporative cooler. Applied Energy. 2018;**217**:126-142. DOI: 10.1016/ j.apenergy.2018.02.120

[24] Jafarian H, Sayyaadi H, Torabi F. Modeling and optimization of dew-point evaporative coolers based on a developed GMDH-type neural network. Energy Conversion and Management. 2017;**143**: 49-65. DOI: 10.1016/j.enconman.2017. 03.015

[25] De Antonellis S, Joppolo CM, Molinaroli L. Simulation, performance analysis and optimization of desiccant wheels. Energy and Buildings. 2010;**42**: 1386-1393. DOI: 10.1016/j.enbuild. 2010.03.007

[26] Dowdy A, NSK PEPD. Experimental determination of heat and mass transfer coefficients in rigid impregnated cellulose evaporative media. ASHRAE Transactions. 1987

[27] 22095:2020 I. International standard iteh standard preview iTeh Standard Preview 2020. 2020

[28] Lin J, Bui DT, Wang R, Chua KJ. On the fundamental heat and mass transfer analysis of the counter-flow dew point evaporative cooler. Appl Energy 2018;

217:126-142. DOI: 10.1016/j.apenergy.2018.02.120

[29] Chai T, Draxler RR, Prediction C. Root mean square error (RMSE) or mean absolute error (MAE)? – Arguments against avoiding RMSE in the literature. 2014:1247-1250. DOI: 10.5194/gmd-7-1247-2014

Section 2

Different Forms of Heat Transfer and Applications in Various Aspects

Chapter 5

Dropwise Condensation and Heat Transfer

R. Yuvaraj

Abstract

The dropwise condensation is obtained on a copper surface by modifying the texture of the bare surface using the thermo-solution immersion method. In this method, the solution of 0.003–0.007 M of ethanol and myristic acid is used, and heating the plate in the solution at 40–65°C for 2–5 h using hot plate apparatus. The heat-transfer coefficient of the dropwise condensation is increased on the prepared superhydrophobic surface that exhibits very low surface energy causing the non-wetting nature of the water droplet on the prepared surface. The contact angle of the water droplet is measured on the obtained superhydrophobic copper surface, giving the average value of 160° ± 2° with a low-inclination angle of 2°. The maximum contact angle of 162° is obtained by adjusting the composition of the solution, the temperature of the solution, and immersion time at 0.005 M, 45°, and 3 h, respectively. Further, the prepared superhydrophobic surface is experimented with for dropwise condensation, which provides a high heat-transfer coefficient of 196 W/m^2 K over the bare surface providing around 186 W/m^2 K. The condensation rate of water droplet fall-off time is about 1 s on the superhydrophobic surface, and 2 s for bare surface is obtained against the mass flow rate of 300 lph.

Keywords: superhydrophobic, copper surface, contact angle, dropwise condensation, heat transfer

1. Introduction

The water vapour normally condenses on a bare surface by film-wise mode of condensation. The larger surface energy of the bare surface results in higher wetting nature of the water on the bare surface and forms a water film throughout the bare surface. This water film causes higher resistance to heat flow across the film. To overcome this, the surface texture is modified to offer lower surface energy, resulting in the non-wetting nature of water on the modified surface and forming many water droplets on the surface to reduce the resistance to heat flow across the surface. For the same temperature difference between the vapour and the surface, dropwise condensation is much more effective than film-wise condensation about more than times. In dropwise condensation, the water droplets initially formed on the nucleation sites and grow up by condensation and coalescence of nearer droplets and then flow downwards, accumulating static droplets below them along the way. By increasing the contact angle of the water droplet, the heat flow resistance between the substrate and the vapour decreases, the heat-transfer coefficient increases, and the condensation rate also increases.

In dropwise condensation, the heat-transfer coefficient is several times larger than that of film-wise condensation. To obtain dropwise condensation, the texture of the surface needs to be modified to offer lower surface energy. This can be implemented by introducing a non-wetting chemical into the vapour, by special physical treatment of the condensation surface/substrate, or by the chemical coating of the solid substrate with a low-surface-energy substance to promote dropwise condensation [1–3]. Among the many processes of promoting dropwise condensation, the chemical method offers superhydrophobic nature on the given surface with a simple process involved. Marto et al. [4] tested several polymer coatings, gold, and silver for sustaining dropwise condensation of steam and reported the heat-transfer coefficients in dropwise condensation as high as six times when compared with film-wise condensation. Zhao et al. [5] reported that the heat-transfer coefficients of dropwise condensation on Langmuir-Blodgett treated surfaces are more than 30 times higher than that of film-wise condensation on bare surfaces. Vemuri et al. [6] experimentally investigated the effects of various chemical coatings and their long-term durability on the dropwise mode of heat transfer. They reported a decrease in heat-transfer coefficient with the elapsed condensation time, suggesting possible leaching of the chemical coating. Rausch et al. [7] reported that the heat-transfer coefficient on an ion-implantation surface is more than five times that of film-wise condensation. In recent years, with the advent of newer coating/manufacturing and nanoscale fabrication techniques, promoting the long-term sustainability of dropwise condensation by chemical coating now holds the considerable prospect for enhancing heat transfer in a variety of industries [2, 6, 8]. An example of enhanced performance of compact steam condensers having chemically coated flow passages of only a few millimetres width is demonstrated by Majumdar and Mezic [9].

In this work, the superhydrophobic copper surface is prepared in a quick time of 2–5 h by using the thermo-solution immersion technique. The maximum contact angle of 162° is obtained by adjusting the composition of the solution, the temperature of the solution, and immersion time at 0.005 M, 45°, and 3 h, respectively. The prepared surface is then experimented with for analysing the nature of dropwise condensation, temperature variation on the substrate, heat-transfer coefficient variation, and condensation time variation. Also, the condensation on the prepared surface is compared with that of the bare copper substrate for the same conditions.

2. Surface preparation

The various methods for the preparation of superhydrophobic surfaces, as shown in **Figure 1**, have been reported, such as electrochemical deposition, plasma method, crystallisation control, chemical vapour deposition (CVD), wet chemical reaction, sol-gel processing, lithography, and solution immersion processes.

Wang et al. [10] conducted an experiment to produce superhydrophobicity on the copper plate using a solution immersion process in which a copper plate was immersed in an ethanol solution of n-tetradecanoic acid (0.01 M) at room temperature for 3–5 days. The immersed copper plate was rinsed with deionised water and ethanol thoroughly and then dried in air. The prepared films were immersed in common solvents (water, acetone, ethanol, and toluene) for 5 days at 25°C. The same

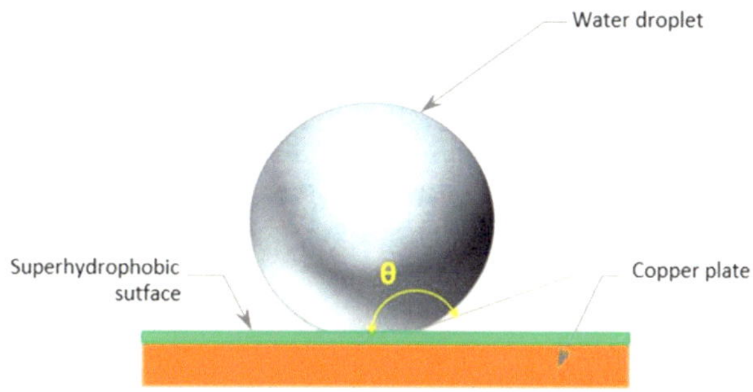

Figure 1.
Water droplet on a superhydrophobic surface.

treatment was also carried out using hot water at 80°C. The contact angle was measured for the dried films. After immersion, the contact angle of the droplet slightly decreases in water, acetone, ethanol, and hot water. The contact angle is slightly increased in solution immersion with toluene. Their experiments suggest that a concentration of around 0.01 M is ideal for the formation of stable flower-like clusters. When the concentration is about 0.001 M, a self-assembled monolayer of tetradecanoic acid is formed on the copper substrate and the contact angle is only about 124°. At above 0.02 M, the microclusters or nanosheets prefer to form in solution rather than on the copper plates. Thus, the concentration of tetradecanoic acid is crucial for the formation of flower-like clusters. They concluded with a concentration of 0.01 M that gives a high contact angle of about 160°. The prepared superhydrophobic surface reliability is checked by immersing the substrate in water, acetone, ethanol, and toluene for 5 days, its superhydrophobic nature remains unaffected.

The drawback of their work is it requires a high concentration of 0.01 M and 5 days to form a superhydrophobic surface on a copper plate compared with our work, which required only 0.003–0.007 M low concentration and less time of around 2–5 h. After immersing the superhydrophobic surface in water, acetone, ethanol, and toluene for 5 days, we obtained an unchanged contact angle of about 162°.

In this work, the superhydrophobic surface is obtained on a copper plate by using the thermo-solution immersion process. **Figure 2** shows the process diagram of preparing a superhydrophobic surface on a copper plate. At first, the plate is cleaned to remove the impurities on the copper plate using an emery sheet. The plate is immersed in acetone for an ultrasonic bath for about 15 min. The plate is washed with deionised water and dried with air. Then, the solution is prepared for the different compositions and the concentration varies from 0.003 to 0.007 M of ethanol and myristic acid. The cleaned copper plate is then immersed in the solution, and the plate within the solution is heated by a hot plate for different temperatures varying from 40° to 65° with atmospheric pressure. After heating, the plate is taken out from the hot solution and cleaned with deionised water and dry air; the self-assembled monolayer is formed on the copper plate that produces superhydrophobic nature.

The copper plate of the required size (25 × 25 × 3 mm) is first rubbed with a rough emery sheet and later with a fine emery sheet. Then, the plate is kept in an ultrasonic bath with acetone for 10–15 min. After being rinsed with deionised water, the plate is dipped in hydrochloric acid for 2–3 min. After being rinsed with deionised water and dried with air, the plate is immersed in a solution of 0.003–0.007 M of ethanol and myristic acid. The solution immersed plate is heated in a hot plate at 40–65°C for 2–5

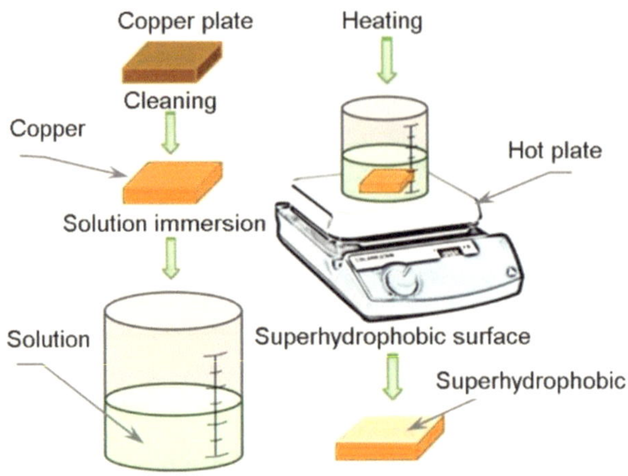

Figure 2.
Surface preparation processes.

h and rinsed with deionised water and ethanol. The superhydrophobic layer is formed on the surface that produces the average contact angle of about 160° ± 2°.

Figure 3 shows scanning electron microscopy (SEM) images of the prepared superhydrophobic copper surface. In **Figure 3(a)**, the SEM image is obtained with ×500 and 50 μm magnification, which shows the ribbon clusters are formed and randomly spread over the copper substrate. In **Figure 3(b)**, the SEM image is obtained with ×10,000 and 1 μm magnification, which shows clusters of a copper compound are formed on the substrate with an air gap between the ribbon clusters. These air gaps will play a vital role to promote the superhydrophobic nature of water droplets on the prepared surface. When the water droplet is dropped on the prepared surface, the water droplet traps the air between the ribbon clusters and reduces the surface energy, which causes an increase in the contact angle of the water droplet. When the contact angle approaches more than 90°, the hydrophobic nature of the prepared surface is formed, and for the contact angle of more than 150°, a superhydrophobic nature of the prepared surface is formed.

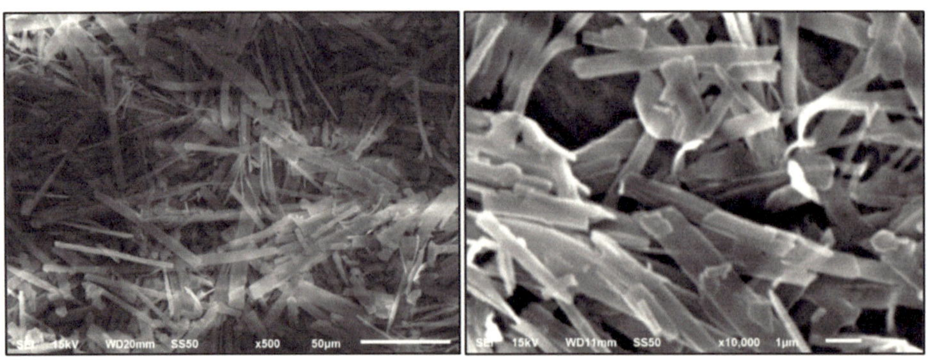

(a) (b)

Figure 3.
(a) SEM image with 50 μm magnification and (b) 1 μm magnification.

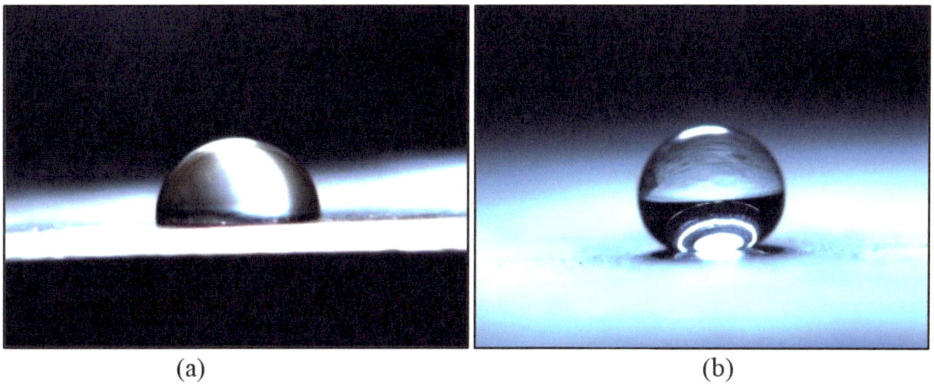

Figure 4.
(a) Droplet on the bare copper surface and (b) droplet on a superhydrophobic surface.

We have carefully monitored this solution immersion process. After immersion into the solution and heating the plate around 45°, a few copper carboxylate nanosheets and small clusters self-assembled from these nanosheets. Also, the surface is very sparsely covered with varying temperatures. Upon increasing the heating time to 60 min, the copper carboxylate nanosheets and self-assembled clusters grow bigger and longer. A further increase in the heating time and temperature leads to an increased surface density of the clusters and nanosheets, and the nanosheets start to grow upwards. For heating times close to 3 h, the ribbon-like clusters grow much bigger, become continuous, and almost completely cover the copper surface, as shown in **Figure 3**. Thus, an interesting continuous coating of ribbon clusters is formed on the copper surface. The ribbon clusters are the morphology that tends to provide the superhydrophobic behaviour.

Figure 4(a) shows the water droplet on the bare copper surface, and **Figure 4(b)** shows the water droplet on the prepared copper surface. On the bare surface, the morphology is not allowed to increase the contact angle due to higher surface energy and exhibits the Wenzel model of a water droplet with a contact angle <90°. Whereas on the prepared copper substrate, the ribbon clusters offer the water droplet to exhibit the Cassie Baxter model for water droplet and offer a superhydrophobic surface with a contact angle >150° due to low surface energy. Further, the contact angle varies for different concentrations, temperatures, and times. The best value obtained is 162° with a concentration of 0.005 M at 45°C for 3 h of heating.

3. Experimental setup

The experimental apparatus was designed and developed to study dropwise condensation under controlled conditions underneath a copper surface shown in **Figure 5**. The setup primarily consisted of the main cylindrical stainless steel vacuum chamber of 180 mm inner diameter and 120 mm length. It was closed from the two ends by specially designed flanges. The lower flange was fitted with an optical viewing window.

In addition, it also had an annular space around this viewing window wherein the working fluid inventory of distilled and de-ionised water was stored. An electric heater is attached at the bottom of the vacuum chamber to heat the water and generate steam inside the vacuum chamber. **Figure 6** shows the cut sectional view of the

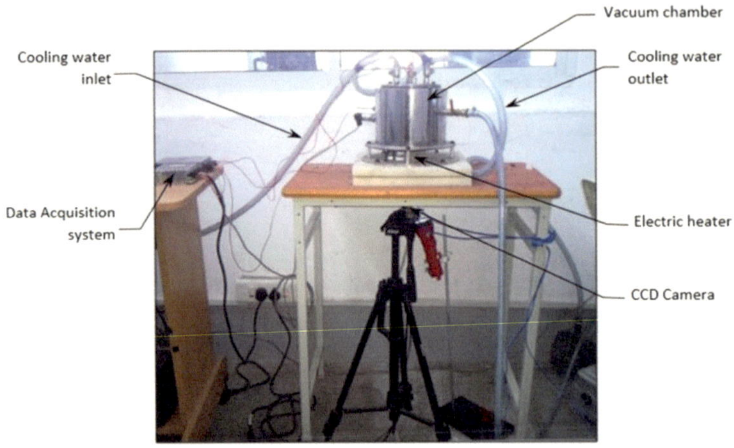

Figure 5.
Experimental setup.

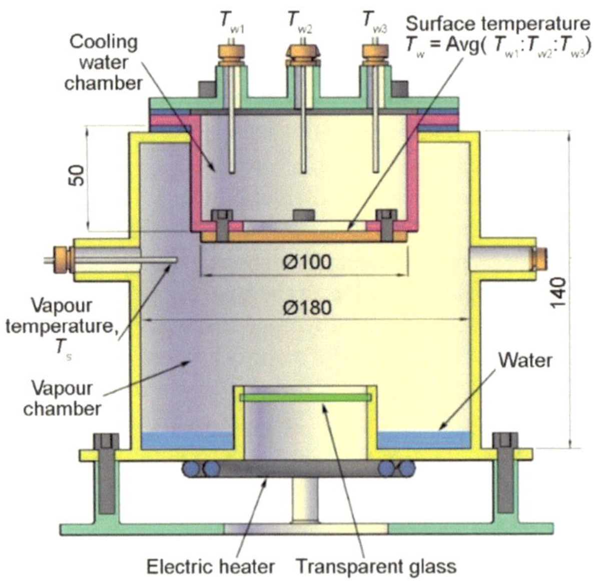

Figure 6.
Condensing chamber cut sectional view.

condensation chamber with a circular heater o.d = 70 mm, i.d = 40 mm was attached outside this annular space to give the necessary heat input. The upper end of the main vacuum chamber was closed with a circular flange with an inbuilt cavity wherein cold water was circulated to maintain constant temperature boundary conditions. Connections for evacuation, pressure transducer, and temperature sensors were provided on the main condensing chamber wall. The temperature of the condensing vapour was measured with one K-type thermocouple Omega, 0.5 mm diameter of the accuracy of 0.2°C after calibration. It was placed centrally in the chamber at a distance of 25 mm from its sidewall. Also, three different thermocouples are placed inside the cooling water cavity to measure the cooling water temperature. The average of these three thermocouple values is treated as the substrate temperature on which the condensation is going to occur. The vacuum is obtained by connecting the vacuum chamber to a vacuum pump, and the vacuum is maintained to improve the condensation process. The data acquisition was carried out with National Instruments. The entire assembly could be tilted to any desired inclination at 0–90 deg. This setup is suitable for the inclined condition after the water is converted to steam by vertical position. After the generation of steam at a vertical position, the setup can be tilted to the required angle to analyse the behaviour of dropwise condensation on the required inclined superhydrophobic surfaces. A colour charge-coupled device CCD video camera Basler with 1024 × 1024 pixels at 100 fps was used to capture the images of the droplets forming on the underside of the superhydrophobic copper plate through StreamPix software. StreamPix software can be used for creating both image and video formats and further converted into required image formats. The LED light source placed under the set of the camera was directed on the substrate from the optical window on the bottom flange to maintain a near-parallel and symmetric beam on the droplets, ensuring a proper contrast level for subsequent edge detection. The volume flow rate of the cooling water and the steam temperature inside the vacuum chamber is modified with different conditions to analyse the dropwise condensation process underneath the prepared copper substrate.

4. Experiment and results

The experiment is conducted on both bare and superhydrophobic surfaces for varying the properties such as vapour temperature (T_s), surface temperature (T_w), power input (I), and cooling water flow rate (lph). The maximum input power that can be set by the setup is 100 W by regulating with the least of 1 W. The cooling water flow can be adjusted up to 500 lph. In this work, three different water flow rates are used for conducting the experiments on both the bare surface and the superhydrophobic surface. The temperatures of the vapour and surface are acquired from the data acquisition system connected with the setup to the computer through LabVIEW software. The surface temperature and the vapour temperature are obtained from the initial state to the steady state by varying the flow rate to 100 lph, 200 lph, and 300 lph, respectively. The images obtained are stored and videos are captured by the camera connected to another computer through StreamPix software with a 100-fps rate. The images shown in **Figure 7** are the bottom view and the droplets are falling downward underneath the bare and the prepared superhydrophobic surface, respectively.

Figure 7(a) shows the condensation of water vapour on the bare copper plate with irregular water particles underneath the horizontal copper plate. The wetting nature of the bare copper surface results in water film formation on the bare substrate, which reduces the contact surface area between the vapour and the substrate. This film increases the resistance to heat flow between the vapour and the copper surface. Also, the fall-off diameter of the droplet is high in bare copper surface condensation. This causes further resistance to heat transfer, a low-heat-transfer coefficient, and a low-condensation rate of water on the bare surface. **Figure 7(b)** shows the dropwise condensation on the prepared superhydrophobic surface with regular complete spheres-like droplets that increase the contact surface area between the vapour and the plate surface. The contact angle of the water droplet on the prepared surface is more than 150°, which offers dropwise condensation on the prepared

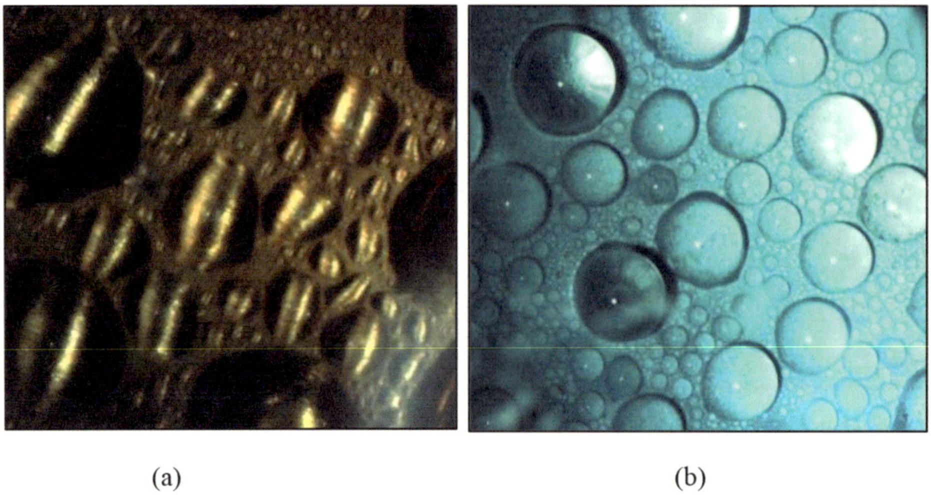

Figure 7.
(a) Condensation on the bare copper plate and (b) condensation on the superhydrophobic surface.

superhydrophobic surface. The water droplets are formed, which results in a decrease in resistance to the flow of heat and the fall-off diameter of the droplet from the superhydrophobic surface. A higher contact angle produces a lower diameter of the falling water droplet, which increases the fall-off frequency of the water droplets. This causes an increase in heat-transfer coefficient and rate of condensation.

4.1 Contact angle

The water contact angle is the angle between the liquid-solid contact surface and the tangential line obtained from the liquid-vapour interface. Wenzel model and Cassie Baxter model are the familiar contact angle model in which Cassie Baxter contact angle model is used in this work for dropwise condensation. In this model, the air

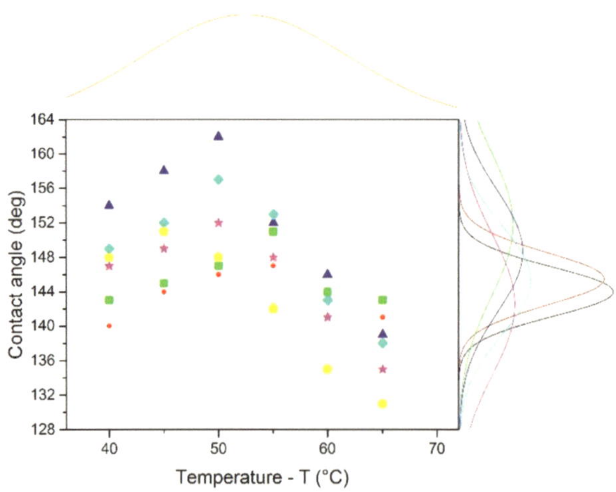

Figure 8.
Contact angle vs. temperature.

is trapped in between the liquid and the solid surface, which offers superhydrophobic nature of the prepared surface with a contact angle of 120–162°. The contact angles obtained for the different temperatures vary from 40° to 65° and times are plotted in the distribution graph shown in **Figure 8**, which reveals that the maximum contact angle of 162° against the temperature of around 50°C for 3 h. The large number of contact angles obtained is in the region 140–152 for the time durations of 2–3 h. The higher contact angles are also obtained in the region of 45–55°. The average contact angle on the prepared surface is obtained as 160°, which is greater than 150° and offers superhydrophobic nature on the prepared surface.

4.2 Vapour and surface temperature comparison

First, the experiment is conducted on a bare copper plate. The vacuum pump is used to obtain vacuum inside the vacuum chamber and the vacuum pump connection is closed after 15 min. Water is poured inside the vacuum chamber and the bare surface is located underneath the cooling water chamber. The power is given to the heater inside the vacuum chamber and regulated with different inputs from 30 to 100 W with a difference of 5 W. The cooling water flow is controlled for three conditions, 100 lph, 200 lph, and 300 lph, respectively. The temperatures are observed for different power inputs for each flow rate. The values of surface temperature T_w and vapour temperature T_s are obtained through the data acquisition system through LabVIEW software. The frequency of water droplet falling is obtained for finding the condensation rate. The experiment is repeated for superhydrophobic copper substrate with the same conditions and obtained the values of surface temperature, vapour temperature, and condensation rate. The heat-transfer coefficient is calculated by Newton's law of cooling and also compared all the parameters for the bare plate with a superhydrophobic surface.

Figure 9 shows the comparison of vapour temperature T_s with the corresponding power input I for the bare and superhydrophobic surfaces for different water flow rates. In both, cases 100 lph conditions give the highest vapour temperature due to the low mass flow rate. Increasing the water flow rate causes decreases in the vapour temperature due to higher heat transfer from the water vapour to the cooling water. **Figure 10** shows the comparison of surface temperature T_w with the corresponding

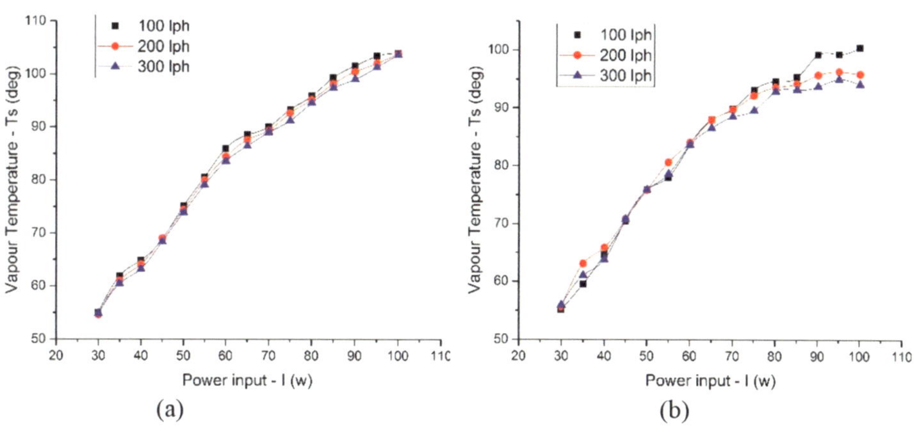

Figure 9.
(a) Vapour temperature variation with bare surface and (b) with the superhydrophobic surface.

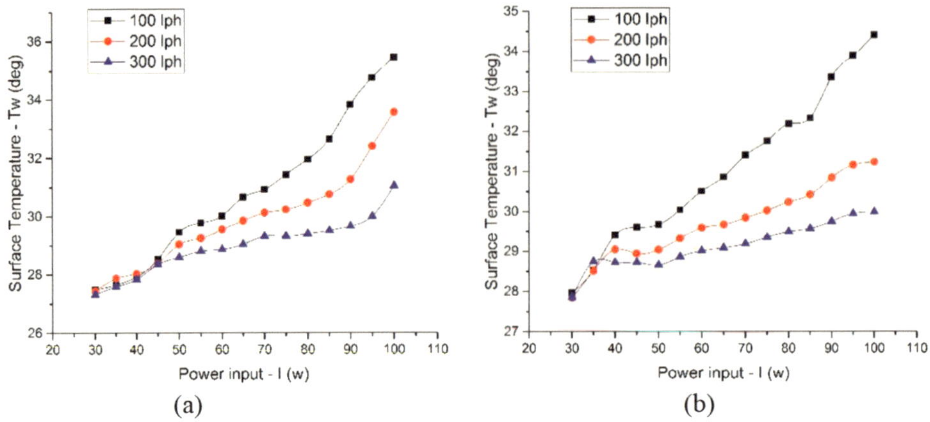

Figure 10.
(a) Surface temperature variation with bare surface and (b) with the superhydrophobic surface.

power input I for the bare and superhydrophobic surface for different water flow rates in which the surface temperature is low with 300 lph of water flow and it is high with 100 lph of water flow rate. Similar to that of vapour temperature, the surface temperature also decreases when the cooling water flow rate is increased. The lower surface is obtained for 300 lph of flow rate for both the bare and the prepared superhydrophobic surfaces.

The difference in temperature $\Delta T = T_w - T_s$ is higher on the bare plate than on the superhydrophobic surface, which increases the heat-transfer coefficient and condensation rate. The temperature difference directly affects the condensation of water particles from the water vapour onto the condensed droplets. The resistance offered by the extra coating layer of the superhydrophobic surface is increased and heat flow from the vapour to the plate is decreased. Due to this, the temperature of the plate is always higher on the superhydrophobic surface than on the bare plate.

Figure 11 shows the comparison of heat-transfer coefficient h obtained through Newton's law of cooling with corresponding heat input I. For bare surface, the

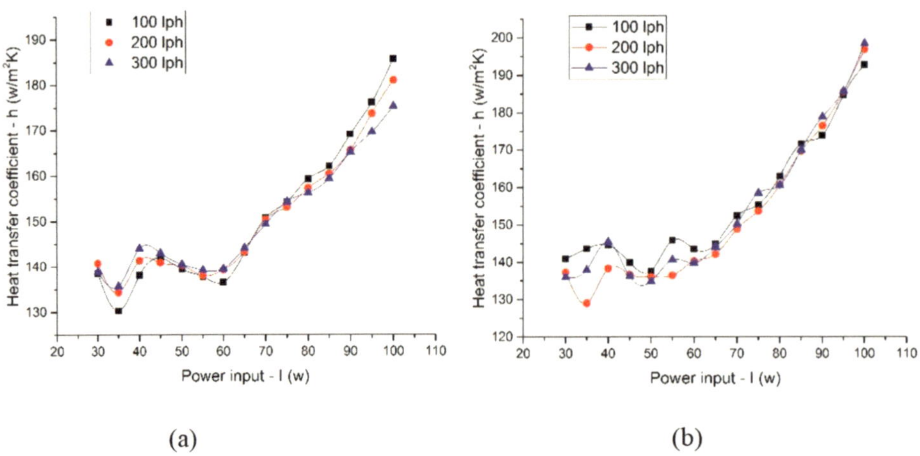

Figure 11.
(a) Heat-transfer coefficient variation with bare surface and (b) with the superhydrophobic surface.

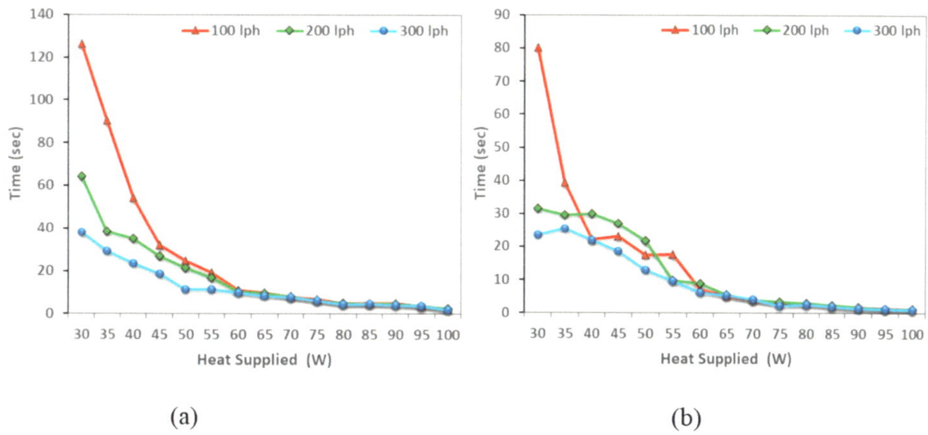

Figure 12.
Condensed droplet falling time on (a) bare surface and (b) superhydrophobic surface.

higher amount of heat-transfer coefficient obtained as 186 W/m^2 K obtained for the cooling water flow rate of 100 lph at the maximum heat input of 100 W. Also, the increase in cooling water flow rate decreases the heat-transfer coefficient, as shown in **Figure 11(a)**. Whereas on a superhydrophobic surface, the maximum heat-transfer coefficient is obtained as 196 W/m^2 K with a cooling water flow rate of 300 lph. In this case, an increase in the cooling water flow rate increases the heat-transfer coefficient, as shown in **Figure 11(b)**. The condensation rate is also higher on the superhydrophobic surface.

Figure 12(a) and **(b)** shows the condensed water droplet falling time on the superhydrophobic and bare surface for different water flow rates of 100 lph, 200 lph, and 300 lph, respectively. It shows that about 120 s is required for the first condensed fall-off water droplet from the bare surface, whereas it takes about 80 s in the case of the prepared superhydrophobic surface. The condensed water droplet fall-off time is about 1 s on the superhydrophobic surface and 2 s on the bare surface against the mass flow rate of 300 lph. It is noted that the condensing rate of falling water droplets is higher on a superhydrophobic surface than on a bare surface.

4.3 Ordinary surface and superhydrophobic surface comparison

In the previous chapter, the surface temperature, vapour temperature, heat-transfer coefficient, and condensation rate are discussed for three different cooling water flow rates, 100 lph, 200 lph, and 300 lph, respectively. The higher heat-transfer coefficient and good condensation rate are obtained on the superhydrophobic surface for the cooling water flow rate of 300 lph. To compare the surface temperature, vapour temperature, heat-transfer coefficient, and condensation rate directly between the ordinary surface and superhydrophobic surface, 400 lph of cooling water flow rate is used and discussed in this chapter. On comparing 300 lph results with 400 lph cooling water flow rate, there is a negligible amount of variation obtained for steam temperature and the surface temperature. Whereas for heat-transfer coefficient, a slight increase in variation is obtained, and for condensation rate, a notable variation is obtained between an ordinary surface and the superhydrophobic surface.

The water vapour temperature, surface temperature, heat-transfer coefficient, and condensation rate are measured for different heat supplies varying from 30 W to

100 W, and the comparisons between ordinary surface and superhydrophobic surface are shown in **Figures 13–16**, respectively. The temperature of the water vapour T_s inside the vacuum chamber is decreased on the superhydrophobic surface compared to that of the ordinary surface when the heat supplied is increased due to thermal resistance offered by the coating on the superhydrophobic surface, as shown in **Figure 13**. In both cases, the vapour temperature varies similarly when the heat supplied is between 40 and 75 W. Further increase in temperature decreases the vapour temperature on the superhydrophobic surface and increases the vapour temperature on the ordinary surface. **Figure 14** shows the comparison of surface temperature T_w with the corresponding power input I varying from 30 to 100 W for the ordinary and superhydrophobic surface for a water flow rate of 400 lph in which the surface temperature of the ordinary surface is always lesser than that of the superhydrophobic surface. This is because there is smooth heat flow across the ordinary surface with lower thermal resistance. Whereas for a superhydrophobic surface, the thin coating offers higher resistance to heat flow across the superhydrophobic coating, resulting in lower heat transfer through the substrate. Similar to that of vapour temperature, the surface temperature also decreases when the cooling water flow rate is increased. A very negligible amount of surface temperature is obtained for both 300 lph and 400 lph of flow rate for both the bare and the prepared superhydrophobic surfaces.

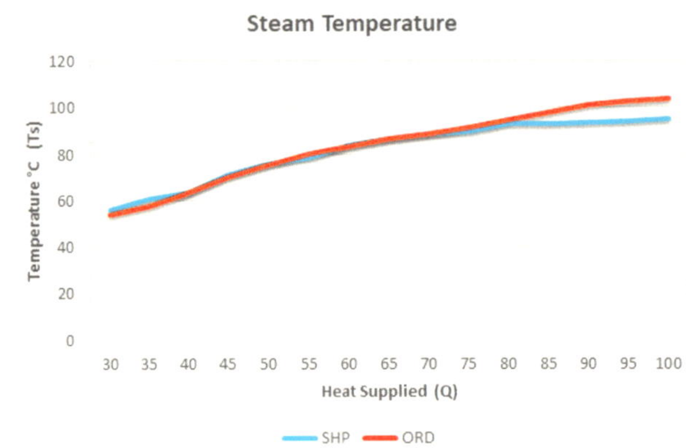

Figure 13.
Steam temperature of ordinary vs. superhydrophobic surfaces.

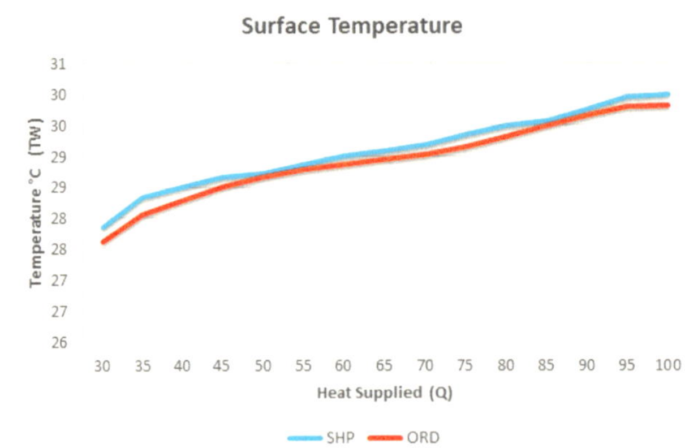

Figure 14.
Surface temperature of ordinary vs. superhydrophobic surfaces.

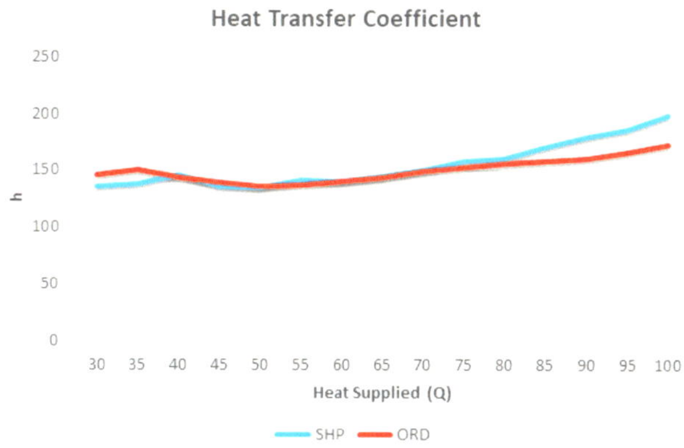

Figure 15.
Heat-transfer coefficient of ordinary vs. superhydrophobic surfaces.

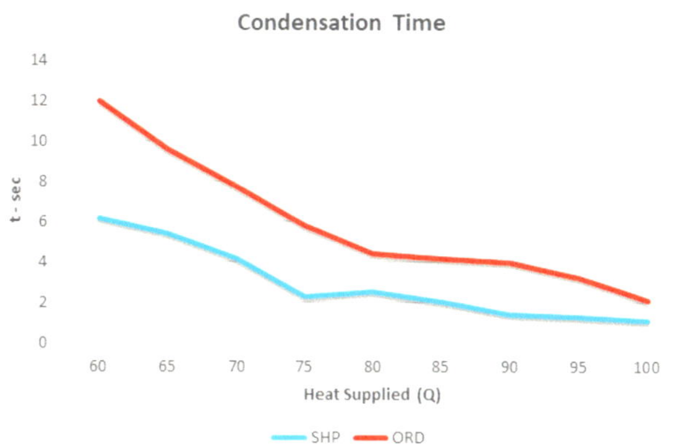

Figure 16.
Condensation time for drop fall off on ordinary vs. superhydrophobic surfaces.

The heat-transfer coefficient is obtained from Newton's law of cooling for both ordinary and superhydrophobic surfaces. The superhydrophobic surface provides a higher heat-transfer coefficient than the ordinary surface when the heat supplied increases from 70 W. This is due to the higher contact angle of the water droplets on the superhydrophobic surfaces. Higher contact water angle increases the surface area contact between the vapour and the solid surface. Whereas in the ordinary surface, the film-wise condensation offers a decrease in direct surface contact between the vapour and the substrate and provides higher thermal resistance across the film, and decreases the heat-transfer coefficient. When the heat supplied is lesser than 40 W, the formation of condensed particles on the ordinary surface is lesser and offers an increase in heat-transfer coefficient. Whereas on a superhydrophobic surface, the nucleation and coalescence of tiny water particles offer more thermal resistance and decrease the heat-transfer coefficient across the substrate.

When the vapour is directly in contact with a solid surface with a temperature lower than that of vapour temperature, the condensation of vapour starts on the solid surface. This offers film-wise condensation on an ordinary surface and dropwise condensation on the superhydrophobic surface. Although the superhydrophobic coating offers higher thermal resistance than that of an ordinary surface, the condensation rate on the superhydrophobic surface increases due to a lower fall-off diameter of the water droplets. **Figure 16** shows the condensation time, that is, water droplet fall-off time, for both ordinary surface and superhydrophobic surface with different heat supplied. When the heat supplied is less than 60 W, the formation of the water droplet on the superhydrophobic surface and condensation of film on the ordinary surface are not occurring at a faster rate. When the heat supplied is increased from 60 W, the condensation of vapour occurs on the ordinary surface as well as the superhydrophobic surface. The first fall-off water droplet takes 6 s on a superhydrophobic surface and 12 s on an ordinary surface. This time is lower than the ordinary surface due to the higher contact angle of the water droplet on the superhydrophobic surface. In both cases, the superhydrophobic surface offers a lower fall-off time and higher condensation rate than the ordinary surface, as shown in **Figure 16**.

5. Conclusion

The superhydrophobic copper surface is successfully prepared by thermo-solution immersion technique with a quick time of 2 h. The highest contact angle of 162° is obtained with the concentration of the ethanol and myristic acid of 0.005 M when the solution is heated at 45–50°C for the time duration of 2–3 h.

The obtained surface is experimented with for dropwise condensation by varying the power input and water flow rate. The complete sphere-like water droplets are obtained on the condensing surface with an increase in contact area that increases the heat-transfer coefficient and condensation rate. The maximum value of temperature difference between the vapour and the surface is 74°C and 64°C for bare and superhydrophobic surfaces, respectively. The maximum heat-transfer coefficient obtained is 196 W/m^2 K for a superhydrophobic surface and the bare surface is 185 W/m^2 K. The condensed water droplet fall-off time is about 1 s on the superhydrophobic surface and 2 s on the bare surface against the mass flow rate of 300 lph. It is noted that the heat-transfer coefficient and condensing rate of falling water droplets on a superhydrophobic surface are higher than that of a bare surface. The superhydrophobic surface always promotes dropwise condensation, and the heat-transfer coefficient is increased by increasing the contact angle of the water droplet.

The experimental analysis is made by conducting a condensation experiment on both ordinary and superhydrophobic copper surfaces. The following conclusions have arrived in the present work:

1. The number of droplets formed is higher on a superhydrophobic surface than on an ordinary surface, which tends to increase the condensation rate.

2. An increase in static contact angle and a decrease in wettability of the substrate reduce the droplet area coverage.

3. A decrease in wettability also results in earlier water droplet fall off.

4. At a higher saturation temperature, the rate of growth of the drop is higher. Heat-transfer coefficient increases with an increase in the degree of sub-cooling of the substrate.

5. The droplets are formed on a superhydrophobic surface and look like perfect circles that decrease the contact area and increase the heat-transfer rate.

Author details

R. Yuvaraj
Sona College of Technology, Salem, Tamilnadu, India

*Address all correspondence to: yuvarajr@sonatech.ac.in

References

[1] Rose JW. Dropwise condensation theory and experiments: A review. Proceedings of the Institution of Mechanical Engineers, Part A. 2002;**216**:115-128

[2] Leipertz A, Fröba AP. Improvement of condensation heat transfer by surface modification. Proceedings of the Seventh ASME, Heat and Mass Transfer Conference, IIT Guwahati, India. 2006;**K7**:k85-k99

[3] Carey VP. Liquid-Vapour Phase-Change Phenomena. New York: Hemisphere; 1992. pp. 342-351

[4] Marto PJ, Looney DJ, Rose JW. Evaluation of organic coating for the promotion of dropwise condensation of steam. International Journal of Heat and Mass Transfer. 1986;**29**:1109-1117

[5] Zhao Q, Zhang DC, Lin JF, Wang GM. Dropwise condensation on L-B film surface. Chemical Engineering and Processing. 1996;**35**:473-477

[6] Vemuri S, Kim KJ, Wood BD, Govindaraju S, Bell TW. Long term testing for dropwise condensation using self-assembled monolayer coating of N-Octadecyl mercaptan. Applied Thermal Engineering. 2006;**26**:421-429

[7] Rausch MH, Fröba AP, Leipertz A. Dropwise condensation heat transfer on ion implanted aluminum surfaces. International Journal of Heat and Mass Transfer. 2008;**51**:1061-1070

[8] Ma X-H, Wang B-X. Life time test of dropwise condensationon polymer-coated surfaces. Heat Transfer - Asian Research. 1999;**28**(7):551-558

[9] Majumdar A, Mezic I. Instability of ultra-thin water film and the mechanism of droplet formation on hydrophobic surfaces. Transactions of the ASME, Journal of Heat Transfer. 1999;**121**:964-970

[10] Wang S, Feng L, Jiang L. One-step solution-immersion process for the fabrication of stable bionic superhydrophobic surfaces. Advanced Materials. 2006;**18**:767-770

Chapter 6

Boiling Heat Transfer on the Micro-Textured Interfaces

Tatsuhiko Aizawa, Naoki Ono and Hiroki Nakata

Abstract

Higher heat flux than its normal criticality from high-power transistors, LIDAR (Laser Imaging Detection and Ranging), stacked CPUs, high-power transistors, and lasers must be efficiently transferred to cooling media through the metallic interface. The micro-/nano-textured aluminum and copper devices were highlighted among several approaches and fabricated to enhance the boiling heat transfer process to the subcooled water. The plasma printing was proposed to fabricate a pure aluminum device with concave micro-textures and to describe the boiling heat transfer behavior with comparison to the bare aluminum plate. A copper device was wet-plated to have convex micro-textures and to discuss the effect of micro-textures on the heat transfer characteristics under the forced water cooling by varying the Reynolds number. The boiling curve on the micro-textured interfaces was newly constructed by improving the boiling heat transfer process by micro-/nano-texturing.

Keywords: boiling heat transfer, micro-/nano-texturing, aluminum, copper, boling curve, critical heat flux (CHF), superheat, micro-bubbles, Reynolds number

1. Introduction

High-performance forced cooling in VLSI (Very-Large-Scale Integrated) circuits [1] became a turning point to reconsider how to improve the capacity to cool down the highly energy-consuming devices and systems. Since then, various efforts were reported on the heat and mass transfer engineering. Micro-fluid channel heat sink in a single phase was proposed to improve the heat transfer from the heated surface to the cooling media. As surveyed in [2–4], numerical simulations were performed to optimize the micro-channel network. Various computational models were developed to control the heat transfer process [5]. The experimental works followed these theoretical discussions to enhance the boiling heat transfer with the use of porous interface [6]. Those studies insisted that thermal spreading with high heat flux could never be installed in practice without innovative modification in heat and mass transfer process.

In general, there are several heat transfer processes with and without phase transformation; e.g., the natural flow convection, the forced flow convection, and the boiling heat transfer. Among them, the boiling heat transfer process provides the highest heat flux condition in practice [7]. Toward the innovation in the boiling heat transfer technologies, the heat and mass transfer mechanism with phase transformation must be reconsidered to modify a boiling curve. In this chapter of book on the heat transfer, several new ideas were discussed to control the heat and mass transfer; e. g., a micro-porous interface design in [6], a micro-structured surface design via the spray cooling [8], a micro-/nano-scale surface modification [9, 10], and wettability-controlled surface design [11]. They provided a possibility to attain higher heat flux than the critical heat flux (CHF) limit [7]. In addition, the nanoscale morphology of heat-dissipating media became important to enhance the heat transfer on the heating surface even with phase transformation [12]. Those studies suggest the importance of the boiling curves on the micro-/nano-textured interface between the heating metallic solid and the cooling media.

After [7], three engineering items must be considered to describe the heat transfer with phase transformation and to discuss how to control this process. At first, the nonlinear relationship between the heat flux (q) and the superheat (ΔT), or the boiling curve, is precisely understood, as illustrated in **Figure 1**. In the natural convection flow heat transfer regime, log (q) or logarithm of heat flux increases linearly with the log (ΔT) or logarithm of ΔT. This linear relation between log (q) and log (ΔT) along A – B changes at B. At this point B, the phase transformation occurs from the liquid to its gaseous phase in local. That is, the liquid water locally transforms to the vapor in this local boiling. Starting from B through C to D, the isolated vapor bubbles grow by themselves, gradually agglomerate in themselves on the heating surface, and take off from the heating interface by buoyancy or by forced flow. In this regime, the heat flux significantly increases with increasing the superheat because of the latent heat by boiling, the mass transfer from liquid water to vapor, and the local convection flow.

Those growing vapor bubbles coalesce with each other to form a vapor film when approaching the D point. Once this film locally forms on the heating interface, the heat transfer deteriorates by itself since the thermal conductance significantly reduces in local. The whole interface is covered by this vapor film at D; the heat flux becomes

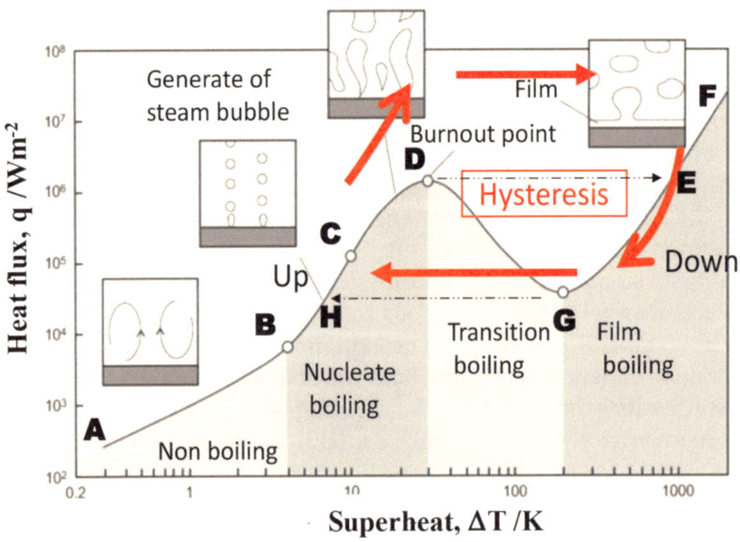

Figure 1.
The boiling curve on the smooth metallic surface [7].

maximum at this point. In the normal heat transfer process, this D point is physically unstable enough to induce the snap-through or the jump of state from the point D to its neighboring stable condition at E even by slightly increasing the heat flux. If the constituent material for heating media had sufficiently high melting temperature not to melt down during this snap-through, the heat transfer process could advance in stable from E to F in **Figure 1**. However, most of metallic heating plates melt down or burn out after this snap-through behavior from D to E.

On the way back by decreasing ΔT from E, this heat transfer process accompanies another nonlinearity. That is, the heat flux decrease from E with decreasing ΔT, and approaches to G where the heat flux minimizes itself under the film boiling regime. At another critical point G, another snap-through takes place from G to H back to the original boiling curve. Every heat exchanging system and device with smooth heating interface has been designed after this boiling curve. The feasible heat flux is limited by the critical heat flux in practice. To be free from this engineering constraint, this relationship must be modified by nontraditional heat transfer process.

To be noticed as the second item, the phase transformation from the liquid water to the vapor bubble commences in local on the heating interface to coolant; each vapor bubble nucleates at the imperfect point on the heating interface. Remember that most of theories on the nonlinear mechanics for heat transfer [7] and buckling mechanics [13] presume the geometric imperfections on the heating interface and on the shell structure, respectively. As summarized in [7], most of classical theories on the boiling heat transfer assumed that a vapor bubble nucleated at the preexisting wedge on the heating interface. Consider that an initial bubble shape is modeled by a half-sphere with its radius of r_c, then, the superheat (ΔT) is inversely proportional to this r_c. When the bubble size is controlled to be small, ΔT increases by itself to start the boiling heat transfer in the early stage of ln (q) – ln (ΔT) relationship in **Figure 1**. In practical situation, few methods were reported to reduce the wedge size, to decrease the wedge depth, to increase the wedge density on the heating interface, to regularize the distribution of wedges, and to control this regularity in the wedge alignment.

The third feature in the boiling process is characterized by the dynamic behavior of vapor bubbles both on the heating interface and in the coolant. After nucleation, one isolated vapor bubble grows, coalesces with each other, and forms a vapor film on the interface. The other bubble takes off from the interface and transports with coolant. The fresh cooling water comes onto the interface to continue this boiling heat transfer process. Unless any bubbles swell on the heated surface, it could be cooled with higher heat flux under constant ΔT. Otherwise, the interface is gradually covered by the vapor bubbles and finally separated from the liquid coolant by the vapor film.

Three engineering items in the above have interaction with each other. The boiling curve in **Figure 1** is governed by the change in vapor bubble morphology. Hence, if the nucleation and growth stages of the vapor bubbles are controlled by micro-texturing the interface, the boiling curve in **Figure 1** is significantly modified in the boiling heat transfer regime. In addition to the studies on controlling the boiling transfer by micro-/nano-texturing [8–12], MEMS (microelectronicmechanical system) technique was utilized to develop the cooling system by impinging the droplets [14] and to investigate the effect of micro-pillared micro-pipe on its thermal performance [15]. Those micro-/nano-texturing methods are useful to demonstrate the improvement of boiling heat transfer in the laboratory scale, but most of them are difficult to be used in the industrial applications. Micro-/nano-textures onto the high conductive metal members must be essential to find a new way in developing the effective cooling devices with linkage to industries.

Three approaches to form these micro-/nano-textures onto the product surfaces and interfaces were developed feasible to industrial and medical applications. In the plasma-oxidation-assisted printing [16–18], the micro-textures were formed onto the carbon-based materials through oxygen plasma etching. In the plasma-nitriding-assisted printing [19–21], the micro-textures were also formed into the metals and alloys by selectively embedding the nitrogen solute. The extremely short-pulse laser micro-/nano-texturing [22–24] is the third approach to directly imprint the tailored fine textures.

In the first approach, carbon-derivative coatings such as DLC (diamond-Like carbon), CNT (carbon nanotube), graphene sheet, and diamond coatings were processed to yield the carbon-based dies with micro-textures [25–27]. The aspect ratio was predetermined by the coating thickness. In second, the selective nitrogen supersaturation process advanced into the unmasked surfaces and interfaces. The ink-jet printing [19–21], the screen printing [28, 29], and the lithography [30, 31] were utilized as an effective masking techniques. In this case, the aspect ratio of micro-/nano-textures was determined by the nitrided layer thickness up to sub-millimeter thickness. In third, the nano-textured ripples superposed on the tailored micro-textured surface profile by the femtosecond laser processing [32]. In particular, the unidirectional nano-textures were simultaneously formed onto the laser-trimmed tool surface [33, 34]. On the basis of these micro-/nano-texturing methods, the surface-topological design is available to control the heat transfer process with phase transformation on the textured interface [35]. Different from the micro-/nano-texturing by the porous structuring, the particle deposition, and the MEMS techniques in [8–12, 14, 15], the regularly or semi-regularly aligned micro-/nano-textures were utilized to discuss the effect of micro-/nano-textures on the aluminum and copper heating devices to the boiling heat transfer in the water vapor system [36, 37].

In the present chapter, two micro-texturing methods are utilized to fabricate the textured aluminum and copper heating units for boiling heat transfer experiments. In the first approach, the plasma-oxidation-assisted plasma printing is used to yield the regular alignment of micro-square pillars on the thick DLC die. This multi-micro-pillared micro-textures on the DLC die are imprinted into the pure aluminum sheet. This concave micro-textured aluminum sheet with regular alignment of micro-cavities is fixed to the copper block to prepare the heating device for heat transfer experiment to describe the vapor bubble nucleation and taking-off behavior. In the presence of micro-cavities on the aluminum interface, the bubble size is reduced by 1/100 or less than that. These super-fine vapor bubbles do not swell on the interface but flow away with the coolant. Due to this change in the vapor-bubble dynamics, the bubble nucleation commences in lower superheat and the heat flux steeply increases with the superheat. In the second approach, the wet plating is utilized to form the acicular Fe–Ni micro-textures onto the oxygen-free pure copper unit for heat transfer experiment to demonstrate the effect of coolant flow velocity or its Reynolds number on the boiling heat transfer. The original boiling curve on the smooth metallic interface is significantly improved by this micro-texturing. The superheat to onset the nucleation of vapor bubbles is reduced by homogeneous and dense generation of bubbles. The heat flux steeply increases with the superheat and exceeds CHF. Through these steps, a physical model is considered to design and fabricate the heat transfer device with $q > q_{cr}$ and $q >> q_{cr}$.

2. Experimental procedure

Two micro-/nano-texturing methods were developed to make concave and convex micro-textures onto the metallic sheets and blocks, respectively. Two experimental setups were prepared to describe the boiling curves, to demonstrate the superiority of micro-textured interface in heat transfer, and to discuss the effect of coolant flow behavior on the boiling heat transfer.

Micro-texturing. The plasma printing method was employed to form the concave micro-textures with regular alignment of square micro-cavities with the specified unit size and pitch onto the pure aluminum sheet. The whole procedure from the micro-texturing design to the CNC imprinting onto the aluminum sheet is depicted in **Figure 2**.

The lithography with ion milling and reactive ion etching was utilized to print the square unit cell pattern with the size of 3.5 μm x 3.5 μm and the pitch of 5 μm onto the DLC-coated AISI420 punch with the coating thickness of 15 μm. Then, the platinum deposit on the nano-carbon film was left as a unit cell. The controlled oxygen plasma etching system was utilized to chemically remove the unmasked DLC regions and to fabricate the micro-pillared DLC coating punch. This punch was inserted into a die set for CNC imprinting of micro-pillar textures into the aluminum sheet. The original two-dimensional micro-texture pattern transformed to the three-dimensional regular alignment of micro-cavities on the aluminum sheet. The wet plating was employed to directly form the convex micro-textures onto the oxygen-free copper specimen with use of the nickel and iron-ionized solution. Each unit cell of acicular Fe-Ni crystals nucleated and grew on the substrate surface as depicted in **Figure 3**. Every acicular unit cell aligned with each other in semi-regular on the substrate. Its root size and height were completely determined by the wet plating conditions including the

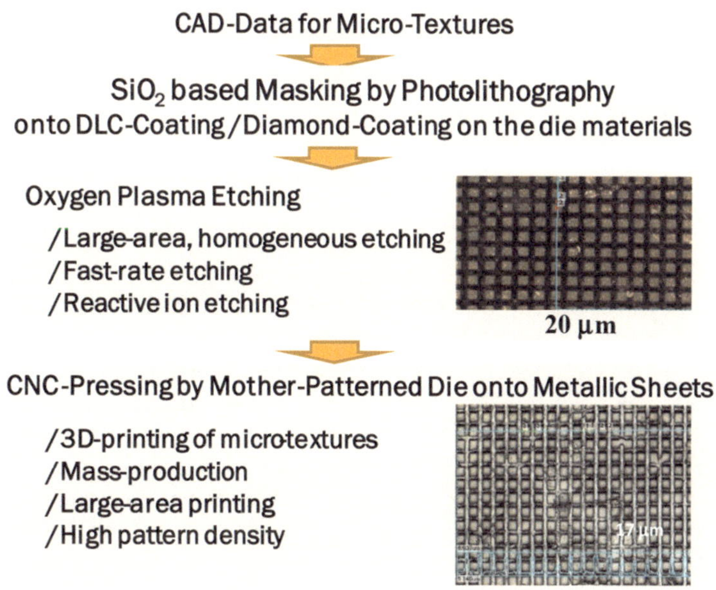

Figure 2.
A plasma printing procedure from the CAD (computer-aided design) of micro-textures through the fabrication of micro-textured DLC die to the CNC (computer numerical control) – Imprinting to metallic sheet.

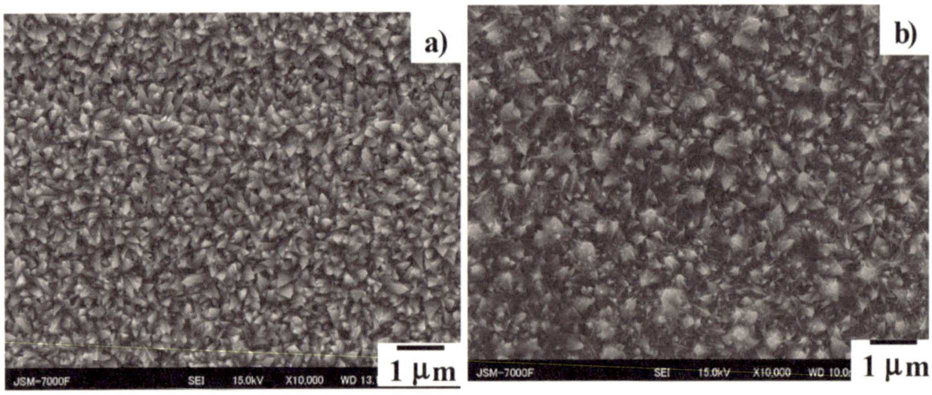

Figure 3.
Nucleation and growth of Fe-Ni acicular unit cells on the substrate during the wet plating. a) τ = 210 s, and b) τ = 630 s.

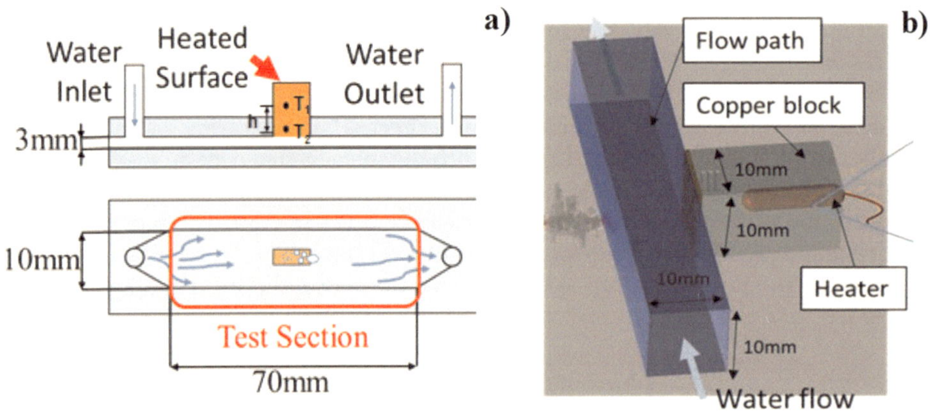

Figure 4.
The boiling heat transfer experimental test sections. a) Narrow, lateral channel section to observe the vapor bubble nucleation and taking-off phenomena with the measurement, and b) square, longitudinal channel section to investigate the effect of coolant flow on the boiling heat transfer.

duration (τ). In **Figure 3a**, when τ is 210 s, the root size (B) of acicular micro-texturing unit cell ranged from 0.1 to 0.4 μm, and its height (H) was 0.5 μm in average. This tiny unit cell significantly grew as compared between **Figure 3a** and **b**. When τ = 730 s, B ranged from 0.2 to 0.7 μm, and H ~ 1 μm.

Heat-transfer test-sections. In the following experiments, two test sections were used to describe the vapor bubble nucleation and taking-off and to investigate the micro-texturing effect on the boiling heat transfer process. In the first setup, the coolant channel with the length of 70 mm is laterally placed so that the subcooled water is heated by the copper block as depicted in **Figure 4a**. Pure aluminum devices with and without micro-textures were joined to the top of this block. The heat flux (q) through these devices was measured by the difference of temperature histories in the block. High-speed video camera was also utilized to visualize the vapor bubble nucleation and its taking-off from the aluminum devices.

Figure 4b shows another setup to measure the heat transfer characteristics with the use of the heating copper block. The micro-textures were directly yielded onto this

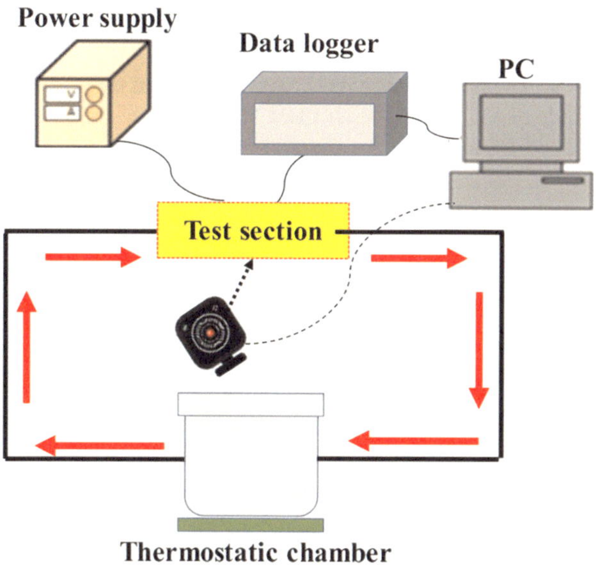

Figure 5.
Experimental setup for measurement and observation of the boiling heat transfer in the test-section.

block. The water channel stood in vertical to control the flow velocity and to investigate the effect of forced coolant velocity on the heat transfer. The heat flux was measured in the similar manner to the setup in **Figure 4a**.

These test sections were respectively set up into the experimental system including the power controllers and the monitoring apparatus, as illustrated in **Figure 5**. The superheat was directly controlled by the applied voltage in the power supply. The thermostatic chamber was utilized to keep the subcooled water temperature by 30 K and to control the coolant velocity. The measured temperature histories were monitored by using three thermocouples, which were embedded into the copper heating block. The surface temperature on the interface was estimated by directly extrapolating the measured temperature depth profile in the block. The heat flux was also calculated from this profile by using the Fourier's law, which was represented by

$$q = -\kappa(\delta T/\delta X), \tag{1}$$

where κ was the thermal conductivity of copper, δT was the difference between two measured temperatures away from each other by the distance of δX. The whole data were transferred and accumulated in PC.

The digital video imaging unit was utilized to record the time history of the boiling and cooling behavior in the inside of channels through the transparent window. The vapor bubble nucleation and growth process were directly monitored to describe the effect of coolant velocity on the boiling behavior.

3. Experimental results

Regular alignment of micro-cavities onto aluminum sheet. Two-dimensional micro-pattern was printed by the lithography onto the DLC film coated on the AISI420 substrate. A square dot with 3.5 μm x 3.5 μm was aligned with the pitch of

5 μm on this DLC coating die. The unprinted DLC films were chemically etched out by using the high-density plasma oxidation system. Through this plasma oxygen etching for 5 ks, a DLC micro-pillared punch array was fabricated by selectively removing the unprinted mesh-lines of DLC films with their width of 1.5 μm. As depicted in **Figure 6a**, the arrayed DLC-punch by plasma oxygen etching had a square micro-pillar head with its area of 3.5 μm x 3.5 μm and its height of 8 μm. The depth profile of DLC punch array was measured by the laser surface profilometer and supposed in **Figure 6a**. DLC micro-pillars with the same head size were regularly aligned with the pitch of 5 μm by the plasma oxidation etching. This regular alignment of micro-pillars onto the DLC-punch assures the regular duplication of micro-textures in inverse to micro-pillars into the metallic sheets or plates by the precise stamping with the use of this DLC-punch.

CNC stamping system was utilized for this imprinting of DLC micro-pillar array into the as-rolled pure aluminum sheet with the thickness of 0.2 mm. As shown in **Figure 6b**, the micro-cavity array was imprinted onto the aluminum sheet surface by indentation of the DLC micro-pillar array in **Figure 6a**. Through this indentation of DLC-punch, the pure aluminum work was extruded in backward into the clearance with the width of 1.5 μm between adjacent DLC pillars. The thin walls of each aluminum micro-cavity were formed to have the thickness of 1.5 μm. That is, each DLC pillar with its head size of 3.5 μm x 3.5 μm was simultaneously indented to shape a bottom of a micro-cavity with the size of 3.5 μm x 3.5 μm. Four clearances surrounding each DLC micro-pillar became four micro-cavity side walls with the width of 1.5 μm and the height of 5 μm. Then, the micro-pillar alignment on the DLC-punch was duplicated to the aluminum sheet as a regular micro-cavity array with its unit cell of 3.5 μm x 3.5 μm x 5 μm and the pitch of 5 μm. This imprinting process incrementally advanced by stamping the DLC-coated AISI420 punch with the head size of 80 mm x 10 mm into the pure aluminum sheet. Each stamped aluminum sheet segment had 3.2 x 10^7 micro-cavities on its area of 80 mm x 10 mm. Then, the density of micro-cavities reached 4 x 10^4/mm^2.

Figure 7 compares the aluminum sheets without and with the micro-cavity textures. The static contact angle of pure water onto the aluminum sheets increased from 92° for smooth surface to 110° by this micro-texturing.

Bubble heat transfer on the concave micro-textured interface. The pure aluminum sheet segment with micro-cavity alignment and the bare aluminum one were

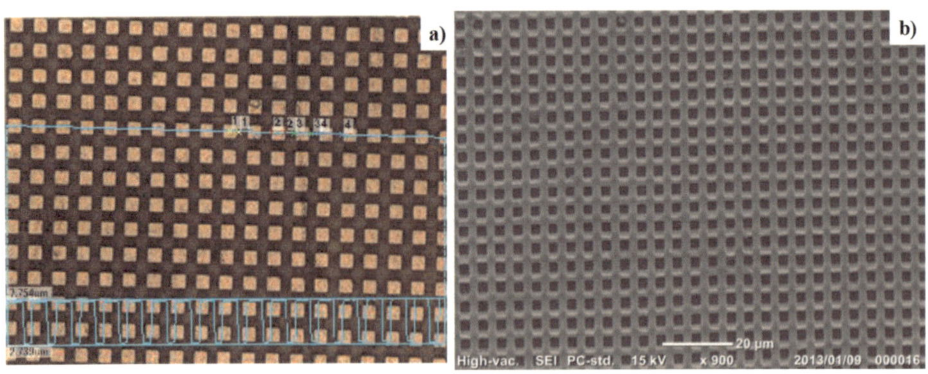

Figure 6.
Comparison of the micro-pillared DLC die surface with the CNC-imprinted aluminum sheet. a) DLC micro-pillar alignment on the die, and b) micro-cavity alignment on the aluminum sheet.

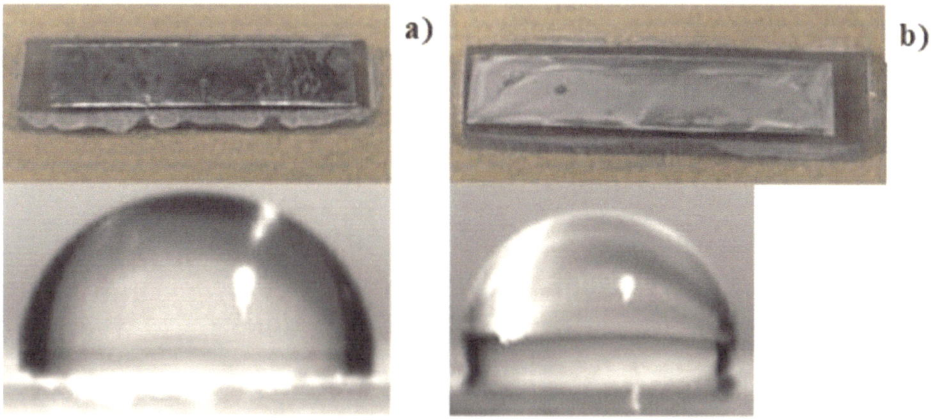

Figure 7.
Comparison of aluminum sheet without and with micro-textures. a) Aluminum sheet without micro-textures (contact angle of pure water on its surface (θ) is 92°), and b) aluminum sheet with micro-textures (θ = 110°).

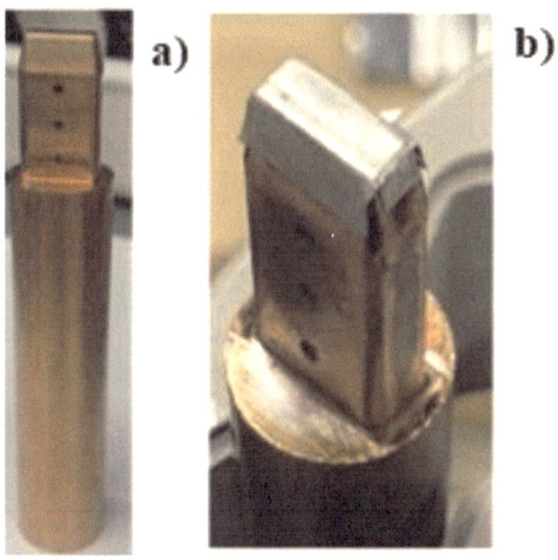

Figure 8.
*A heating block in the setup in **Figure 4a**. a) a bare copper heating block, and b) heating block with the joined aluminum sheets in **Figure 6** by using the nanoparticle silver paste.*

respectively joined to the copper block by using the nanoparticle silver paste. **Figure 8** compares the heating copper block before and after joining the aluminum sheet segment. Except for the thermal gap in conductance by this silver paste, this heating block was suitable to experimental analysis on the effect of micro-textured metallic sheet on the boiling heat transfer. This heating block was fixed into the test section in **Figure 4a**. In this setup, the phase transformation from the liquid water to the vapor bubble took place on the surface of aluminum device with and without the micro-cavity textures.

Figure 9 depicts the nucleation and growth stage of vapor bubbles on the aluminum sheet without and with micro-cavities. As shown in **Figure 9a**, when the micro-textures were absent on the aluminum sheet, large vapor bubbles only nucleated and swelled on the interface and grew by themselves. No taking-off of bubbles was observed by the video imaging.

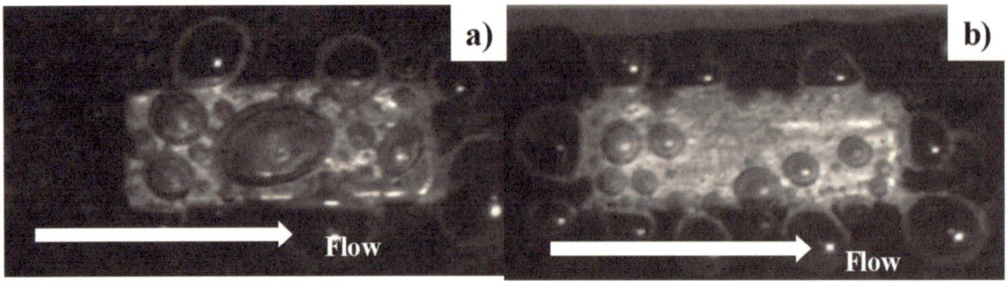

Figure 9.
A bubble nucleation on the aluminum sheet a) without the micro-textures and b) with the micro-texture.

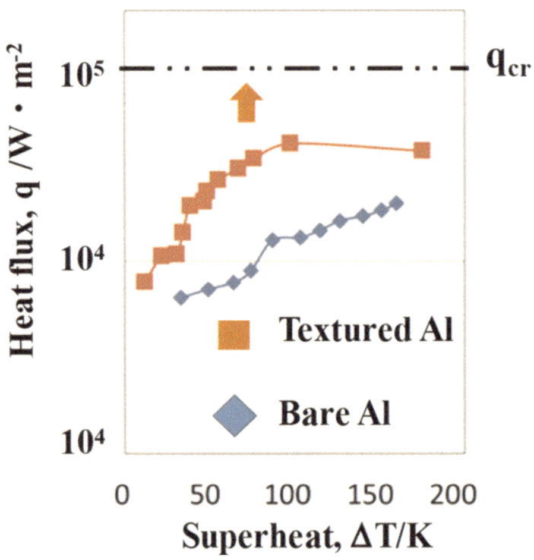

Figure 10.
Comparison of the boiling curves in the heat transfer through the aluminum sheets with and without the micro-textures.

On the other hand, as depicted in **Figure 9b**, only fine bubbles with the average size of 5–10 μm nucleated on the interface without significant growth. Most of them flew away together with the coolant and the medium-sized bubbles. As compared between **Figure 9a** and **b**, the heating interface with micro-textures was cleared by this flow-away of fine vapor bubbles. This easiness of vapor bubble taking-off from the heating interface enabled new liquid coolant to directly come onto the interface for further heat transfer. Owing to this micro-texturing effect on the bubble nucleation and growth mechanism, the heat transfer on the aluminum interface between the heating block and the coolant was much improved in the presence of micro-textures.

Figure 10 depicts the relationship between the measured heat flux q and the superheat ΔT with and without micro-textures. When using the bare aluminum interface, the heat flux gradually increased with ΔT and the onset superheat to nucleate the boiling was retarded in correspondence to the normal heat transfer mechanism in **Figure 1**. On the other hand, in the presence of micro-textures on the aluminum interface, the heat flux steeply increased with ΔT at the beginning of superheating. The onset of superheat for bubble nucleation started earlier than the normal boiling nucleation. Although this increasing heat flux was suppressed around ΔT = 60 K due to the insufficient coolant flow volume, the micro-textures enhanced the increase of heat flux with ΔT so that q approaches to the critical heat flux, q_{cr}.

This heat transfer through the micro-cavity arrayed aluminum device teaches:

1. the bubble size is reduced by decreasing the wedge size,

2. the onset superheat of ΔT_{onset} to nucleate a mass of bubbles is also reduced, and

3. the heat flux q increases steeply toward the critical heat flux, q_{cr}.

After the classical theory, the effect of the wedge size on the heat transfer is considered by

$$\Delta T \sim 2\sigma T_{sat}/[\rho_v \bullet \Delta h_v \bullet r_c \bullet (1 - r_c/\delta)] \quad (2)$$

In this relation, σ denotes the thermal conductivity, T_{sat} is the saturated temperature, ρ_v is the mass density of coolant, Δh_v is the latent heat, r_c is the radius of wedge, and δ is the thermal boundary layer thickness. From Eq. (2), the superheat increases by decreasing r_c. On the other hand, ΔT_{onset} decreased with decreasing the micro-cavity size in **Figure 8**. This contradiction between the classical model and the experimental results is attributed to vapor bubbling mechanism on the interfaces without and with the regular micro-cavity alignment.

The classical knowledge by Eq. (2) presumes that each vapor bubble nucleates at the preexisting wedge on the flat interface through the phase transformation of liquid coolant in local. This nucleation and growth process advances independently in each bubble, which is isolated from the neighboring bubbles. In this low bubble density nucleation and growth process, higher superheat is needed to stimulate the nucleation and growth step for each bubble till it takes off from the wedge. In the heat transfer across the micro-textured interface, many bubbles nucleate simultaneously at the regularly aligned wedges with the specified distance between adjacent wedges. Under this high bubble density nucleation from regularly distancing wedges by 5 μm in **Figure 6b**, lower superheat was enough to nucleate a fine bubble in the surrounding coolant flows. **Figure 9b** reveals that a mass of fine bubbles easily moves with the coolant flow and that the interface is cooled down to restart the bubble-nucleation stage under the fresh coolant. That is, the generated bubble size is much small enough to drive the routine of bubble nucleation, growth, and taking-off promptly and repeatedly. This tiny bubbling routine on the heating interface with regularly aligned concave micro-texture modifies the original boiling heat transfer mechanism in **Figure 1** and pushes up the heat flux even at the lower superheat.

Two engineering items are considered to further control the boiling heat transfer by micro-texturing. In the first item, the micro-textured interface property is taken into account together with the regular alignment of micro-cavities or wedges in **Figure 7b**.

Since the metal surface is usually hydrophilic with the static contact angle of 60°–70°, this surface condition is controlled by micro-/nano-texturing to be more hydrophilic or to be hydrophobic, as an engineering policy. In second item, the coolant design is surveyed to improve the q – ΔT relationship or the boiling curve. Among several parameters, the coolant flow velocity is employed as an important item. After the classical treatise on the boiling heat transfer [7], its mechanism was considered to be invariant to the coolant flow velocity. There is a possibility to improve the q – ΔT relationship with increasing the flow velocity or the Reynolds number of coolant on the micro-textured interface.

Semi-regular alignment of micro-acicular textures onto copper. The wet plating process was employed to build up the acicular iron-nickel (Fe–Ni) alloy micro-textures directly onto the copper heating block surface. The acicular Fe–Ni pyramids nucleated and grew onto the copper surface with increasing the duration in the wet plating, as stated before. In particular, as depicted in **Figure 11a**, the root size and height of acicular pyramids were controlled by the wet-plating conditions.

Since no pastes were used in this micro-texturing, no thermal gap conductance was present between the micro-textured layer and the heating copper block. As shown in **Figure 11b**, this acicular Fe–Ni layer was thermally well-contact to the top surface of copper heating block. In the previous experiment, the micro-textured aluminum sheet was joined to the copper block by using the solder paste. Then, the measured q – ΔT relationship was biased by its thermal gap conductance. In this second experimental setup, the micro-textured layer was directly deposited onto the copper heating block head surface without the use of paste. The heat flux in the copper block was directly conducted to the wet-plated Fe–Ni layer without loss of thermal conductance. The acicular micro-textures influenced on the surface property and on the vapor bubble nucleation and growth during the boiling process.

Two Fe–Ni acicular layers were formed onto the copper block by varying their unit cell size to understand the micro-texture size effect on the boiling heat transfer mechanism. The microstructure of the first Fe–Ni micro-textured film (specimen-1) is shown in **Figure 12a**. Three-dimensional profilometer was utilized to define the average height (H) of acicular unit cell, the average spacing (D) between adjacent unit cells, and the aspect ratio (H/B) of H to the average bottom size (B) of unit cells. This specimen-1 has H = 1.6 μm, D = 3.9 μm, and H/B = 0.42. As depicted in **Figure 12b**, this textured surface becomes hydrophilic with the contact angle of 20–30°. A bare copper block is also used as a reference to compare the q –ΔT relationship with two micro-textured specimens. The mechanically ground copper interface has the average roughness of Ra = 0.6 μm and Rz = 1.2 μm. Due to this surface roughness, the contact angle reaches 100–110°; the copper interface is hydrophobic.

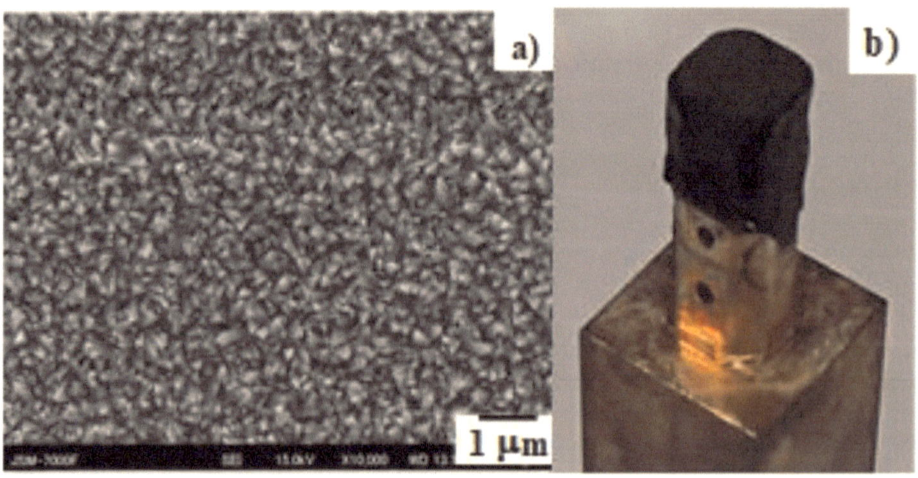

Figure 11.
An acicular Fe–Ni alloy film, wet-plated onto the copper heating block. a) SEM image on the wet-plated Fe – Ni alloy film, and b) copper heating block with the Fe–Ni layer on the top of block.

Figure 12.
The first Fe–Ni micro-textured heating block specimen-1. a) Its microstructure in the plain view, and b) its wettability.

Figure 13.
The second Fe–Ni micro-textured heating block specimen-2. a) Its microstructure in the plain view, and b) its wettability.

Remembering that a smooth metallic surface is usually hydrophilic with the contact angle of 40–70°, this high contact angle implies that original roughness of copper block surface influences the measured boiling curve. In particular, this micro-textured surface of specimen-1 has nearly the same geometric features as the bare copper surface.

The second specimen or specimen-2 was prepared by the wet plating with longer duration. **Figure 13a** shows the SEM image of the Fe–Ni acicular microstructures on the copper block. The micro-textures in **Figure 13a** have a self-similar morphology to the specimen-1 in **Figure 12a**. This Fe–Ni acicular unit cell has H = 4.8 μm, D = 2.2 μm, and H/B = 0.46. After the wettability testing, this specimen is also hydrophilic with the same contact angle.

By using two specimens in the above, the boiling heat transfer experiments in **Figure 3b** are performed to investigate the effect of the acicular micro-textures on the q – ΔT relationship and the heat transfer mechanism.

Boiling heat transfer on the convex micro-textured interface. Two types of experiments are performed in the following. In the former experiment, the acicular micro-texturing effect on the heat transfer is investigated. In the latter, the effect of coolant velocity or Reynolds number (Re) to the heat transfer mechanism is precisely discussed. In the first experiment, the coolant velocity was held constant by V = 1.9 m/s or Re = 460. The coolant was completely degassed and its temperature was also controlled to be 347 K or subcooled by 30 K. The copper block was gradually heated by increasing the applied voltage in every 10 V.

In correspondence to the boiling curve in **Figure 1**, q increases gradually with increasing ΔT in case of copper block without the micro-textures. When using the specimen-2 with its microstructure in **Figure 14**, the heat flux rapidly increases with ΔT after the superheat of ΔT = 2 K, and approaches the critical heat flux, q_{cr} at ΔT = 17 K. Remember that q = 0.9 x 10^6 W/m^2 at ΔT = 17 K when using the copper block without the micro-textures. The heat flux becomes two and a half times higher at the same superheat of 17 K by using this hydrophilic micro-texture.

On the other hand, in case of the specimen-1 in **Figure 12**, its q – ΔT relationship becomes nearly the same as q – ΔT for non-textured specimen. This is because the micro-textured interface of specimen-1 has nearly the same topological aspects as the bare copper interface. The difference between two specimens implies that the boiling heat transfer behavior becomes sensitive to the acicular micro-texture morphology under the laminar coolant flow. To be noticed, q_{cr} on the micro-textured surfaces seems to be always more than q_{cr} on the smooth surface.

In the second experiment, the coolant flow velocity was increased to change the laminar flow to the turbulent flow and to investigate the sensitivity of boiling heat transfer mechanism to the flow pattern change. The specimen-1 was employed in this experiment, while the bare copper specimen was also used as a reference. **Figure 15** compares the variation of q – ΔT curves with increasing the Re between the specimen-1 and the normal copper block without micro-textures.

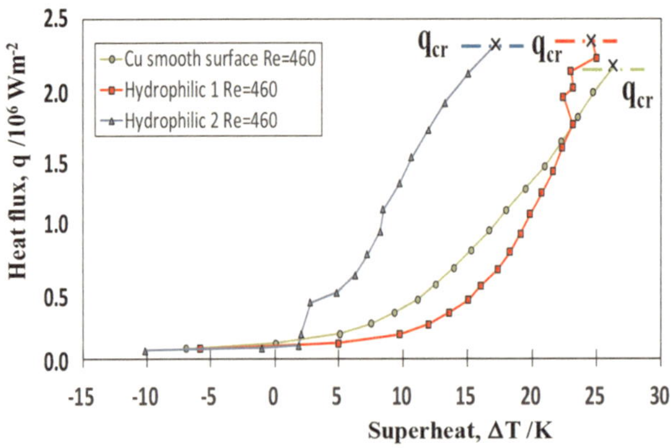

Figure 14.
Comparison of the boiling curves under the laminar flow condition among the copper smooth surface and two micro-textured copper surfaces.

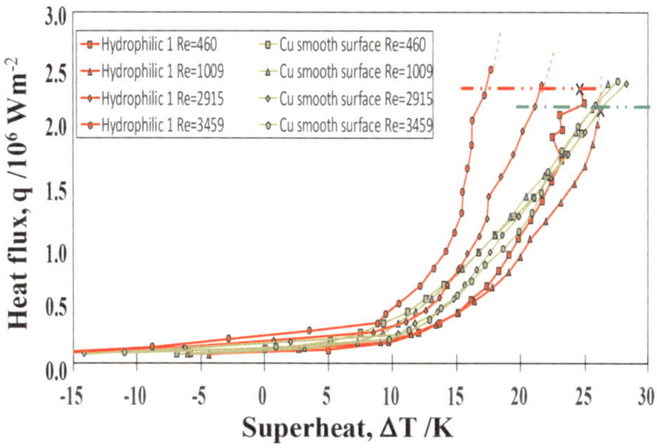

Figure 15.
Q – ΔT relationships for normal copper block without micro-textures and for the specimen-1 with increasing Re.

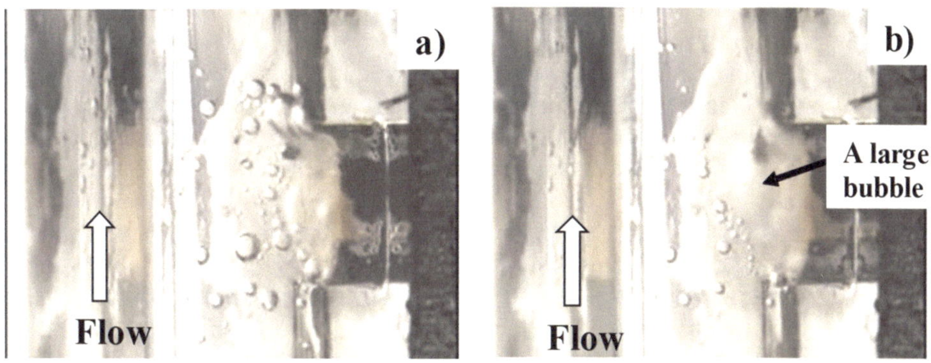

Figure 16.
High-speed camera flame on the nucleation boiling process on the bare copper block head without micro-textures in the coolant channel at Re = 3459 with increasing the applied power (P) for heating. a) P = 110 W, and b) P = 180 W.

When using the bare copper heating block, four measured q – ΔT relationships in **Figure 15** are nearly the same among them and insensitive to the coolant velocity even by increasing Re from 460 to 3459, or by changing the laminar flow to the turbulent one. This insensitivity of boiling heat transfer process to the Reynolds number was just stated in the classical treatise [7]. The coolant flow pattern has nothing to do with the boiling heat transfer process or the bubble nucleation and growth process in the classical common knowledge. This is because each wedge on the bare copper is isolated from each other and a bubble nucleates and grows mainly in the function of superheat.

In case of the micro-textured specimen-1, the onset of superheat to nucleate the bubbles is reduced and the gradient of dq/d(ΔT) becomes steeper with increasing Re. CHF also seems to increase with increasing Re. This enhancement of q – ΔT relationship with Re implies that a mass of vapor bubbles is transported by the coolant flow and that the fresh coolant comes onto the heating interface.

The video imaging often helps to describe the boiling behavior with increasing the electrical power (P). In the following, an overall boiling and flow behavior is observed during the heat transfer experiment by using two snap-shots in the video imaging. **Figure 16** depicts the boiling and flow processes in the coolant channel under Re = 3459. When P = 110 W, the coolant was uniformly boiled in down- and upper streams, surrounding the heating copper block head as shown in **Figure 16a**. The small-sized bubbles were seen in this boiled coolant, but the boiled coolant region widened symmetrically in both streams of coolant. A fresh coolant did not approach the heating copper block head for further forced cooling. When increasing the power to P = 180 W, the boiled coolant pattern in **Figure 16b** was nearly the same as seen in **Figure 16a**. The boiling behavior did not change itself by increasing the applied power to the heating unit. Larger bubbles were seen in **Figure 16b**; the bubbles nucleated and easily agglomerated themselves to form a larger bubble. This appearance of large bubbles teaches a signal of risk where these large bubbles promptly coalescence to a film at the vicinity of interface, and the nucleate boiling mode changes to the film boiling mode. That is, the nucleate boiling behavior in **Figure 16b** continues by itself till the burn-out point when the heat flux approaches the critical one.

The difference of q–ΔT relationship with and without micro-textures in **Figure 15** predicts that the nucleation boiling process in **Figure 16** with increasing the power changes itself by the micro-texturing on the heating copper block head.

In using the same experimental setup, the nucleation boiling behavior on the micro-textured copper block is analyzed by high-speed camera. When P = 110 W, the boiling coolant region was seen only in the downstream and became narrow near the micro-textured interface between the heating copper block and coolant in **Figure 17a**. When P = 180 W, the bubbled coolant flew away in the downstream together with the main coolant as seen in **Figure 17b**. To be noticed, a single-phase or liquid-phase coolant entered into the channel inlet, mixed with the two-phase, turbulent flow from the interface, and flew away from the channel outlet in **Figure 4b**. This observation on the flow pattern at Re = 3459 reveals that the coolant flow has a significant interaction with the boiling heat transfer on the micro-textured interface. Two-phase mass with fine bubbles moves away from the vicinity of interface to the downstream of coolant. The micro-textured interface is cooled down by a new coolant from the upper stream. The mixing process of two flows and the mass transfer of fine bubbles to coolant flow work independently to sustain the higher heat transfer through the micro-textured interface. This results in the steep increase of heat flux to CHF and above CHF even at lower superheat.

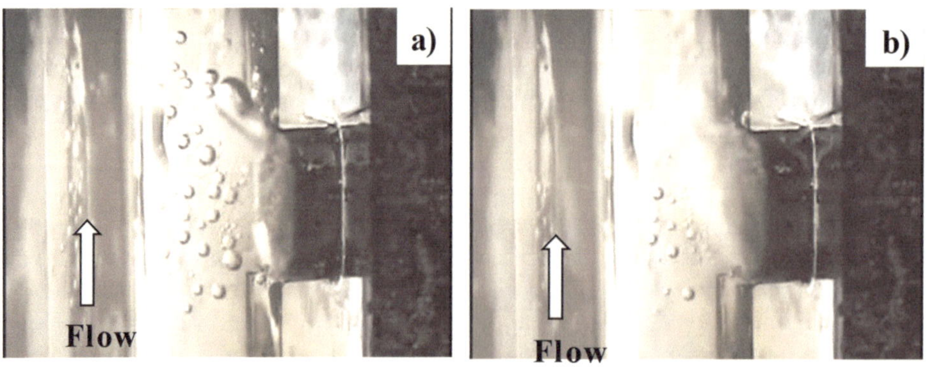

Figure 17.
High-speed camera flame on the nucleation boiling process on the micro-textured copper block head in the coolant channel at Re = 3459 with increasing the applied power (P) for heating. a) P = 110 W, and b) P = 180 W.

Micro-texturing effect on coolant flow and heat transfer. A micro-texture on the interface has a possibility to influence both on the flow behavior and the heat transfer. Various studies have been reported on the effect of micro-texture to the coolant turbulent flow in the literature [38]. After those results, the turbulent flow behavior at the vicinity of the channel wall is affected by the viscous effect. In this viscous sub-layer, its friction velocity u* is estimated by the following equation with the Blasius equation for friction factor λ,

$$u^* = \sqrt{\frac{t_0}{\rho}}, t_0 = \frac{1}{8}\lambda\rho u^2, \lambda = 0.3164\, Re^{-\frac{1}{4}}, \tag{3}$$

From this friction velocity, let us calculate the wall coordinate (y+) at the height of the micro-texture. In the above experiment, Re = 3459; then, u* = 0.0103 m/s. Since the height of micro-texture is k_s = 2.2 μm or 2.2 x 10^{-6} m, and the kinetic viscosity of coolant is ν = 4.13 x 10^{-7} m^2/s, y + = u* k_s/ν = 0.04. This y + is much lower than the critical number of ~5. The surface angulation by this micro-texturing is identified to be included into the viscous sub-layer. That is, this microstructure has nothing to do with the coolant turbulent flow, and it can be regarded as a hydraulically smooth surface.

Next let us consider the micro-texturing effect on the heat transfer coefficient. The turbulent flow heat transfer on the flat interface without the micro-textures is described by the Nusselt number (Nu). After [39], this Nu is expressed by the function of the Reynolds number and the Prandle number (Pr) in the following:

$$\mathrm{Nu} = 0.023\ \mathrm{Re}^{0.8}\mathrm{Pr}^{0.4} \tag{4}$$

Since the heat transfer coefficient (h) is proportional to Nu, h ~ $Re^{0.8}$ on the flat interface. On the other hand, the measured h (= q/ ($T_w - T_L$)) on the micro-textured interface is estimated from **Figure 15** to be

$$\mathrm{h} \sim \mathrm{Re}^{\mathrm{n}} \tag{5}$$

Here, n is a power exponent, which is given by n = 2.71 at ΔT = 9 K, n = 2.79 at ΔT = 15 K, and n = 3.31 at ΔT = 17 K. This result with n > 0.8 proves that the boiling heat transfer is much enhanced by the coolant turbulent flow on the micro-textured interface. Remembering that micro-textures have no influence on the turbulent flow profile, this enhancement is induced by the vapor-bubble transportation with high coolant flow velocity even in the viscous sub-layer.

4. Discussion

Boiling curves on the micro-textured interface. The micro-/nano-texturing onto the heating device surface has a significant influence on the boiling heat transfer. First, the heat flux to superheat relationship in **Figure 1** is controlled by the micro-/nano-texturing. The superheat to start the nucleation of vapor bubbles is much reduced, the heat flux gradient by ΔT or the heat penetration factor (K_q) becomes steep along B – C – D curve, and the heat flux can exceed the critical heat flux (CHF).

In second, the classical model on the nucleation and growth of vapor bubbles must be exchanged with a new physical model, where superfine vapor bubbles nucleate on the micro-/nano-textured interface and flow away with coolant flow. In particular, the nucleation and growth of vapor bubbles are dependent on the local geometric topology of micro-/nano-textures. The bubble density, nucleating on the textured interface, much increases even in the early stage on B – C regime.

In third, higher heat flux is attained even at lower superheat with increasing the coolant flow velocity or its Reynold number. In the absence of textures on the heating interface, the boiling heat transfer has nothing to do with the flow velocity change from the laminar flow to the turbulent flow. This is just corresponding to the classical theory in [7]. In the presence of micro-textures, most of vapor bubbles densely nucleate and easily take off from the heating interface at the early stage of superheating. Since the local mass in two phase with lots of fine bubbles moves away with main turbulent flow of coolant, higher heat flux is attained even at the lower superheat.

Standing on these engineering items, a new boiling curve is proposed for the boiling heat transfer process through the micro-/nano-textured heating surface.

The heat flux to superheat relationship on the micro-textured interface is schematically depicted in **Figure 18** with comparison to the boiling curve on the flat interface. The vapor nucleation starts at B′ or at $\Delta T = \Delta T_i$; the point B′ becomes a turning point from the natural or forced convection heat transfer along A' – B′ to the nucleation boiling process. This intrinsic superheat (ΔT_i) is determined by the minimum thermal conductance of the micro-/nano-textured layer; this ΔT_i is much smaller than ΔT in **Figure 1**. Along the line B′ to C′, the heat flux increases with the steep gradient of K_q even under the laminar coolant flow. This heat penetration rate, K_q, is strongly dependent on the geometric topology of micro-/nano-textured layer. In case of the present acicular micro-textures, their unit cell size, height, aspect ratio, and their alignment have essential influence on K_q. At C′, the heat transfer process is enhanced by the coolant flow pattern change from the laminar flow to the forced turbulent flow. Under the continuous mass transfer of two-phase coolant with dense fine bubbles to the main flow, the heat flux exceeds the CHF and increases monotonously with ΔT above CHF.

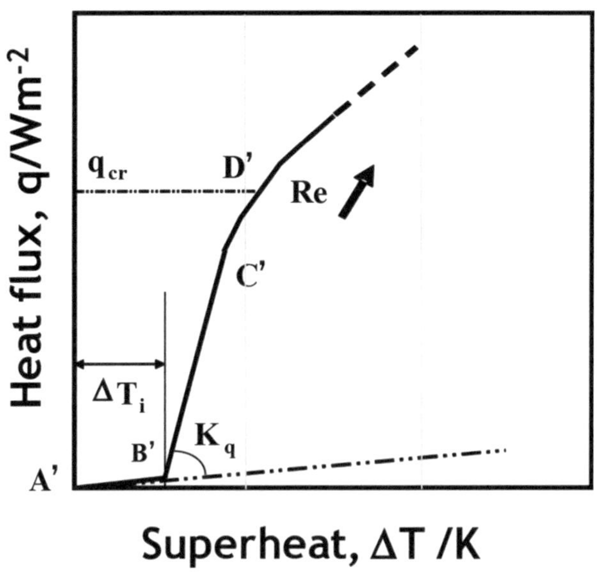

Figure 18.
A new boiling curve for boiling heat transfer on the micro-textured interface between the coolant and the heating solid.

From **Figures 10** and **15**, this ΔT_i was estimated to be less than 10 K and $K_q \sim 0.2 \times 10^6$ W/(m^2•K). The onset temperature of bubble nucleation, ΔT_{onset} reduced to ΔT_i in these Figures. The bubble nucleation at the single wedge on the flat surface is retarded so that $\Delta T_{onset} > \Delta T_i$. On the other hand, the phase transformation to vapor bubbles commences at every spot near the aligned micro-textures at the same time. That is, the micro-/nano-texturing onto the heating surface stimulates the onset of bubble nucleation at the network of micro-cavities in **Figure 10** and at the micro-pillars in **Figure 15**. This difference of onset superheat reveals that the mode change from the convection heat transfer process to the boiling heat transfer process is triggered in the very early stage of superheat by the simultaneous nucleation of fine bubbles in mass.

The least onset superheat of ΔT_i is needed for thermal conductance of micro-/nano-textured layer. In case of the heat transfer on the flat interface, the difference of ($\Delta T_{onset} - \Delta T_i$) is needed to trigger the phase transformation to vapor at each isolated wedge on the interface. This retardation is reduced, or, $\Delta T_{onset} \rightarrow \Delta T_i$ by simultaneous nucleation of fine vapor bubbles due to the micro-/nano-texturing. The simultaneous nucleation reflects on the increase of bubble density even at lower superheat.

The steep branch at B′ with high gradient of K_q in **Figure 18**, implies that the fine bubbles nucleating at the micro-textures easily take off from the heating interface to the main stream of coolant. Until this nucleation and taking-off continues even in increasing the superheat, the heat flux increases even under the laminar coolant flow. After the nucleation mode changes to the growth mode of bubbles, the increase of heat flux by ΔT becomes redundant and slow. In particular, $q/\Delta T$ in the forced turbulent coolant flow regime is still large enough to sustain the flow interaction between the generated two-phase local flow and the main forced coolant flow.

Bubble density nucleating on the micro-textured interface. Nucleation of super-fine vapor bubbles and their taking-off with coolant is essential to start the boiling heat transfer at lower superheat and to sustain high heat flux condition. The unit-cell size and its pitch play an important role to reduce the vapor-bubble size and to increase the bubble density. Through the image processing of frame pictures in **Figures 9b**, **17a**, and **b**, the vapor-bubble size was still larger than the unit-cell size; e.g., the unit-cavity size of micro-textures was 3.5 μm x 3.5 μm but the bubble size was larger than 10 μm. In other words, the local phase transformation process is redundant to nucleate the vapor bubble at the selected unit cell among many nucleation sites. This low bubble density suppresses the actual heat flux to be slightly higher than CHF.

Micro-/nano-texture design for enhancement of heat transfer. In the present study, two types of micro-textures were employed to investigate the interface conditions on the boiling transfer process; e.g., the concave micro-textures with hydrophobicity in **Figure 7**, and the convex micro-textures with hydrophilicity in **Figures 12** and **13**. The controllability of wettability by micro-texturing has been intensely discussed through previous studies in the literature [32]. Let us describe the effect of surface properties on the micro-textured interface to the heat transfer. High surface energy surface is preferable to cover the whole surface by coolant liquid and to easily release the nucleating vapor bubbles. That is, the micro- and nano-textures might well be redesigned to consider their multiple functions in the boiling heat transfer mechanism. The measured $q - \Delta T$ relationships in **Figures 10** and **15** suggest that the interfacial condition seems to have little influence on the boiling heat transfer process. Other micro-/nano-texturing features might have more importance on the improvement of boiling heat transfer process; e.g., the multidimensional texturing with self-similarity, the super-hydrophilic texturing with high aspect ratio and the super-hydrophobic texturing with higher spatial frequency ratio.

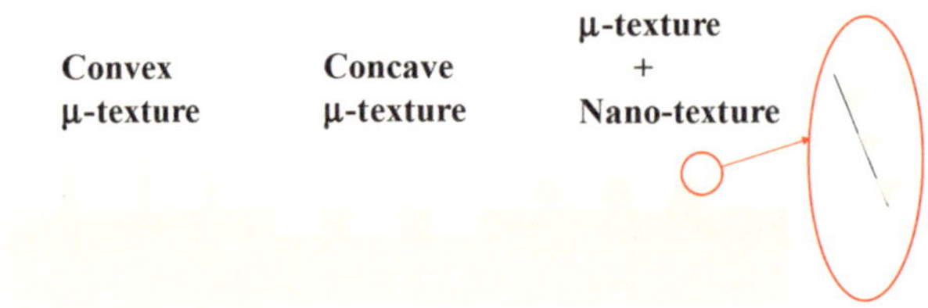

Figure 19.
Feasible micro-/nano-texturing onto the heating interface.

As illustrated in **Figure 19**, various microstructures are available in CAD to modify the heating surface conditions. Various nano-textures are also superposed onto each specified microstructure surface. These micro-/nano-textures are aligned with different regularity in their topological design. For an example, the micro-/nano-textures are considered in trial to improve the coefficient of K_q at B′ in **Figure 18**. The hydrophilic or super-hydrophilic micro-textures with high peak-to-valley intensity ratio are suitable to attain much higher heat penetration rate by improving the surface properties of specimen-2.

Manufacturing to boiling heat transfer devices. In addition to the scientific understanding on the effect of micro-/nano-textures to the boiling heat transfer, how to make mass production of the textured heat-transfer devices must be also taken into account. In this chapter, two methods were proposed to build up the micro-textured devices. In the former method, a thick DLC-coating with more thickness than 20 μm was employed as a mother die to build up the tailored micro-/nano-textures by the plasma-oxidation-assisted printing. This die with the aligned micro-pillared punches was indented into the metallic work sheet to form the micro-cavity micro-textures on it. This plasma-oxidation-assisted technique is exchanged with other approaches. As demonstrated in [40–43], the nitrogen supersaturated stainless steels as well as this DLC coating are also utilized to make micro-/nano-texturing via the pico-/femto-second laser micromachining. The LIPSS (laser-induced periodical surface structured) nano-textures are precisely imprinted to metal plates and sheets by CNC-stamping. Through this precise stamping, the mother topology of micro-/nano-structures is imprinted onto the aluminum and copper device surfaces together with their engineering functions such as their grating in colors and surface plasmonic brilliance.

The latter method is a coating procedure including the dry and wet plating to form the micro-/nano-textures directly onto the metallic substrates. As shown in the acicular micro-texture formation in **Figures 12** and **13**, the anisotropic deposition of Fe–Ni alloy layers or the etching of deposited layers played a key process to control the unit cell size of the convex and concave micro-textures. Their topological alignment of each unit cell is also controllable in this approach. In application of these micro-textured devices to practical heat transferring system, they must be post-treated to have high hardness and strength against the erosion by the vapor bubble attack. Once these micro-/nano-texture layers are hardened and strengthened, the heat-transferring device surface can be functionally decorated by the tailored coating with complex topology.

5. Conclusion

The boiling heat transfer process with two-phase coolant flow has been utilized in various engineering fields including the heat exchanging facilities, the heat-spreading devices of waste heats, or the high-cooling-rate equipment. Their design base stands on the classical knowledge of heat transfer mechanism through the flat metallic interface. The present study experimentally demonstrates that micro- and micro-/nano-texturing on the metallic interface can control the boiling heat transfer between the flowing coolant and the heated solid. The ultra-fine bubbles nucleate and take off with the coolant flow on the pure aluminum sheet and plate with regular alignment of micro-cavities. In addition to fine vapor bubbling in nucleation, these fine bubbles flow away with coolant without swelling onto the interface. Owing to this easy and prompt taking-off of bubbles, the heating interface is continuously wetted by fresh coolant to sustain the high heat flux through the micro-textured interface. This finding teaches the important role of fine bubble density and its dynamics in the nucleation step of boiling.

The heat flux to superheat relationship or the boiling curve is essential in the boiling heat transfer. It is much improved by the present micro-/nano-texturing. The onset superheat for transition from the convection heat transfer to the boiling nucleation is much reduced by homogeneous nucleation of bubbles with high density. With increasing the superheat over this onset, the heat flux increases in steep gradient. In addition, this steep increase of heat flux is much enhanced with increasing the coolant velocity. In classics, the boiling heat transfer has been believed never to be dependent on the coolant velocity. The high-speed camera observation in the present experiments discovers that the mass of two-phase local flow with fine bubbles takes off from the textured interface and moves away to the downstream of forced turbulent coolant flow. Since the fresh coolant flows onto the textured surface, the higher heat flux is sustained on the textured heating interface. This significant change of physical models in the boiling curve by micro-texturing suggests that multi-scaled physics are necessary to describe the microscopic heat transfer on the bubble nucleation, to analyze the mesoscopic interaction of vapor bubbles and fresh coolant flow, and to modify the macroscopic relationship between the heat flux and the superheat.

The present study started to consider the micro-/nano-textures tailored for each boiling heat transfer design. Much more scientific idea and engineering effort is still necessary to find a way to attain the much higher heat flux than the CHF in the heat exchanger, the heat pipes, the heat spreaders, and the thermal device to efficiently release the waste heat. A graphene with much higher thermal conductivity becomes a candidate to be working instead of metals for small-scaled cooing devices. The micro-/nano-textured sheet is near-net-shaped to a channel or a pipe with textured inner surfaces for highly efficient heat transferring and exchanging systems. The unit cell size and geometric topology are much modified to improve the convection heat transfer together with the boiling transfer [44]. The regularity in the alignment of unit cells in micro-/nano-texturing is further controlled to discuss the simultaneous nucleation and taking-off at the early stage of bubble nucleation.

Acknowledgements

The authors would like to express their gratitude to Dr. T. Yamada, Mr. M. Tasaka, and Mr. S. Suwa (Graduate School of Science and Engineering, SIT), and Mr. S. Kurozumi (Nano-Film Coat, llc.) for their help in experiments.

Conflict of interest

Authors declared no conflict of interests.

Nomenclature

q	Heat flux
q_{cr}	Critical heat flux (CHF)
ΔT	Superheat, the difference between wall temperature and saturation temperature
ΔT_{onset}	Superheat to onset for nucleation
ΔT_i	Intrinsic superheat
K_q	Heat penetration factor
h	Heat transfer coefficient
P	Power applied to heater
Re	Reynolds number
Pr	Prandle number
Nu	Nusselt number
u^*	Friction velocity
λ	Friction factor
u	Velocity of coolant
ρ	Mass density of coolant
ν	Viscous velocity
$y+$	Wall coordinate
κ	Thermal conductivity
θ	Static contact angle

Author details

Tatsuhiko Aizawa[1*], Naoki Ono[2] and Hiroki Nakata[3]

1 Surface Engineering Design Laboratory, Shibaura Institute of Technology, Tokyo, Japan

2 Shibaura Institute of Technology, Tokyo, Japan

3 Ebina Denka Kogyo, Co., Ltd., Tokyo, Japan

*Address all correspondence to: taizawa@sic.shibaura-it.ac.jp

References

[1] Tickerman DB, Pease RFW. High-performance heat sinking for VLSI. IEEE Electron Device Letters. 1981;**EDL-2**(5):126-129

[2] Gong L, Zhao J, Huang S. Numerical study on layout of micro-channel heat sink for thermal management of electric devices. Applied Thermal Engineering. 2015;**88**:480-490

[3] Huang S, Zhao J, Gong L, Duan X. Thermal performance and structure optimization for allotted microchannel heat sing. Applied Thermal Engineering. 2017;**115**:1266-1276

[4] Li P, Luo Y, Zhang D, Xie Y. Flow and heat transfer characteristics and optimization study on the water-cooled microchannel heat sinks with dimple and pin-fin. International Journal of Heat and Mass Transfer. 2018;**119**:152-162

[5] Milica L, Milan P, Vladimir S. Boiling heat transfer modeling: A review and future prospectus. Thermal Science. 2018;**23**:249-272

[6] Chang JY, You SM. Enhanced boiling heat transfer from micro-porous surfaces: Effects of a coating composition and method. International Journal of Heat and Mass Transfer. 1997;**40**(18):4449-4460

[7] Kattoh Y. Fundamentals in Heat Transfer. Yokendo; 1964

[8] Sodtke C, Stephan P. Spray cooling on micro structured surfaces. International Journal of Heat and Mass Transfer. 2007;**50**:4089-4097

[9] Hendricks TJ, Krishnan S, Choi C, Chang C-H, Paul B. Enhancement of pool-boiling heat transfer using nanostructure surface on aluminum and copper. International Journal of Heat and Mass Transfer. 2010;**53**:3357-3365

[10] Ahn HS, Lee C, Kim H, Jo H, Kang S-H, Kim J, et al. Pool boiling CHF enhancement by micro/nanoscale modification of zircaloy-4 surface. Nuclear Engineering and Design. 2010;**240**:3350-3360

[11] Jo H, Ahn HS, Kang S-H, Kim MH. A study of nucleate boiling heat transfer on hydrophilic, hydrophobic and heterogeneous wetting surfaces. International Journal of Heat and Mass Transfer. 2011;**54**:564305652

[12] Jung SM, Preston DJ, Jung HY, Deng Z, Wang EN, Kong J. Porous Cu nanowire aerosponges from one-step assembly and their applications in heat dissipation. Advanced Materials. 2016;**28**:1413-1419

[13] Simitses GJ, Hodges DH. Fundamentals of Structural Stability. Butterworth Heinemann; 2006

[14] Amon CH, Murthy J, Yao SC, Narumachi S, Wu C-F, Hsieg C-C. MEMS-enabled thermal management of high-heat-flus devices EDIFICE; embedded droplet impinging for integrated cooling of electronics. Experimental Thermal and Fluid Science. 2001;**25**(5):231-242

[15] Hamidnia M, Luo Y, Li Z, Wang X. Capillary and thermal performance enhancement of rectangular grooved micro heat pipe with micro pillars. International Journal of Heat and Mass Transfer. 2020;**153**:119581

[16] Aizawa T, Tamaki M, Fukuda T. Large area micro-texture imprinting

onto metallic sheet via CNC stamping. Journal of Procedia Engineering. 2014; **81**:1427-1432

[17] Aizawa T, Fukuda T. Pulsewise-motion controlled stamping for micro-texturing onto aluminum sheet. Journal of Micro- and Nano Manufacturing. ASME. 2016;**4**:014502-014511

[18] Aizawa T. Micro-Texturing for Tribology and Surface Engineering in Manufacturing Process. Proceedings of the 4th ISAST Conference, Indonesia; 2016. pp. 1-17

[19] Aizawa T, Suga H, Yamaguchi T. Plasma-nitriding assisted micro-texturing into stainless steel molds. Proceedings of the 4th ICNFT. 2015: 9002/1-9002/6

[20] Aizawa T, Takashima T, Shiratori T. Plasma printing to fabricate the micro-piercing dies for miniature metal products. Proceedings of the 8th AWMFT. 2015:J1/1-J1/6

[21] Nagata T, Aizawa T. Plasma-nitriding assisted modification of stainless steels for micro surgery knives. Proceedings of the 11th ICOMM. 2016; **59**:1-5

[22] Aizawa T, Inohara T. Pico- and femtosecond laser micromachining for surface texturing. Ch. 1. In: Micromachining. London UK: IntechOpen; 2019. pp. 1-24

[23] Hasegawa T, Aizawa T, Inohara T, Yoshihara S-I. Mold-stamping of optical glasses by micro/nano-textured die to transcript the hydrophobicity. J. Japan Society of Technology of Plasticity. 2019; **60**:23-27

[24] Aizawa T, Inohara T, Wasa K. Femtosecond laser micro-/nano-texturing of stainless steels for surface property control. Journal of Micromachines. 2019;**10**(512):1-10

[25] Aizawa T, Fukuda T. Oxygen plasma etching of diamond-like carbon coated mold-die for micro-texturing. Surface and Coating Technology. 2013;**215**:364-368

[26] Aizawa T. Micro-texturing onto amorphous carbon materials as a mold-die for micro-forming. Applied Mechanics and Materials. 2014;**289**:23-37

[27] Yunata EE, Aizawa T. Micro-grooving into thick CVD diamond films via hollow cathode. Manufacturing Letters. 2016;**4**:17-22

[28] Shiratori T, Aizawa T, Saito Y, Wasa K. Plasma printing of an AISI316 micro-meshing punch array for micro-embossing onto copper plates. Journal of Metals. 2019;**9**(396):1-11

[29] Aizawa T, Saito Y, Hasegawa H, Wasa K. Fabrication of optimally micro-textured copper substrates by plasma printing for plastic mold packaging. International Journal of Automation Technology. 2020;**14**(2):200-207

[30] Aizawa T. Development of micro-manufacturing by controlled plasma technologies. J. Japan Society of Technology of Plasticity. 2017;**58**: 1064-1068

[31] Aizawa T, Yoshihara S-I. Microtexturing into AISI420 dies for fine piercing of micropatterns into metallic sheets. J. Japan Society of Technology of Plasticity. 2019;**60**:53-57

[32] Aizawa T, Inohara T, Wasa K. Fabrication of super-hydrophobic stainless steel nozzles by femtosecond laser micro-/nano-texturing. International Journal of Automation Engineering. 2020;**14**(2):159-165

[33] Aizawa T, Shiratori T, Yoshino T, Inohara T. Femtosecond laser trimming of CVD-diamond coated punch for fine embossing. Materials Transactions. 2020;**61**(2):244-250

[34] Aizawa T, Shiratori T, Kira Y, Inohara T. Simultaneous nano-texturing onto CVD-diamond coated piercing punch with femtosecond laser trimming. Journal of Applied Sciences. 2020; **10**(2674):1-13

[35] Aizawa T, Ono N. Boiling heat transfer control by micro–/nano-texturing of metallic heat-spreading devices. In: Proc. 3rd WCMNM-2021. Mumbai, India; September 21st, 2021 (in press)

[36] Aizawa T, Wasa K, Tamagaki H. A DLC-punch array to fabricate the micro-textured aluminum sheet for boiling heat transfer control. Journal of Micromachines. 2018;**9**(147):1-10

[37] Aizawa T, Gregorius A, Ono N. Boiling heat transfer on the micro-textured copper interface. Journal of Micromachines. 2022 (in press)

[38] Tani I. The flow close to the wall of a turbulent boundary layer. Japan Journal of Aerospace and Aeronautics Society. 1973;**21**(228):39-43

[39] JSME. Heat Transfer Data Handbook. JSME; 2009

[40] Aizawa T, Yoshino T, Inohara T. Micro–/nano-texturing of aluminum by precise coining for functional surface decoration. Journal of Metals. 2020; **10**(1044):1-10

[41] Aizawa T, Inohara T, Wasa K. Nano-texturing onto tool-surface by the femtosecond laser processing. In: Proceedings of the WCMNM-2021. India, Mumbai; September, 2021 (in press)

[42] Aizawa T, Yoshino T, Shiratori T, Inohara T. Femtosecond laser printing of micro–/nano-textures into DLC dies for functional decoration of light metals. Journal of Nanomaterials. 2021 (in press)

[43] Aizawa T, Yoshino T, Suzuki Y, Komatsu T, Inohara T. Micro–/nano-texture surface decoration of metals and metallic alloys via femtosecond laser processing and precise imprinting. In: Proceedings of the 13th AFGS. China, Hong Kong; 2021. pp. 79-86

[44] Aizawa T, Nakata H, Nasu T. Manufacturing and characterization of acicular Fe-Ni micro-textured heat-transferring sheets. In: Proceedings of the 5th WCMNM Conference. Leuven, Belgium; September, 2022 (in press)

Chapter 7

Multi-Track Overlapping by Laser-Treated and Its Effects on the Microstructural Behavior of Al-Fe Alloy Assessed by FEM

Moises Meza Pariona

Abstract

In this work, a numerical simulation by finite element method (FEM) and optimized multigrid technology was applied. In order to study the influence of multi-track overlapping on the microstructure processed by laser surface remelting (LSR) and so, the validation of the experimental and the numerical simulation results of the multi-track overlapping were accomplished, which was the goal. It was verified in this work, multi-track overlapping and thermal cycling allied to high-speed cooling of liquid metalinfluence the liquid flow, mechanism of heat transport, and microstructure evolution in the molten pool of laser-treated alloy, promoting so a nanostructure characteristic. Thus, FEM is capable of accurately simulating the multi-track overlapping of the workpiece LSR-treated. Results of the overlapping ratio, as well as, the depth where the phase transformation occurs from liquid to solid was very nearby, therefore, simulation and experimental results agree quite well. This type of laser-treated alloy has very special characteristics and it is of innovative character, then, in aerospace, aeronautical, and automobile industries can be applied.

Keywords: heat transfer, laser surface remelting, multi-track overlapping, microstructure, FEM, numerical simulation

1. Introduction

When multiple laser tracks are used, the latter track causes the tempering of the previous tracks and hardness reduction in multiple laser tracks, which leads to non-uniform hardness in the surface of components, this was argued by Li et al. [1]. Furthermore, because of the rapidity of thermal cycle, a series of complicated phenomena such as heat and mass transfer, phase transition, material properties, laser beam absorption, reflection, and radiation occur in a very short time, this study was corroborated too by Guan et al. [2]. In addition, the later authors argued that the numerical model suggests that Marangoni convection plays a predominant role in determining the solidification microstructure, and it increases significantly with the overlapping rate. However, Pariona et al. [3] investigated this effect and verified by

FEM, which the Marangoni effect is inversely proportional to the fluid flow velocity and directly proportional to the thermal gradient, hence, this effect controls the quality, morphological characteristic, and geometry of the molten pool. On the other hand, the study carried out by Karbalaian et al. [4], pointed out that FEM reduces the costs related to experimental measurement. Moreover, with simulation, it is easier to understand results analyze, and optimize the process.

Although Pariona et al. [5, 6] analyzed, during LSR-treatment in Al alloy, the melted zone was constituted of metastable phases by LAXRD analysis and it revealed the presence mainly of Al_2O_3 and AlN phases. These authors emphasized that these phases contributed to the microstructural modification, favored the characteristics of high hardness and corrosion resistance of LSR-treated workpiece in sulfuric acid.

Together with COMSOL Multiphysics based in heat transfer modeling and the experimental approaches by LSR technique were utilized to perform this work. This research was accomplished the numerical simulation by finite elements method of the isothermal temperature field for multi-track, thermal cycles for the multi-track overlapping on the workpiece surface and in the workpiece cross-section. Experimentally was studied the microstructure of Al-2% Fe alloy was with multi-track overlapping. Finally, the experimental validation of numerical simulation of the multi-track overlapping laser was accomplished.

2. Computational model

The thermal analysis of this work involves studying the transient thermal history according to from heat transfer. The following will be presented the theoretical aspects of numerical simulation.

2.1 Mathematical model

2.1.1 3D Mathematical model of transport phenomena

In the past decade, many mathematical models have been developed to study the transport phenomena in laser-treated alloys. Until now, the most recent and complete 3D mathematical models of transport were performed with the evolution of software and hardware. The authors [7] have already discussed this subject.

The main equations of the model of transport phenomena are shown in the following. Eq. (1) represents the transient heat conduction equation, which it was proposed by Yilbas et al. [8]:

$$\rho\frac{\partial(CpT)}{\partial t} = \frac{\partial}{\partial x}\left(k\frac{\partial T}{\partial x}\right) + \frac{\partial}{\partial y}\left(k\frac{\partial T}{\partial y}\right) + \frac{\partial}{\partial z}\left(k\frac{\partial T}{\partial z}\right) + (1 - rf)I_0\,exp\left(-\frac{x^2+y^2}{a^2}\right)exp\,(-\delta z) \tag{1}$$

where *x*, *y*, and *z* are the vertical, depth and horizontal coordinates, respectively, ρ is the density, c_p is the specific heat, *k* the thermal conductivity, r_f the surface reflectivity, I_o is the laser peak intensity,$_\delta$ is the absorption depth, *t* is time, and *a* is the diameter of the laser beam, assumed as Gaussian form (Gaussian parameter).

The laser beam moving on surface is considered as thermal loads for the simulate LSR process, which it has continuous scanning that describes the moving heat source and it has the Gaussian-like distribution form (1), which is considered as a volumetric heat source and that determine the beam penetration depth. The boundary conditions have been established. The heat exchange with the air at room temperature surrounding the treated material is described by Eq. (2).

$$k.\nabla T = h(T_{amb\_}T) \tag{2}$$

Where h indicates the heat flux caused by convective loss and T_{amb} is room temperature.

The losses of energy in form of radiation released from the hot surfaces of material are included in Eq. (3).

$$k.\nabla T = \varepsilon\sigma\left(T^4_{amb} - T^4\right) \tag{3}$$

The parameter ε indicates the emissivity of the material and the term σ represents Stefan-Boltzmann constant for radiation, σ = 5.67e^{-8}W/m^2 K^4).

2.2 Multigrid method

In the numerical simulation works, it is usually necessary to solve large systems through direct or iterative methods, for this, requests a large amount of processing time by Central Processing Unit (CPU), however, Fedorenko [9] proposed Multigrid method (MG) that dramatically reduce the processing time. According to Briggs et al. [10], this method consists, of the transference of information among a refined grid to coarse auxiliary grids, so, which the numerical smoothers are more efficient that optimizing the process. Besides that, Brandt and Trottenberg et al. [11, 12] investigated several parameters types, which can to modified the Multigrid method optimizes process and details and discussion about this method is given by Pariona et al. [13].

3. Materials and methods

3.1 Experimental characterization

The material tested was Al-2.0 wt. % Fe alloy, this alloy was prepared with pure raw materials with chemical composition in mass % as Al: 99.76, with impurity as Fe:0.09, Si: 0.06, Cu: 0.06, Ni: 0.03 and Fe: 99.97, with impurity Si: 0.01, Cu: 0.01, Ni:0.01. The casting assembly used in the solidification experiments consists of water-cooled mold, being the heat extracted was done only in the bottom, and it promotes a vertical upward directional solidification. The laser surface treatment was performed with a 2 kW Yb-fiber laser (IPG YLR-2000S) and its intensity at initial moment was I (0) = 1.81×10^9 W.m^{-2}. The power density was 4.8×10^5 W.cm^{-2} with multi-phase distribution of energy and with an approximately Gaussian profile, and the laser beam velocity fixed was 40 mm/s. The laser treatment without an assisting gas jet was executed, with the purpose of creating alumina and aluminum nitride phases in the environment at high temperature, being that these phases presented high resistant and corrosion mechanic resistant, which were studied by Pariona et al. [5].

Furthermore, for metallographic characterization in the cross-section, small samples were cut, sanded, and polished with colloidal silica. Micrographs were recorded by optical microscopy (OM, Olympus-BX51) and by an SEM, Shimadzu SSX-550 microscope. Besides, a chemical attack with 0.5% HF was also made on these samples for analysis microstructure.

3.2 Aspects of the numerical simulation by FEM technique

The simulations were carried out with COMSOL Multiphysics software^{TM}, in a microcomputer with Intel i7 2.8 GHz processor with 32 GB RAM and Linux operational system. For the simulation procedure, the geometry and mesh were constructed. For carrying out the simulation process, the proposed equations of chapter 2.1 were used.

Initial and boundary conditions were applied in the geometry, and the material's thermophysical properties were considered dependent on temperature, presented by Pariona et al. [14], and these properties were found through [15]. A moving heat source was established in the x-axis with a scan velocity of 40 mm.s^{-1}. The laser parameters used in this work (Eq. 1) were shown in previous work, such as given by Pariona et al. [3, 13].

Corresponding the heat flux and radiation expressions (Eqs 2 and 3) were applied as boundary conditions, where: h_{up} is 12.25 W.m^{-2}K^{-1}; h_{down} is 6.25 W.m^{-2}K^{-1}, these values were used of the literature [16], because, these values were not determined experimentally; meanwhile, the surface emissivity (ε) equaled to 0.33 and the room temperature was 300 Kelvin.

For the execution of solution process, Multigrid method was applied for optimization procedure. Hence, the following parameters were established in his method: in Solver mode was selected the geometric multigrid, the number of iterations was 2, the multigrid cycle was V cycle, the hierarchy generation method, the number of multigrid levels was 1, and mesh coarsening factor was 2.

4. Results and discussion

In this work was performed the numerical simulation by finite elements method of multi-track overlapping by means of RSL-treated technique. Simulation was conducted based in **Figure 1**, in this figure was accomplished a schematic diagram of the multi-track overlapping, therefore, to carry out this work, four tracks were considered, thereby, the multi-track overlapping and distance between consecutive tracks are displayed. So, the results of the thermal cycles were shown on the workpiece surface and in the cross-section for the four tracks.

4.1 Isothermal temperature field for multi-track overlapping

Isothermal temperature field for the multi-track by numerical simulation study using FEM was performed for the four tracks (**Figure 1**), whose tracks are schematized in **Figure 1** and at this figure are shown the coordinates of the points: I, II, III, and IV. The arrow indicates way that follows the laser beam. Whereas, after executing one of the tracks by the laser beam and soon to pass to the next track was calculated at 0.01 s the delay time between two tracks, as also it was displayed on the schematic. According to the scheme of **Figure 1** for the laser beam velocity v = 40 mm/s that was considered, then, the total time it takes to run the four tracks was calculated at 2.04 s.

In **Figure 1**, the isothermal temperature field has the form of an ellipse, nevertheless, they are deformed, this fact was confirmed by [7, 17]. These last authors affirmed

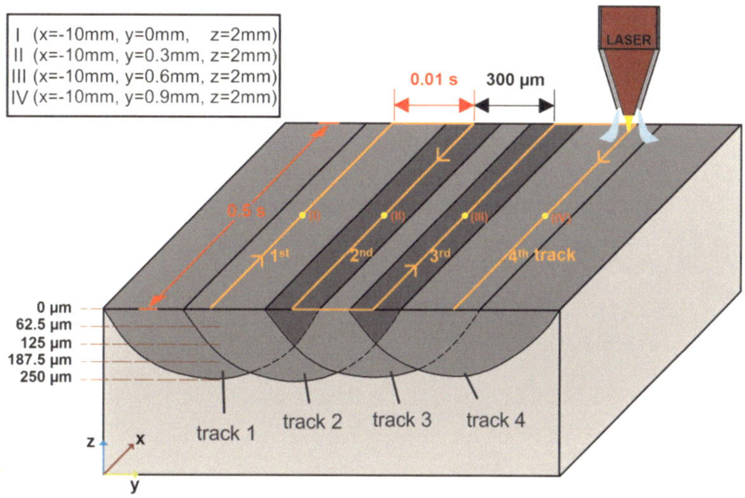

Figure 1.
Schematic diagram of the multi-track overlapping, where, the laser move occurred in a parallel line fashion. Hence the designation I, II, III, and IV are midpoints of each track.

that the ellipse of temperature field is not symmetrical by the center of the facula, but form the deviation on one side of the cladding, still, they argued, the phenomenon that the multi-track cladding forms the partial ellipse, is induced by influence of the former cladding on the latter one.

In **Figure 2a** the isothermal temperature field for two tracks on the workpiece surface was presented and it can be observed in 3D, where, the directions of the laser beam are indicated by arrows in figure. The first track was presented at instant 0.4 s, which corresponds to the forward motion of the laser beam and for the fourth track at instant 1.9 s was displayed, which corresponds to the return of the laser beam. In this figure, the solidification conditions ahead for the first track, presents an enlargement in the direction of the movement of the laser beam, though, the solidification conditions ahead for the fourth track is deformed, due to the fact that it reached the final part of its movement, where the thermal waves exhibited an aspect as if it were a rebound. While, to elucidate with more detail, in **Figure 2b** the isothermal is displayed in 2D, where all the phenomena mentioned are observed again. Still, the authors [1] obtained similar results to **Figure 2**, however, these authors did their research for two tracks, meanwhile, and this research was done for four tracks.

4.2 Thermal cycles for multi-track overlapping

In the next will be presented a result of analysis of thermal cycles, so much, on the workpiece surface and in the workpiece cross-section.

The result of thermal cycles on the workpiece surface is shown in **Figure 3**. It was based on **Figure 1**, where thermal cycles on the surface were raised at the midpoints of each track and those points were represented by: I, II, III, and IV. e.g., the first peak (or point I) of thermal cycle corresponds at the instant 0.25 s; the second peak (or point II) at 0.76 s; the third peak (or point III) at 1.27 s, and the quarter peak (or point IV) at 1.78 s. Fits to highlight here, the travel total time of the laser beam along the way is indicated by arrows, such as indicated in **Figure 1**.

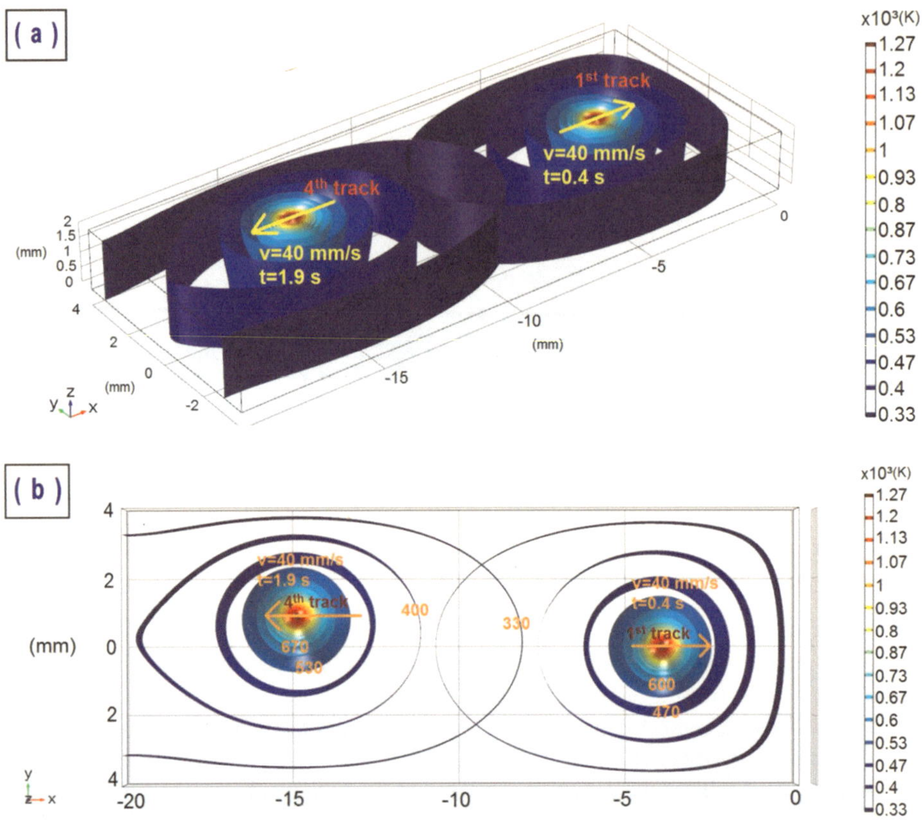

Figure 2.
The isothermal temperature field for multi-track on the workpiece surface. (a) Viewed in 3D and (b) Viewed in 2D.

As we can see in **Figure 3a**, the temperature evolution of point I of the 1st track versus time and influences of other tracks on the first track is shown. So, the first peak in **Figure 3a** is the temperature rise induced by the 1st track, it reached a temperature of around 1300 K. When the laser beam moves away, the temperature of point I decrease to room temperature at t = 0.5 s. The second peak shows the temperature rise of point I when performing the second track. Therefore, the second peak is the influence of the 2nd track at point I, this peak is above the solidification temperature (933 K) and it is related to phase transformation. The third peak did not reach the phase transformation, but its intensity is relatively high, this peak displays the temperature elevation of point I when passing the third track. Finally, the fourth peak, which is in solid-state temperature of the alloy, likewise, it was thermal influence of the 4th track on the 1st track (point I), thereby, whose influence is insignificant.

Furthermore, temperature field characteristic for the second track can be seen in **Figure 3b**, it is the temperature rise induced by the 2nd track of the laser beam, over again, an intense peak appears so equal to peak of the first track (**Figure 3a**); but here is a difference in relation to the first peak that was described previously. In this figure, it can be observed that the temperature peaks of the first neighbors around the second track are similar, as well as in intensity. In addition, both are above of solidification point of alloy and still correspond to phase transformation, thus, influence of temperature around the first neighbors (the first and third peaks) at the 2nd peak, it is very

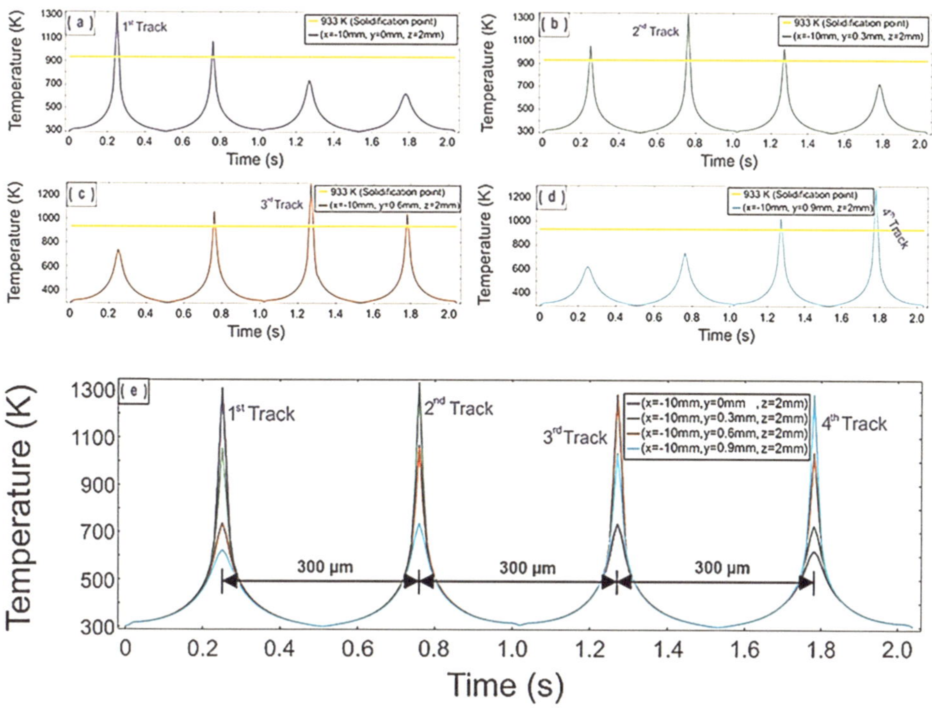

Figure 3.
Thermal cycles for the multi-track overlapping on the workpiece surface. (a) influences of other tracks on the first track, (b) influences of other tracks on the second track, (c) effects of other tracks on the third track and (d) effects of other tracks on the fourth track, and (e) summarized all four cases previously treated.

considerable. However, influence of the temperature field of the fourth track at the second is negligible.

Moreover, intensity of the temperature field of the third track (**Figure 3c**), where the characteristic is of form similar to the temperature field of the second track (**Figure 3b**), therefore, the temperature peaks of the first neighbors at the third track have considerable influence, however, the effect of the temperature field of the 1st track at the 3 rd track is depreciable. Nevertheless, the behavior of the temperature field of the fourth track (**Figure 3d**) is similar to the temperature field of the first track (**Figure 3a**), but in inverse mode.

In **Figure 3e** is summarized all four cases previously treated, where intensities are denoted with different colors as shown in this figure. In addition, the distances between tracks are displayed, as well as the coordinates of each track are presented. The time interval between each peak was 0.5 s.

Result of the thermal cycles for multi-track overlapping on the workpiece surface by numerical simulation studied using FEM was shown in **Figure 3**, this result gives an idea when a track was performed by the laser beam, the temperature of treated track reaches a high magnitude, and soon decreases to room temperature, as was observed, also, showed the temperature rise due to the influence of the other neighboring tracks. Certainly, this result could influence the microstructural characteristic and electrochemical behavior of the workpiece.

Besides, **Figure 4** shows behavior of the thermal field profile of thermal cycles in the workpiece cross-section as a function of depth, as shown schematically in

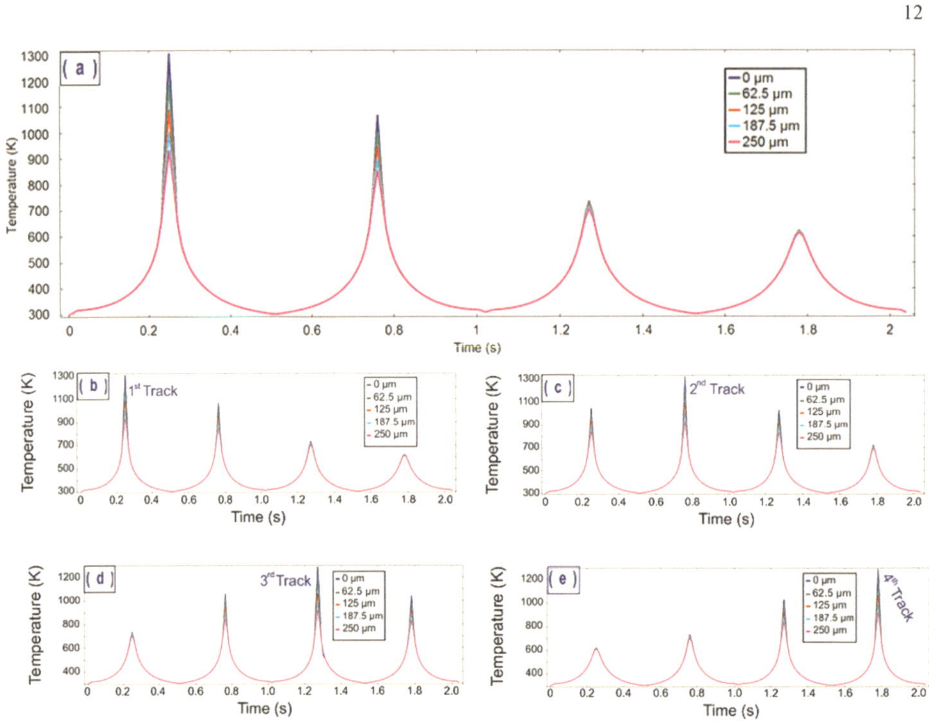

Figure 4.
Shows behavior of the thermal field profile of the thermal cycles at the workpiece cross-section as a function of depth. (a) thermal profile of each track at the point I coordinates, (b) same that (a), (c) thermal profile of each track at the point II coordinates, (d) thermal profile of each track at the point III coordinates, and (e) thermal profile of each track at the point IV coordinates.

Figure 1, where the depths with different colors are displayed. For each track (**Figure 1**), the same depths were considered; in this work only the z-axis (depth) was varied. It should be remembered that the depth at the phase transformation from liquid to solid occurs is 250 μm. E.g., the most intense peak of **Figure 4a** corresponds when the laser beam passed through the first track, which corresponds point I of **Figure 1**. This track will be considered as a reference to remark the influence of other tracks in this one. Therefore, at point I, it can be seen the variation of the laser beam intensity as a function of depth in different colors, the most intense belongs to the molten pool zone of alloy. However, the second peak in this same figure corresponds to when the laser passed through the second track and whose effect is signaled, at point II can be noted the thermal profiles as a function of depth and whose effect of this peak is intense. The third peak of **Figure 4a** connotes influence of the thermal field at the point I coordinates, when the laser beam passed through the third track, whose effect is reasonable. Finally, the fourth peak is when the laser beam passed through the fourth track, whose effect is shown at the point I coordinates with negligible characteristics. Nevertheless, **Figure 4a** represents magnification of **Figure 4b**.

Therefore, **Figure 4c** represents the thermal profile of thermal cycles of each track referred to at the point II coordinates of **Figure 1**. Where, the most intense peak in this figure corresponds to when laser passed by the second track, where thermal profiles are shown as a function of depth. The first peak corresponds to the first track and third peak corresponds to the third track, they have an appreciable and a similar influence at the point II coordinates, whose intensities are above the solidification temperature. The fourth peak represents a negligible influence at the point II coordinates.

Figure 4d represents the thermal profile of thermal cycles of each track referred to at the point III coordinates of **Figure 1**. Where, the most intense peak corresponds to when the laser beam passed through the third track, in this peak can be appreciated the thermal profiles as a function of depth. The thermal profile corresponds to the first peak and whose influence is presented at the point III coordinates; however, it has a negligible characteristic. Nevertheless, the third and fourth peak has a reasonable influence on the referred point.

Finally, **Figure 4e** is the thermal profile of thermal cycles of each track referred to at the point IV coordinates of **Figure 1**. It means that the most intense peak corresponds to the fact that the laser beam passed through the fourth track; also, thermal profiles can be noticed as a function of depth at the same point IV. Besides, the third peak most intense is the influence of the third track at the point IV coordinates, since it has an appreciable influence. However, other peaks have negligible influences at the point IV coordinates.

4.3 Experimental validation of the numerical simulation of multi-track overlapping

For this goal, multi-track overlapping was simulated by finite element technique and finally, these results were confronted and the validation of multi-track overlapping effects on the microstructure. Since this type of study was done for Al–2.0 wt.%Fe alloy and the laser beam velocity was 40 mm/s.

The multi-track overlapping effect on the microstructure is shown in **Figure 5a** for Al–2.0 wt.%Fe alloy, where the multi-track overlapping can be clearly observed. To clarify, the overlapping lines were dashed at blue, where the three tracks are shown within the melted zone. In reference to the first track, whose diameter was 760.7 μm, within this area there are two overlapping due to the influence of the two tracks of first neighbors. For the first neighbor of the first track whose distance of overlapping at adjacent of the two tracks, was 472.2 μm, nonetheless, for the second neighbor was 203.3 μm. As it can be perceived, due to influence of the first neighbors tracks at area of the first track, therefore, there are three regions in the molten pool. Thus, this region showed a homogeneous characteristic, with a higher porosity concentration at the first two regions from left to right, but at the latter region, porosity concentration was lower due to multi-passage of the laser beam, therefore, this leads to greater mechanism of heat transport at the molten pool of the latter region, thus avoiding lower porosity concentration.

The overlapping ratio R, proposed by the authors Li et al. [18], it is denoted as:

$$R = d/D \times 100\% \tag{4}$$

Where d and D are the distances of overlapping of the adjacent of the two tracks and the diameter of the laser beam (=0.3 mm), respectively.

According to **Figure 5a**, the overlapping ratios R was calculated for two overlapping of the first neighbors.

For the first neighbor, R = 62% and for the second neighbor, R = 26.7%.

This means that the first overlapping ratio has a contribution of 62% and the second overlapping ratio has 26.7% at area of the first track. Then we can be stated,

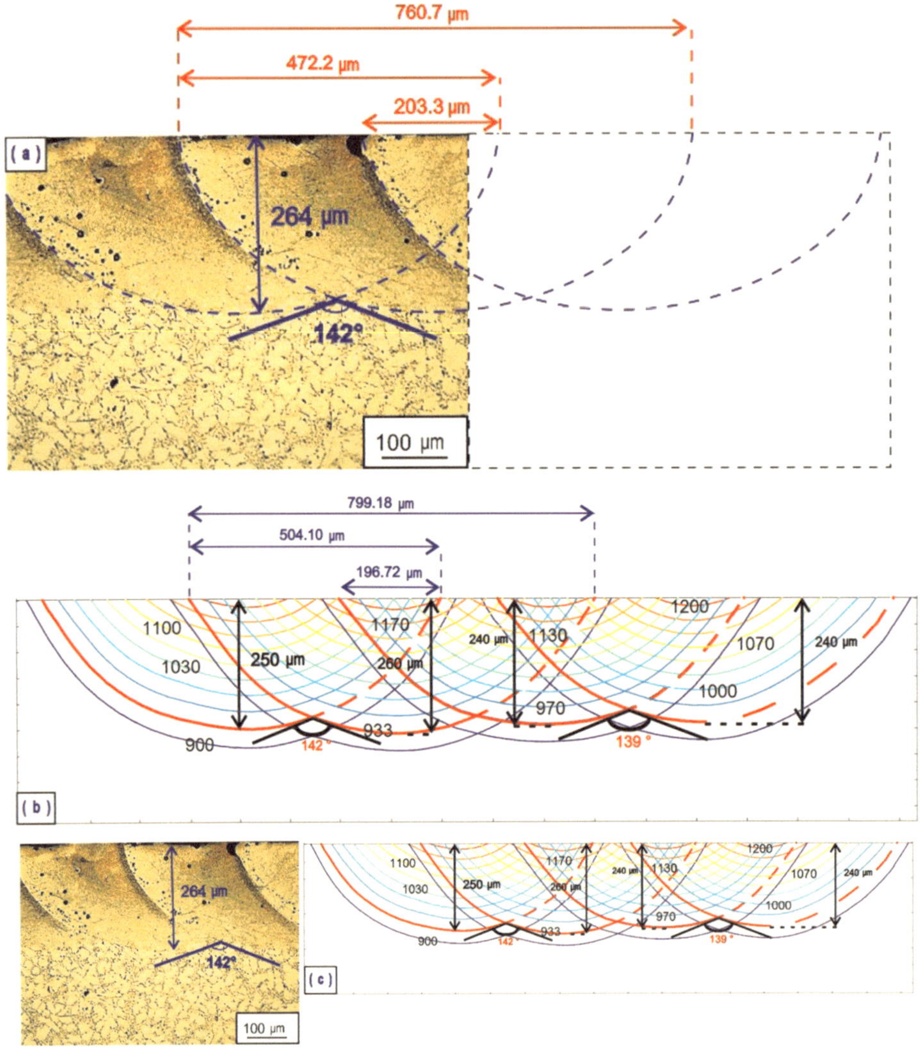

Figure 5.
Experimental validation of numerical simulation of the multi-track overlapping laser. (a) microstructure of Al-2% Fe alloy with multi-track overlapping, (b) numerical simulation of the multi-track overlapping, and (c) confront of the numerical simulation and experimental results.

the overlapping ratio depends on various parameters, among them, type of alloy, distance between tracks, laser beam speed, and laser beam power.

In addition, the overlapping was simulated for all tracks and the result is shown in **Figure 5b**. While, the quantitative data of the overlapping ratio, using Eq. (4) were:

For the first neighbor, R = 63% and for the second neighbor, R = 24.6%

Figure 5a measured the intersection angle between melt tracks, this measure was 142° and for [19] was 128°, since, this author carried out an in-depth study on microstructures of aluminum alloys. Well then, we can state, this angle depends on alloy type, distance of overlapping adjacent of two tracks, laser beam speed and laser power, and all these parameters also influence the alloy properties.

In order to illustrate better this work, a numerical simulation of multi-track overlapping was performed using finite element technique and optimized by the multigrid technique. To carry out the simulation, a schematic diagram is shown in **Figure 1**, where four tracks are shown. The results of this simulation are presented in **Figure 5b**. For each track, the thermal field in form of isotherms was presented, besides, some isotherms are presented with the corresponding temperatures. Furthermore, the depth where phase transformation occurs from liquid to solid are also shown, e.g., for the first track, the depth was 250 μm; for the second it was 260 μm; for the third, it was 240 μm and for the latter, it was 240 μm. However, experimentally the depth was 264 μm (**Figure 5a**), therefore, this small variation of depth can be attributed due to variation of thermal gradient, or owing to the Marangoni effect [3] and also, can be as a consequence of variation of thermophysical properties, given by [13], thus, experimental and simulated results were very close. Furthermore, results of overlapping ratio (**Figure 5**) for both experimental and simulated cases were quite close, as well as, the depth where phase transformation occurs from liquid to solid, also, were very nearby, both the results of the simulation and the experimental part agreeing quite well. Therefore, FEM is capable of accurately simulating the multi-track overlapping, to illustrate better this comparison, in **Figure 5c** these details are shown. According to results of the numerical simulation of the thermal cycles for the multi-track overlapping by LSR-treated, so much, on the workpiece surface and in the cross-sections. Then, we can be stated, distance between tracks, the laser beam velocity, the laser beam characteristic, and alloy type influence the numerical simulation result and consequently also, the microstructural characteristic of alloy.

The following will be presented a discussion about the thermal cycles, the overlapping of laser beam tracks by different authors. Cordovilla et al. [20] argued that effect of preheating induced in the material by previous tracks on kinetic of current track and so on; the thermal cycles induced in the material by a laser surface hardening process are the driving force for all the metallurgical transformations and they performed the numerical simulation of overlapping only for two tracks. On other hand, Li et al. [1] reported, melted zones created by multiple tracks should be uniform and which provides condition to compose fine grains and a high hardness melted layer, in addition, these same authors observed, that temperature change rate in overlapping area increases with increase in overlapping ratio, in agreement with experimental ones. Also, they stated, to predict the hardness distribution under two consecutive parallel tracks carried out with realistic thermal cycles, the transformation maps must be validated that will serve as a tool to confront the overlapping process, this study was supported by Cordovilla et al. [20].

Guan et al. [2] verified, which numerical model suggests that Marangoni convection plays a predominant role in determining the solidification microstructure, and it increases significantly with the overlapping rate and however, it may play a significant role in influencing final surface properties of laser-treated materials, however, Guan et al. [21] analyzed that overlapping is important in determining corrosion resistance due to microstructure in-homogeneities in the molten pool and this study was also remarked by Reitz and Rawers [22]. As described by He et al. [23], overlapping tracks affect heat transfer and liquid flow, microstructure evolution, and these authors from a well-tested numerical heat transfer and fluid flow model analyzed the mechanism of heat transport in the molten pool.

Liu and Qi [7] point out that the overlapping ratio was the key parameter to ensure the continuity and consistency of the epitaxial columnar dendrite growth. Still, they vary the deposition parameters and the alternating scanning method, the epitaxial growth of columnar dendrite microstructure can be achieved in a multi-track and in multi-layer deposit. Moreover, Besides, Lakhkar et al. [24] focused their study on the development of a numerical model to predict the back tempering in multi-track laser hardening of AISI 4140 steel and the predictions of the multi-track by numerical simulation, it was validated in a satisfactory way with the resulted experimental of the microstructure. Similar result was given by Yao et al. [25] that analyzed the structure of the overlapping zone in the microstructure in alloyed steel by SEM.

With all this, overlapping of laser beam tracks, thermal cycling, and allied to high-speed cooling influences the microstructural characteristics of laser-treated alloy, promoting a nanostructure characteristic in the molten pool, where metastable phases are formed, as well as in this case, alumina and nitride phases were produced by environment during laser-treatment, this subject was discussed trough low-angle x-ray diffraction analysis, by Pariona et al. [3, 26]. Furthermore, these authors confirmed, characteristics of overlapping, as well as microstructure in each region and distribution of nanopores, therefore, they must influence mechanical properties and corrosion resistance in corrosive media of the LSR-treated alloy with multi-track.

4.4 Workpiece microstructure by the multi-track overlapping by laser treated

The microstructural result in the workpiece cross-section was already presented in the previous section, where was discussed the analysis result of thermal cycles, the multi-track overlapping by LSR-treated through means of numerical simulation, and finally, the numerical simulation and experimental results were confronted. This section will be presented and discussed of microstructure of the as-received alloy on the workpiece surface, with the purpose of analyzing the multi-track overlapping influence in the microstructure characteristics.

Figure 6 illustrates the morphology of hypereutectic Al-2.0 wt.% Fe alloy laser-treated, analyzed through OM and FESEM, showing the multiple laser tracks

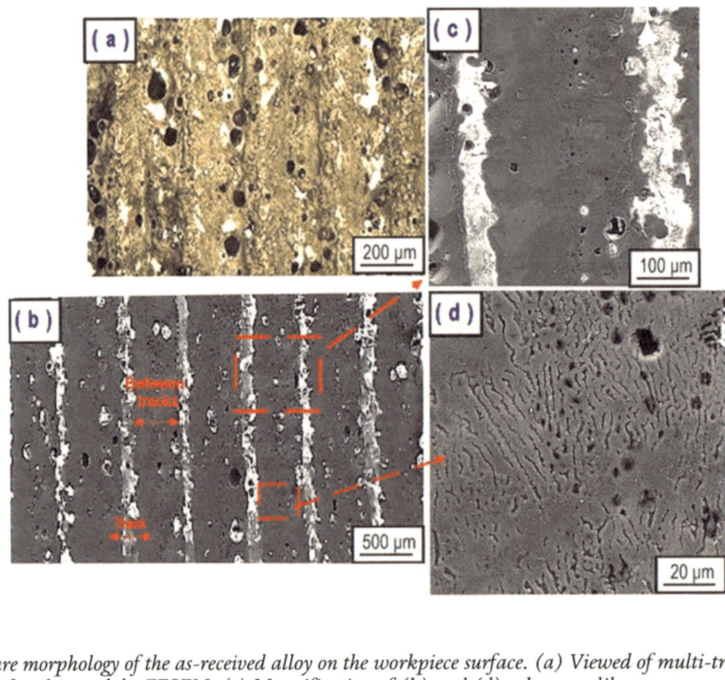

Figure 6.
Microstructure morphology of the as-received alloy on the workpiece surface. (a) Viewed of multi-truck by OM, (b) Viewed of multi-truck by FESEM, (c) Magnification of (b) and (d) columnar-like structure growth on the molten pool.

characteristics formed during laser treatment. OM image in **Figure 6a** shows the surface morphology, where multi-track is observed, while FESEM image in **Figure 6b** shows the morphology in more detail of region on the track and between the tracks, it corresponds to multi-track overlapping region, according to **Figure 5**.

Meanwhile, in **Figure 6**, as can be seen, the region on the track contains a higher concentration of defects than at region between the tracks, therefore, Zhang et al. and Kalita [19, 27] reported a similar result. In **Figure 6b**, the distance between the tracks is approximately 300 μm, in this region we can note presence of several nanopores, which may be attributed to volatilization of inclusions or vaporization of the substrate itself, caused by hydrogen and moisture in atmospheric air, which they are absorbed in laser-treated region, favoring thus the formation of pores, these results are consistent such as reported by Yilbas et al. and Pariona and Micene [5, 8]. The micrograph in **Figure 6c** shows a region on the track and region between the tracks under higher magnification, showing the concentration of defects with more detail. **Figure 6d**, also viewed under higher magnification, shows the region between the tracks, revealing, a uniform morphology with grains varied the form from columnar-like structure to irregular, confirming thus that the molten pool zone showed a fine microstructure due to high quenching rates. Nevertheless, [3] also observed these structures in Al-1.5 wt% Fe and [1], the last authors confirmed that Al-Co-Ce alloys contain Al-rich eutectic regions and whose structure was similar to Al-2.0wt.% Fe alloy. Consequently, due to peculiar characteristics of the microstructure shown in **Figure 6**, which presented highly improved properties, such as hardness, corrosion, and wear resistance, which are a result of dissolution of precipitates and formation of metastable phases, this subject was extensively discussed by Pariona and Micene [5]. Furthermore, several authors have reported similar findings, among them, [6, 18, 28].

The study is done by Pariona et al. [14], analyzed the hypoeutectic Al-1.5 wt.% Fe alloy LSR-treated and they observed the presence of nano-cracks between the tracks. However, this phenomenon in this study was not observed at the hypereutectic Al-2.0 wt% Fe alloy, as can be verified in **Figure 6c** and **d**, because, of the absence of microcracks was expected. Nevertheless, according to [29] stated that the formation Al-Fe alloys is impaired when the material contains coarse Al_3Fe particles, which tend to produce microcracks and reduce formability, whereas, this does not occur in the presence of Al_6Fe finely dispersed at the Al-2.0 wt.% Fe alloy. Meanwhile, Gremaud et al. [30] reported that increasing cooling rate of hypereutectic alloys containing up to 9 wt.% of Fe, which suppresses the formation of the stable Al_3Fe phase and is replaced by Al_6Fe phase, this fact, was confirmed in our result.

Other similar works were discussed by several authors, the following will be displayed [19, 31], these authors confirmed that the overlapping of laser beam tracks has a significant influence on surface quality of laser-treated materials. In addition, these authors showed how overlapping tracks affect heat flow, the solidification microstructure, and the electrochemical behavior, where the melt zone showed fine homogenized microstructure, due to high quenching rates during laser melting. Cordovilla et al. [20] remarked, which each track affects the microstructures produced by previous one and which when two consecutive tracks take place in an overlapping process. Meanwhile, all the studies cited by different authors confirm our results consistently.

So then, the hypereutectic Al-2.0 wt.% Fe alloy laser-treated is very peculiar and which this alloy has very special characteristics, still it has an innovative character and which can be applied in aerospace, aeronautical, and automobile industries. Karbalaian et al. [4], they pointed out that FEM reduces the costs related to experimental measurement. Moreover, with the simulation is easier to understand, analyze the result and optimize the process.

5. Conclusion

Will be pointed out the most important results of experimental validation of multi-track overlapping through the numerical simulation by FEM and whose influence in the microstructure of laser-treated workpiece.

1. The induced preheating effect by two overlapping of the first neighbors tracks on the kinetic of current track in the workpiece was verified through numerical simulation of the multi-track overlapping. Thus, FEM is capable of accurately simulating.

2. The results of overlapping ratio, as well as, the depth where the phase transformation occurs from liquid to solid were very nearby, both for the experimental and numerical simulation results.

3. The multi-track overlapping was simulated by the finite element technique and these results were confronted with the microstructure, being that the results were very close, however, the geometric dimensions of the molten pool and the intersection angle between melt tracks also were very close.

4. The multi-track overlapping and thermal cycling allied to high-speed cooling, influence, the liquid flow, mechanism of heat transport, and microstructure evolution in the molten pool of laser-treated alloy, promoting so a nanostructure characteristic.

5. The melt zone showed fine microstructure due to high quenching rates during laser melting and thus producing a fine homogenized microstructure at the melted region.

6. The multi-track overlapping influence on the surface quality of laser-treated materials.

7. This type of laser-treated alloy has very special characteristics and is of innovative character, then, it can be applied in the aerospace, aeronautical, and automobile industries

Acknowledgements

This work was entirely financed by CNPq (Brazilian National Council for Scientific and Technological Development), FundaçãoAraucária (FA), CAPES (Federal Agency for the Support and Evaluation of Postgraduate Education), and FINEP (Research and Projects Financing Agency). Also, thanks to LABMU–UEPG.

Author details

Moises Meza Pariona
Department of Engineering and Materials Science, State University of Ponta Grossa (UEPG), Ponta Grossa, PR, Brazil

*Address all correspondence to: mmpariona@uepg.br

References

[1] Li C, Wang Y, Zhang Z, Han B, Han T. Influence of overlapping ratio on hardness and residual stress distributions in multi-track laser surface melting roller Steel. Optics and Lasers in Engineering. 2010;**48**:1224-1230

[2] Guan YC, Zhou W, Li ZL, Zheng HY. Influence of overlapping tracks on microstructure evolution and corrosion behavior in laser-melt magnesium alloy. Materials and Design. 2013;**52**:452-458

[3] Pariona MM, Taques AF, Woiciechowski LA. The Marangoni effect on microstructure properties and morphology of laser-treated Al-Fe alloy with single track by FEM: Varying the laser beam velocity. International Journal of Heat and Mass Transfer. 2018; **119**:10-19

[4] Karbalaian HR, Yousefi-Koma A, Karimpour AM, Mohtasebi SS. Investigation on the effect of overlapping laser pulses in laser shock peening with finite element method. Procedia Materials Science. 2015;**11**: 454-458

[5] Pariona MM, Micene KT. The alumina film nanomorphology formed to improve the corrosion resistance of Al–2.0 wt.%Fe alloy as result of the laser surface melting technique applied. Advances in Chemical Engineering and Science (ACES). 2017;7:10-22

[6] Pariona MM, Teleginski V, dos Santos K, dos Santos ELR, de Lima AAOC, Riva R. AFM study of the effects of laser surface remelting on the morphology of Al Fe aerospace alloys. Materials Characterization. 2012a;**74**:64-76

[7] Liu Z, Qi H. Mathematical modeling of crystal growth and microstructure formation in multi-layer and multi-track laser powder deposition of single-crystal superalloy. Physics Procedia. 2014;**56**: 411-420

[8] Yilbas BS. Analytical investigation into laser pulse heating and thermal stresses. Optics & Laser Technology. 2009;**41**:132-139

[9] Fedorenko RP. On the Speed of Convergence of an Iteration Process. USSR Computational Mathematics and Mathematical Physics. 1964;**4**:227-235

[10] Briggs WL. Henson VE, Mccormick SF. A Multigrid Tutorial, second ed., SIAM, Philadelphia, USA. 2000

[11] Brandt A. Multi-level adaptive solutions to boundary-value problems. Mathematical Computing. 1977;**31**:333-390

[12] Trottenberg U, Oosterlee CA, Schüller A. Multigrid. Germany: St. Augustin, Academic Press; 2001

[13] Pariona MM, de Oliveira F, Teleginski V, Machado S, Pinto MAV. Simulation and optimization of al-fe aerospace alloy processed by laser surface remelting using geometric multigrid solver and experimental validation. Heat Mass Transfer. 2016;**52**:1037-1049

[14] Pariona MM, Teleginski V, dos Santos K, de Lima AAOC, Zara AJ, Micene TM, et al. Influence of laser surface treated on the characterization and corrosion behavior of Al–Fe aerospace alloys. Applied Surface Science. 2013;**276**:76-85

[15] THERMO CALC Software. 2015. Stockholm, Sweden

[16] COMSOL. Release notes 5.0, Friction Stir Welding. Stockholm, Sweden. 2017

[17] Hong-yun Z, Hong-tao Z, Chun-hua X, Xian-qun Y, Hong-yun Z, Hong-tao Z, et al. Temperature and stress fields of multi-track laser cladding. Transactions of Nonferrous Metals Society of China. 2009;**19**:495-501

[18] Bertelli F, Meza ES, Goulart PR, Cheung N, Riva R, Garcia A. Laser remelting of Al–1.5wt.% Fe alloy surfaces: Numerical and experimental analyses. Optics and Lasers in Engineering. 2011;**49**: 490-497

[19] Kalita SJ. Microstructure and corrosion properties of diode laser melted friction stir weld of aluminum alloy 2024 T351. Applied Surface Science. 2011;**257**:3985-3997

[20] Cordovilla F, García-Beltrán A, Sancho P, Domínguez J, Ruiz-de-Lara L, Ocaña JL. Numerical/experimental analysis of the laser surface hardening with overlapped tracks to design the configuration of the process for Cr-Mo steels. Materials and Design. 2016;**102**: 225-237

[21] Guan YC, Zhou W, Zheng HY, Li ZL. Solidification microstructure of AZ91D Mg alloy after laser surface melting. Applied Physics A. 2010;**101**: 339-344

[22] Reitz W, Rawers J. Effect of laser surface melted Zirconium alloys on microstructure and corrosion-resistance. Journal of Material Science. 1992;**27**: 2437-2443

[23] He X, Elmer JW, DebRoy T. Heat transfer and fluid flow in laser microwelding. Journal of Applied Physics. 2005;**97**:084909

[24] Lakhkar RS, Yung SC, Krane MJM. Predictive modeling of multi-track laser hardening of AISI 4140 steel. Materials Science and Engineering A. 2008;**480**: 209-217

[25] Yao C, Xu B, Huang J, Zhang P, Wu Y. Study on the softening in overlapping zone by laser-overlapping scanning surface hardening for carbon and alloyed steel. Optics and Lasers in Engineering. 2010;**48**:20-26

[26] Pariona MM, Teleginski V, Santos K, Machado S, Zara AJ, Zurba NK, et al. Yb-fiber laser beam effects on the surface modification of Al–Fe aerospace alloy obtaining fillet weld structures, low fine porosity and corrosion resistance. Surface and Coatings Technology. 2012b;**206**:2293-2301

[27] Zhang X, She J, Li S, Duan S, Zhou Y, Yu X, et al. Simulation on deforming progress and stress evolution during laser shock forming with finite element method. Journal of Materials Processing Technology. 2015;**220**:27-35

[28] Majumdar JD, Pinkerton A, Liu Z, Manna I, Li L. Microstructure characterization and process optimization of laser assisted rapid fabrication of 316L stainless steel. Applied Surface Science. 2005;**247**:320-327

[29] Mondolfo LF. Aluminum Alloys: Structure and Properties. 9th ed. London: Butterworths; 1976

[30] Gremaud M, Carrard M, Kurz W. The microstructure of rapidly solidified Al-Fe alloys subjected to laser surface treatment. Acta Metallurgia et Materialia. 1990;**38**:2587-2599

[31] Cao S, Gu D, Shi Q. Relation of microstructure, microhardness and underlying thermodynamics in molten pools of laser melting deposition processed TiC/Inconel 625 composites. Journal of Alloys and Compounds. 2017; **692**:758-769

Chapter 8

A Review on Condensation Process of Refrigerants in Horizontal Microfin Tubes: A Typical Example

Thanh Nhan Phan

Abstract

Heat transfer performance of refrigerant on the condensation process is very important in the designing of condensation equipment, especially in air conditioning and refrigeration systems. The outstanding advantages of microfin tubes are reducing the weight and size of condensers and also reducing the amount of refrigerant on the system. Reviewing the general concept of condensation and detailing the formation of flow pattern map that is also the procedure to determine heat transfer coefficient and pressure drop during condensation process of refrigerant inside the horizontal microfin tubes would be considered. Also, a typical example will be presented to illustrate a detailed procedure to calculate the value of heat transfer coefficient and pressure drop during the condensation process in horizontal microfin tubes. The data results show that microfin tube J60 with 60 number of fins inside 8.96 mm inner diameter, 0.2 mm height of fin, 40° of apex angle and 18° of helix angle, the condensation procedure of R1234ze at 35°C, heat flux of 8.62 kW/m^2, and mass flux of 222 kg/m^{2}s that could be estimated with heat transfer coefficient would be reduced from 8160.4 [W/m^2 K] to 1413.8 [W/m^2 K] follow with quality x changes from 0.99 to 0.01, and the maximum pressure drop of this process is 3173.8 [Pa/m].

Keywords: condensation, microfin tube, flow pattern, heat transfer coefficient, pressure drop

1. Introduction

The condensation procedure of refrigerant is an important issue in the refrigeration industry. It is strongly supported to the optimization of equipment by using the right refrigerant with the very good thermodynamics properties and friendly environment and also by applying the microfin tube to conceive, design, implement, and operate in order to improve the heat transfer performance of system and to reduce the size, weight of the system, and the amount of refrigerant used in the system. The best performances of microfin tubes are consequences of the increase of the heat transfer and the increase of the pressure drop. To understand this phenomenon, many researchers these days are demonstrating to analyze and create the new models for flow pattern map, heat transfer, and pressure drop.

The group research of Thome and Cavallini with their colleagues [1–8] is concentrated in condensation phenomena from the past to now. They presented so much information from experiments to empirical correlation for these fields, condensation on tubes (smooth, microfin tubes, micro channel, etc.) to consider about the formation of flow regime and also the penalization terms for heat transfer and pressure drop. With the group of Muzzio, Colombo and Lucchini [9–12], they have been considering about an experimental facility to determine the flow patterns map and heat transfer performance of condensation on the difference geometries of microfin tubes with vary of refrigerant from R22 to R134a, and now they keep working with new refrigerant R1234ze and R1234yf. Besides, the group research of Sunden and Wu [13–15] is also focused on two-phase phenomena on microfin tube with some difference geometries for the current refrigerants by using their experimental data as well as collected data results from the others group to make an overview for their purposes.

More specific, about the flow pattern maps, until now is no new update of the maps for microfin tubes. Those are still presented the flow pattern map of condensation procedure of refrigerant on smooth horizontal tube, with the maps for condensation in microfin tubes are come from experimental map, it is not represented to build for other conditions. So that, apply the map on smooth tubes to estimate and get the stepping stone of the behavior and the changing of two-phase flow in the tube in circumstances. The map of Breber [16] was introduced in 1980 after 2 years the map of Tandon [17] was published, and in 2003 the map of Hajal [1], the group research of Thome and Cavallini, was presented.

About the performance of heat transfer, it cannot be denied that the outstanding performance of heat transfer in microfin tubes for condensation is normative to compare with smooth tubes. A huge number of experimental researches have been done to contribute to a lot of databases for establishing the predictive and empirical models to determine the heat transfer coefficients and pressure drop. General correlations for heat transfer of condensation processes on microfin tubes are published and used to estimate data results with an acceptable value to design condensers in refrigeration systems. Those are Yu and Koyama [18] Kedzieski and Goncalves [19], Han and Lee [20], Cavallini et al. [4]. Also, for the pressure drop of two-phase flow, some correlations were built for both evaporation and condensation procedure of Choi et al. [21], Goto et al. [22], and the others used for separated purposes. Some particular pressure drop correlations could be mentioned for condensation procedure of Kedzieski and Goncalves [19], Haraguchi et al. [23], and Cavallini et al. [24].

This chapter is focused on the condensation procedure of refrigerants in horizontal microfin tubes. Summary, the phenomenon, the effectiveness of fluid dynamics on the formation of flow pattern map, heat transfer coefficient, and pressure drop during condensation process under the empirical correlations are the examples to calculate condensation parameters.

2. Flow pattern map

With the aim of understanding the trend of increasing and decreasing heat transfer coefficient and pressure drop during the changing phase of refrigerants during condensation processes, the formation and transient of flow regime during the changing phase of condensation should be considered.

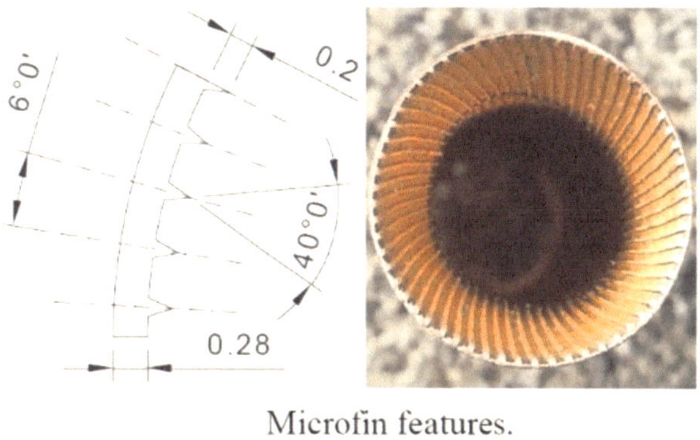

Figure 1.
A geometrical parameter of microfin tube J60 [9].

In order to draw a map of flow mechanism, there needs to be a classification of different regimes of flow in microfin tube, which relies on flow patterns such as slug, fully stratified, stratified wavy, helix, annular, and some regimes which are the combination of two or three flow pattern simultaneously happened. Many parameters have to define as an input parameter such as geometries of tube, refrigerants (consider about thermophysical and thermodynamic of refrigerant) heat flux, and mass flux that could take microfin tube from Colombo et al. as an example as shown in **Figure 1**, and the geometrical of microfin tube consists of number of fins, fin height, apex angle, and helix angle besides the inner and outer diameters of tubes.

It is could be considered about an experimental result of flow pattern map in microfin tube from the group of Colombo et al. [10], and they classified the pattern into five regimes: slug flow, stratified wavy flow, intermittent (slug, stratified wavy, and helix), helix flow, and annular flow. They showed their map as an indifferent dot of data points and compared it with the map of Hajal et al. as presented in **Figure 2**.

2.1 The flow pattern map of Breber et al.

Breber et al. [16] published their map in 1980, and the map could be used for smooth tube that divided into four quadrant zones based on two parameters: dimensionless gas velocity J_G– and turbulent-turbulent Martinelli parameter X_{tt} (annular and mist-annular flow, wavy and stratified flow, slug and plug flow, and bubble flow) and some transition zones between each other.

Dimensionless gas velocity J_G is given by:

$$J_G = \frac{G.x}{\sqrt{g.d.\rho_G(\rho_L - \rho_G)}} \quad (1)$$

Turbulent-turbulent Martinelli parameter X_{tt} is given by:

$$X_{tt} = \left(\frac{1-x}{x}\right)^{0.9}\left(\frac{\rho_v}{\rho_l}\right)^{0.5}\left(\frac{\mu_l}{\mu_v}\right)^{0.1} \quad (2)$$

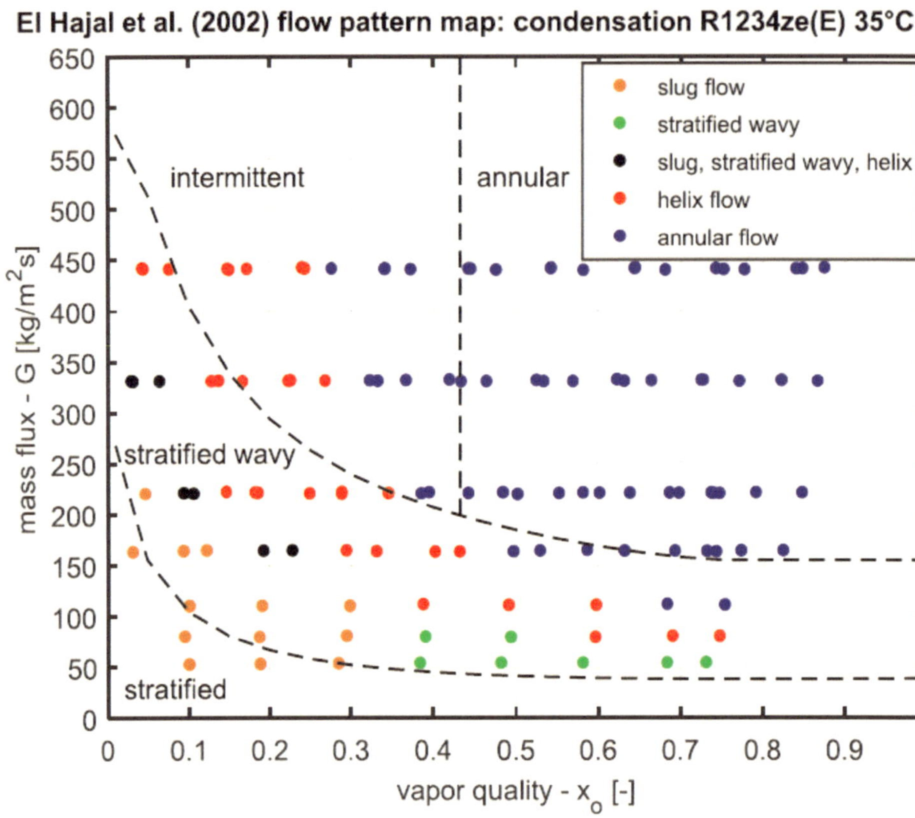

Figure 2.
Flow pattern results observed from experiment of Colombo et al. [10] draw on the map of Hajal [9].

Transition criteria for flow regimes on the map:

Zone 1	Zone 2	Zone 3	Zone 4
Annular flow	Wavy and stratified	Slug flow	Bubble (at high reduced pressure)
$J_G > 1.5$ $X < 1.0$	$J_G < 0.5$ $X < 1.0$	$J_G < 1.5$ $X > 1.5$	$J_G > 1.5$ $X > 1.5$

2.2 The flow pattern map of Tandon et al.

The map of Tandon et al. [17] is presented in five zones of patterns: spray, annular and semi-annular, wavy, slug, and plug. They used two dimensionless numbers to draw their map: gas velocity J_G and $(1-\alpha)/\alpha$ with α is the void fraction.

Dimensionless gas velocity J_G is given by:

$$J_G = \frac{G.x}{\sqrt{g.d.\rho_G(\rho_L - \rho_G)}} \tag{3}$$

DOI: http://dx.doi.org/10.5772/intechopen.105875

Void fraction is defined from Smith's equation (1970) is given as follows:

$$\alpha = \left\{ 1 + \left(\frac{\rho_G}{\rho_L}\right)\left(\frac{1-x}{x}\right)\left[0.4 + 0.6\sqrt{\frac{\frac{\rho_L}{\rho_G} + 0.4\left(\frac{1-x}{x}\right)}{1 + 0.4\left(\frac{1-x}{x}\right)}}\right]\right\}^{-1} \tag{4}$$

Transition criteria for flow regimes on the map.

Spray	Annular and semi-annular	Wavy	Slug	Plug
$J_G \geq 6$ $(1-\alpha)/\alpha \leq 0.5$	$1 \leq J_G \leq 6$ $(1-\alpha)/\alpha \leq 0.5$	$1 \leq J_G$ $(1-\alpha)/\alpha \leq 0.5$	$0.01 \leq J_G \leq 0.5$ $(1-\alpha)/\alpha \geq 0.5$	$J_G \leq 0.01$ $(1-\alpha)/\alpha \geq 0.5$

2.3 The flow pattern map of Hajal et al.

The map of Hajal et al. [1] introduced the logarithmic mean void fraction instead of homogeneous void fraction to define the stratification of liquid and vapor and is based on the transient line between mass velocity and vapor quality to classify the regimes of patterns. The procedure to calculate step by step is shown as follows:

- Step 1: Calculate logarithmic mean void fractions ε based on homogeneous and nonhomogeneous void fractions:

  Homogeneous: ε_h

$$\varepsilon_h = \left[1 + \left(\frac{1-x}{x}\right)\left(\frac{\rho_V}{\rho_L}\right)\right]^{-1} \tag{5}$$

  Nonhomogeneous: ε_{ra}

$$\varepsilon_{ra} = \frac{x}{\rho_V}\left([1 + 0.12(1-x)]\left[\left(\frac{x}{\rho_V}\right) + \left(\frac{1-x}{\rho_L}\right)\right] + \frac{1.18(1-x)[g\sigma(\rho_L - \rho_V)]^{0.25}}{G{\rho_L}^{0.5}}\right)^{-1} \tag{6}$$

  Logarithmic mean void fractions: ε

$$\varepsilon = \frac{\varepsilon_h - \varepsilon_{ra}}{ln\left(\frac{\varepsilon_h}{\varepsilon_{ra}}\right)} \tag{7}$$

- Step 2: Determine the stratified angle θ_{strat} and geometrical parameters for two-phase flow in a circular tube:

  + The stratified angle θ_{strat}:

$$\theta_{strat} = 2\pi - 2\left\{\pi(1-\varepsilon) + \left(\frac{3\pi}{2}\right)^{1/3}\left[1 - 2(1-\varepsilon) + (1-\varepsilon)^{\frac{1}{3}} - \varepsilon^{\frac{1}{3}}\right] - \frac{1}{200}(1-\varepsilon)\varepsilon[1 - 2(1-\varepsilon)]\left[1 + 4\left((1-\varepsilon)^2 + \varepsilon^2\right)\right]\right\} \tag{8}$$

+ Geometrical parameters for two-phase flow in a circular tube:

$$A_{LD} = \frac{A_L}{D^2} = \frac{A(1-\varepsilon)}{D^2} \tag{9}$$

$$A_{VD} = \frac{A_V}{D^2} = \frac{A\varepsilon}{D^2} \tag{10}$$

$$h_{LD} = 0.5\left(1 - cos\left(\frac{2\pi - \theta_{strat}}{2}\right)\right) \tag{11}$$

$$P_{iD} = sin\left(\frac{2\pi - \theta_{strat}}{2}\right) \tag{12}$$

- Step 3: Use the value from step 1 and 2 to determine the transient lines:

+ Transient line of stratifine wavy: G_{wavy}:

$$G_{wavy} = \left\{\frac{16A_{VD}^3 gD\rho_L\rho_V}{x^2\pi^2\left(1-(2h_{LD}-1)^2\right)^{0.5}}\left[\frac{\pi^2}{25h_{LD}^2}\left(\frac{We}{Fr}\right)_L^{-1.023}+1\right]\right\}^{0.5} + 50 - 75e^{-\left(x^2-0.97\right)^2/x(1-x)} \tag{13}$$

G_wavy_min occur at x_min.
then, for x > x_min → G_wavy = G_wavy_min.
+ Transient line of fully stratified: G_{strat}:

$$G_{strat} = \left\{\frac{(226.3)^3 A_{LD}A_{VD}^2\rho_V(\rho_L-\rho_V)\mu_L g}{x^2(1-x)\pi^3}\right\}^{1/3} + 20x \tag{14}$$

+ Marttineli parameter: X_IA:

$$X_{IA} = \left\{\left[0.2914\left(\frac{\rho_V}{\rho_L}\right)^{-1/1.75}\left(\frac{\mu_L}{\mu_V}\right)^{-1/7}\right]+1\right\}^{-1} \tag{15}$$

+ Transient line of mist flow: G_{mist}:

$$G_{mist} = \left\{\frac{7680A_{VD}^2 gD\rho_L\rho_V}{x^2\pi^2\xi}\left(\frac{Fr}{We}\right)_L\right\}^{0.5} \tag{16}$$

$$\xi = \left[1.138 + 2\,log\left(\frac{\pi}{1.5A_{LD}}\right)\right]^{-2} \tag{17}$$

G_mist_min occur at x_min.
then, for x > x_min → G_mist = G_mist_min.
+ Transient line of bubbly flow: G_{bubbly}:

$$G_{bubbly} = \left\{\frac{256A_{VD}A_{LD}^2 D^{1.25}\rho_L(\rho_L-\rho_V)g}{0.3164(1-x)^{1.75}\pi^2 P_{iD}\mu_L^{0.25}}\right\}^{1/1.75} \tag{18}$$

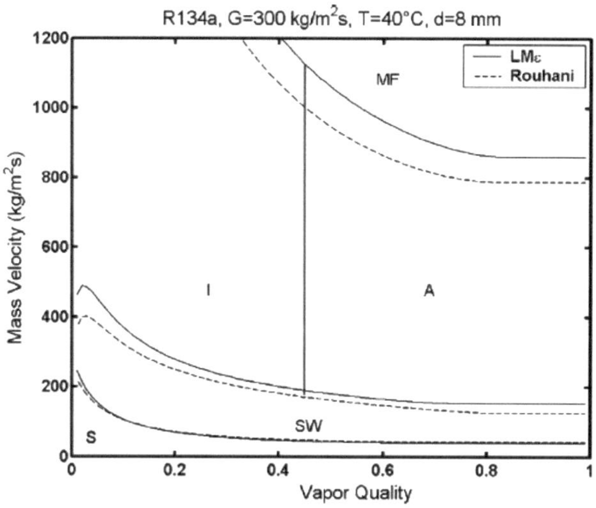

Figure 3.
Flow pattern map of Hajal et al. [1].

- Step 4: Draw the flow pattern map of condensation in 2D with the transient lines of mass velocity and vapor quality. The result of Hajal flow pattern map of condensation R134a in a smooth tube of 8 mm diameter at 40°C temperature with initial value of mass velocity 300 kg/m^2s is shown in **Figure 3**.

3. Two-phase heat transfer coefficient in condensation

3.1 The correlation of Cavallini et al.

The correlation of Cavallini et al. [4] was published in 2009 which derived 558 data points, experimental results, and validated databases from other researches. The working range of microfin tubes: h/D < 0.04 (fin height/fin tip diameter) and helix angle is from 0° to 30°.

D: fin tip diameter; n_g: fins number; h: fin height; β: spiral angle; γ: apex angle.

Heat transfer coefficient of two phases could be determined from two zones as follows:

$$\alpha = \left[\alpha_A^3 + \alpha_D^3\right]^{0.333} \tag{19}$$

Heat transfer coefficient for the ΔT independent zone α_A is given as follows:

$$\alpha_A = \alpha_{AS}.A.C \tag{20}$$

$$\alpha_{AS} = \alpha_{LO}\left[1 + 1.128x^{0.817}\left(\frac{\rho_L}{\rho_G}\right)^{0.3685}\left(\frac{\mu_L}{\mu_G}\right)^{0.2363}\left(1 - \frac{\mu_G}{\mu_L}\right)^{2.144} Pr_L^{-0.1}\right] \tag{21}$$

$$\alpha_{LO} = 0.023\frac{\lambda_L}{D} Re_{LO}^{0.8} Pr_L^{0.4} = 0.023\frac{\lambda_L}{D}\left(\frac{GD}{\mu_L}\right)^{0.8} Pr_L^{0.4} \tag{22}$$

$$A = 1 + 1.119Fr^{-0.3821}(Rx - 1)^{0.3586} \tag{23}$$

$$Fr = \frac{G^2}{gD(\rho_L - \rho_G)^2} \tag{24}$$

$$Rx = \left\{\frac{2h.n_g[1 - sin\,(\gamma/2)]}{\pi D cos(\gamma/2)} + 1\right\}\frac{1}{cos\beta} \tag{25}$$

$$C = 1 \;\; if \;\; (n_{opt}/n_g) \geq 0.8 \tag{26}$$

$$C = (n_{opt}/n_g)^{1.904} \;\; if \;\; (n_{opt}/n_g) < 0.8 \tag{27}$$

$$n_{opt} = 4064.4D + 23.257 \tag{28}$$

Heat transfer coefficient for the ΔT dependent zone α_D is given as follows:

$$\alpha_D = C\left[2.4x^{0.1206}(Rx - 1)^{1.466}C_1^{0.6875} + 1\right]\alpha_{D,S} + C\left(1 - x^{0.087}\right)Rx.\alpha_{LO} \tag{29}$$

$$\alpha_{D,S} = \frac{0.725}{1 + 0.741\left(\frac{1-x}{x}\right)^{0.3321}}\left[\frac{\lambda_L^3.\rho_L(\rho_L - \rho_G)g.h_{LG}}{\mu_L.D.\Delta T}\right]^{0.25} \tag{30}$$

$$C_1 = 1 \;\; if \;\; J_G \geq J_G^* \tag{31}$$

$$C_1 = J_G/J_G^* \;\; if \;\; J_G < J_G^* \tag{32}$$

$$J_G = \frac{xG}{[gD\rho_G(\rho_L - \rho_G)]^{0.5}} \tag{33}$$

$$J_G^* = 0.6\left\{\left[\frac{7.5}{4.3X_{tt}^{1.111} + 1}\right]^{-3} + 2.5^{-3}\right\}^{-0.3333} \tag{34}$$

$$\Delta T = T_s - T_w \tag{35}$$

3.2 The correlation of Han et al.

Heat transfer coefficient correlation of Han et al. [20] was proposed in 2005, which was formed from a database of three various refrigerant R134A, R22, and R410A and four different inner diameter of microfin tubes di = 8.92, 6.46, 5.1, 4 mm. Some hypotheses were used to build this correlation as the uniform of liquid film thickness and turbulent flow in vapor core, neglect the radial temperature gradients and temperature in the vapor, and assume that the saturation temperature is constant at temperatures in the vapor core and at liquid-vapor interface.

Heat transfer coefficient of two-phase flow is given as follows:

$$h = \frac{\rho c_p u_t}{0.904e^{+0.592}Pr^{0.729} + 2.1335\,ln\,(\delta/e)} \tag{36}$$

Liquid film thickness is given as follows:

$$\delta = \frac{(1 - \lambda)D_i}{4} \tag{37}$$

Void fraction from Yashar λ is given as follows:

$$\lambda = \left[1 + \left(\frac{1-x}{x}\right)\left(\frac{\rho_g}{\rho_l}\right)\left\{1 + 1.3\,Re_{lo}^{-0.19}\left(\frac{\rho_l}{\rho_g}\right)^{0.72}\left(\frac{x}{1-x}\right)^{0.5}\right\}\right]^{-1} \tag{38}$$

With Re_{lo} is Reynolds of liquid only, e^+ is roughness Reynolds number and u_t is turbulent friction (or shear velocity) and is given as follows

$$e^+ = \frac{eu_t}{v} \tag{39}$$

$$u_t = \left(\frac{\tau}{\rho}\right)^{0.5} = \left\{\frac{D_i}{4\rho}\left(\frac{dP}{dz}\right)_{fr}\right\}^{0.5} \tag{40}$$

where e is the fin height, $p = \pi D_i/n\tan\beta$ is the axial fin pitch, Di is the inside tube diameter, n is the number of fins, and β is the spiral angle.

If liquid film thickness is less than fin height, the term of logarithm ln(δ/e) is neglected.

3.3 The correlation of Kedzieski and Goncalves

Kedzieski and Goncalves in 1999 [19] collected 1704 data points from their experimental database to regress correlation. Their experimental data recorded in various working conditions as from four different refrigerants R134a, R410A, R125, and R32 in one microfin tube that has 8.91 mm root diameter, 0.2 mm fin height, and 50°apex angle, and the model was derived as follows:

$$Nu = \frac{h_{2\varnothing}D_h}{K_l} = 2.256\,Re^{\beta 1}Ja^{\beta 2}Pr^{\beta 3}\left(\frac{P_r}{P_c}\right)^{\beta 4}\left(-\log_{10}\frac{P_r}{P_c}\right)^{\beta 5}Sv^{\beta 6} \tag{41}$$

$\beta 1 = 0.303$; $\beta 2 = 0.232x_q$; $\beta 3 = 0.393$; $\beta 4 = -0.578x_q^2$; $\beta 5 = -0.474x_q^2$; $\beta 4 = 2.531x_q^{2.}$

The authors also supplied the simpler form with larger uncertainty:

$$Nu = \frac{h_{2\varnothing}D_h}{K_l} = 4.94\,Re^{\beta 1}Pr^{\beta 3}\left(\frac{P_r}{P_c}\right)^{\beta 4}\left(-\log_{10}\frac{P_r}{P_c}\right)^{\beta 5}Sv^{\beta 6} \tag{42}$$

$\beta 1 = 0.235$; $\beta 3 = 0.308$; $\beta 4 = -1.16x_q^2$; $\beta 5 = -0.887x_q^2$; $\beta 4 = 2.708x_q^2$

$$Ja = \frac{i_{fg}}{C_{pr,l}\Delta T_s} \tag{43}$$

$$\Delta T_s = T_s - T_w \tag{44}$$

$$Sv = \frac{v_v - v_l}{v} \tag{45}$$

$$v = x_q v_v + (1 - x_q)v_l \tag{46}$$

$$Re = GD_h/\mu_{r,l} \tag{47}$$

where Dh is the hydraulic diameter.

3.4 The correlation of Yu and Koyama 1998

The correlation of Yu and Koyama [18] for microfin tubes was updated from the correlation of Haraguchi et al. for smooth tubes and was introduced in 1998. The procedure to determine heat transfer coefficient by Nusselt number is presented as follows:

The Nusselt number of two-phase flow is given by:

$$Nu = \left(Nu_F^2 + Nu_B^2\right)^{1/2} \tag{48}$$

$$Nu_F = 0.152\left[0.3 + 0.1Pr_L^{1.1}\right](\Phi_v/X_{tt})\, Re_L^{0.68} \tag{49}$$

$$\Phi_v = 1.1 + 1.3\left[\frac{GX_{tt}}{\sqrt{gd_i\rho_v(\rho_l - \rho_v)}}\right]^{0.35} \tag{50}$$

$$Nu_B = \frac{0.725}{\eta_A^{\frac{1}{4}}}.H(\xi)\left(\frac{Ga.Pr_L}{Ph_L}\right)^{1/4} \tag{51}$$

$$H(\xi) = \xi + A\sqrt{\xi}\left(1 - \sqrt{\xi}\right) \tag{52}$$

$$A = 10(1 - \xi)^{0.1} - 8.0 \tag{53}$$

where di is the mean inner diameter of tube, η_A is the enlargement ration of heat transfer area, and Ph_L is the phase change number.

$$\xi = \left[1 + \frac{\rho_v}{\rho_l}\left(\frac{1-x}{x}\right)\left(0.4 + 0.6\sqrt{\frac{\frac{\rho_v}{\rho_l} + 0.4\frac{1-x}{x}}{1 + 0.4\frac{1-x}{x}}}\right)\right]^{-1} \tag{54}$$

$$Re_L = \frac{G(1-x)d_i}{\mu_L} \tag{55}$$

$$Ga = \frac{g\rho_L^3 {d_i}^3}{{\mu_L}^2} \tag{56}$$

$$Ph_L = \frac{Cp_L(T_{sat} - T_{wi})}{\Delta h_{vl}} \tag{57}$$

4. Pressure drop

4.1 The correlation of Han et al.

Han et al. [20] not only presented the correlation for heat transfer coefficient as aforementioned but also introduced a new correlation to define pressure drop. It is could be considered as follows:

Friction pressure gradient of two phases is given as follows:

$$\left(\frac{dP}{dz}\right)_{fr} = \Phi_l^2\frac{f_l[G(1-x)]^2}{2D_i\rho_l} \tag{58}$$

$$f_l = 0.193\left[\frac{G(1-x)D_i}{\mu_l}\right]^{-0.024}\left(\frac{p}{e}\right)^{-0.539} \quad (59)$$

$$\Phi_l^2 = 2.684X_{tt}^{-1.946} \quad (60)$$

$$X_{tt} = \left(\frac{1-x}{x}\right)^{0.9}\left(\frac{\rho_v}{\rho_l}\right)^{0.5}\left(\frac{\mu_l}{\mu_v}\right)^{0.1} \quad (61)$$

4.2 The correlation of Kedzieski and Goncalves

In 1999, Kedzieski and Goncalves [19] modified the correlation of Pierre 1964 with confirmation by their own experimental data from four refrigerants such as R134a, R410A, R125, and R32 in their microfin tube that has 8.91 mm root diameter, 0.2 mm fin height, and 50°apex angle, and presented the correlation as follows:

$$\frac{\Delta p}{L} = \frac{\Delta p_f}{L} + \frac{\Delta p_m}{L} = G^2\left[f\frac{\left(v_{tp,out} + v_{tp,in}\right)}{d_h} + \frac{\left(v_{tp,out} - v_{tp,in}\right)}{L}\right] \quad (62)$$

$$d_h = 4A_c cos\beta/\left(n.S_p\right) \quad (63)$$

$$f = 0.002275 + 0.00933\,exp^{\left[e/\left(-0.003d_{fr}\right)\right]}\,Re_{h,LO}^{-1/\left(4.16+532e/d_{fr}\right)}K_f^{0.211} \quad (64)$$

$$Re_{h,LO} = Gd_h/\mu_l \quad (65)$$

$$K_f = \frac{\Delta x.h_{lv}}{gL} \quad (66)$$

where d_h is the hydraulic diameter, d_{fr} is the fin root diameter, and e is the fin height.

4.3 The correlation of Harraguchi

Harraguchi et al. [23] presented their correlation for two-phase friction pressure drop from their experimental results with R134a, R123, and R22 in a horizontal smooth tube and a horizontal microfin tube in 1993.

Two-phase friction pressure drop is given as follows:

$$\left(\frac{dP}{dz}\right)_f = \Phi_v^2\left(\frac{dP}{dz}\right)_v = \Phi_v^2 2f_{e,v}(Gx)^2/(\rho_v d_e) \quad (67)$$

$$d_e = \sqrt{\frac{4A_c}{\pi}} \quad (68)$$

$$f_{e,v} = 0.046\,Re_{e,v}^{-0.2} \quad (69)$$

$$\Phi_v = 1.1 + 1.3\left\{X_{tt}G/[gd_e\rho_v(\rho_l - \rho_v)]^{0.5}\right\}^{0.35} \quad (70)$$

$$Re_{e,v} = Gxd_e/\mu_v \quad (71)$$

$$X_{tt} = \left(\frac{1-x}{x}\right)^{0.9}\left(\frac{\rho_v}{\rho_l}\right)^{0.5}\left(\frac{\mu_l}{\mu_v}\right)^{0.1} \quad (72)$$

5. Example

The condensation of refrigerant R1234ze inside horizontal microfin tube at 35°C, heat flux of 8.62 kW/m^2, and mass flux of 222 kg/m^{2}s using the microfin tube J60 with the geometrical parameters: 60 number of fins, 8.96 mm inner surface diameter, 0.2 mm fin height, 18° of helix angle, and 40° of apex angle is considered. Heat transfer coefficient and pressure drop during condensation process are determined. Saturation to wall temperature difference is kept constant (5°C).

Solve:

Thermal properties of refrigerant R1234ze at saturation temperature t_{sat} = 35°C is listed in **Table 1**.

Refrigerant: R1234ze, molar mass: M = 114 g/mol, $p_{critical}$ = 3.64 MPa

Reduce pressure: $p_r = p_{sat}/p_{critical}$ = 0.66756/3.64 = 0.1834

Heat flux: q = 15.3 kW/m^2

Mass flux: G = 220 kg/m^2 s

Microfin tube: N = 60 number of fins, dr = 8.96 mm, h = 0.2 mm, β = 18°; γ = 40°

5.1 Heat transfer coefficients

Quality x = 0.5 could be taken as an example to present the procedure to calculate heat transfer coefficient and pressure drop of boiling refrigerant inside horizontal microfin tube, with the data point of quality x from 0 to 1 that can be determined with the same method.

Applied the correlation of Cavallini et al. to calculate heat transfer coefficient as below:

- Heat transfer coefficient of two phases could be determined from two zones as follows:

$$\alpha = \left[\alpha_A^3 + \alpha_D^3\right]^{0.333}$$

- Heat transfer coefficient for the ΔT independent zone α_A is give as follows:

$$\alpha_A = \alpha_{AS}.A.C$$

Property	Unit	Value	Property	Unit	Value
T_{sat}	[°C]	35	P_{sat}	[MPa]	0.66756
ρ_L	[kg/m^3]	1129	λ_L	[W/m-K]	7.098E−02
ρ_V	[kg/m^3]	35.32	λ_V	[W/m-K]	1.452E−02
ν_L	[m^3/kg]	8.857E−04	μ_L	[Pa-s]	1.772E−04
ν_v	[m^3/kg]	2.831E−02	μ_V	[Pa-s]	1.269E−05
h_{LV}	[kJ/kg]	159.86	Pr_L	[−]	3.548
Cp_L	[kJ/kg-K]	1.421	Pr_V	[−]	0.904
Cp_V	[kJ/kg-K]	1.034	σ	[N/m]	7.578E−03

Table 1.
Properties of R1234ze.

+ Calculate α_{AS}:

$$\alpha_{LO} = 0.023\frac{\lambda_L}{D}\left(\frac{GD}{\mu_L}\right)^{0.8} Pr_L^{0.4} = 0.023\frac{7.098.10^{-2}}{8.96.10^{-3}}\left(\frac{220.(8.96.10^{-3})}{1.772.10^{-4}}\right)^{0.8} 3.548^{0.4}$$

$$\alpha_{LO} = 521.87$$

$$\alpha_{AS} = \alpha_{LO}\left[1 + 1.128x^{0.817}\left(\frac{\rho_L}{\rho_G}\right)^{0.3685}\left(\frac{\mu_L}{\mu_G}\right)^{0.2363}\left(1-\frac{\mu_G}{\mu_L}\right)^{2.144} Pr_L^{-0.1}\right]$$

$$\alpha_{AS} = 521.87\left[1 + 1.128(0.5^{0.817})\left(\frac{1129}{35.32}\right)^{0.3685}\left(\frac{1.772.10^{-4}}{1.269.10^{-5}}\right)^{0.2363}\left(1 - \frac{1.269.10^{-5}}{1.772.10^{-4}}\right)^{2.144} 3.548^{-0.1}\right]$$

$$\alpha_{AS} = 2199.82$$

+ Calculate A:

$$Fr = \frac{G^2}{gD(\rho_L - \rho_G)^2} = \frac{220^2}{9.81.(8.96.10^{-3})(1129 - 35.32)^2} = 0.46035$$

$$Rx = \left\{\frac{2h.n_g[1 - sin\,(\gamma/2)]}{\pi D cos(\gamma/2)} + 1\right\}\frac{1}{cos\beta}$$

$$Rx = \left\{\frac{2.(2.10^4).60.[1 - sin\,(40/2)]}{\pi.(8.96.10^{-3}).\,cos\,(40/2)} + 1\right\}\frac{1}{\cos 18} = 1.79$$

$$A = 1 + 1.119Fr^{-0.3821}(Rx - 1)^{0.3586}$$

$$A = 1 + 1.119(0.46035^{-0.3821}).(1.79 - 1)^{0.3586} = 2.383$$

+ Calculate C:

$$n_{opt} = 4064.4D + 23.257 = 4064.4(8.96.10^{-3}) + 23.257 = 59.7$$

$$C = 1 \quad if \quad (n_{opt}/n_g) \geq 0.8$$

$$C = (n_{opt}/n_g)^{1.904} \quad if \quad (n_{opt}/n_g) < 0.8$$

In this case:

$$\frac{n_{opt}}{n_g} = \frac{59.7}{60} = 0.995 \geq 0.8$$

- C = 1

Heat transfer coefficient for the ΔT independent zone α_A is given as follows:

$$\alpha_A = \alpha_{AS}.A.C = (2199.82).(2.383).(1) = 5242.213$$

- Heat transfer coefficient for the ΔT dependent zone α_D is given as follows:

+ Calculate C1:

$$J_G = \frac{xG}{[gD\rho_G(\rho_L - \rho_G)]^{0.5}} = \frac{(0.5).(220)}{[(9.81).(8.96.10^{-3}).(35.32).(1129 - 35.32)]^{0.5}}$$

$$J_G = 1.8878$$

Martinelli number is given as follows:

$$X_{tt} = \left(\frac{1-x}{x}\right)^{0.9}\left(\frac{\rho_G}{\rho_L}\right)^{0.5}\left(\frac{\mu_L}{\mu_G}\right)^{0.1} = \left(\frac{1-0.5}{0.5}\right)^{0.9}\left(\frac{35.32}{1129}\right)^{0.5}\left(\frac{1.772.10^{-4}}{1.269.10^{-5}}\right)^{0.1} = 0.2302$$

$$J_G^* = 0.6\left\{\left[\frac{7.5}{4.3X_{tt}^{1.111}+1}\right]^{-3} + 2.5^{-3}\right\}^{-0.3333}$$

$$J_G^* = 0.6\left\{\left[\frac{7.5}{4.3(0.2302^{1.111})+1}\right]^{-3} + 2.5^{-3}\right\}^{-0.3333} = 1.3994$$

$$C_1 = 1 \quad if \quad J_G \geq J_G^*$$
$$C_1 = J_G/J_G^* \quad if \quad J_G < J_G^*$$

In this case: $J_G = 1.8878 > J^*_G = 1.3994$ => C1 = 1
+ Calculate $\alpha_{D,S}$:

$$\Delta T = T_s - T_w = 5$$

$$\alpha_{D,S} = \frac{0.725}{1+0.741\left(\frac{1-x}{x}\right)^{0.3321}}\left[\frac{\lambda_L^3.\rho_L(\rho_L - \rho_G)g.h_{LG}}{\mu_L.D.\Delta T}\right]^{0.25}$$

$$\alpha_{D,S} = \frac{0.725}{1+0.741\left(\frac{1-0.5}{0.5}\right)^{0.3321}}\left[\frac{(7.098.10^{-2})^3.(1129).(1129-35.32).(9.81).(159.86)}{(1.772.10^{-4}).(8.96.10^{-3}).5}\right]^{0.25}$$

$$\alpha_{D,S} = 226.31$$

=> Heat transfer coefficient for the ΔT dependent zone α_D is given as follows:

$$\alpha_D = C\left[2.4x^{0.1206}(Rx-1)^{1.466}C_1^{0.6875}+1\right]\alpha_{D,S} + C(1-x^{0.087})Rx.\alpha_{LO}$$

$$\alpha_D = 1.\left[2.4(0.5^{0.1206}).(1.79-1)^{1.466}(1^{0.6875})+1\right].(226.31) + 1.(1-0.5^{0.087}).(1.79).(521.87)$$

$$\alpha_D = 634.51$$

Heat transfer coefficient of condensation at x = 0.5 is given as follows:

$$\alpha = \left[\alpha_A^3 + \alpha_D^3\right]^{0.333}$$

$$\alpha = \left[5242.213^3 + 634.51^3\right]^{0.333} = 5200.58\ W/m^2K$$

5.2 Pressure drop

5.2.1 The correlation of Harraguchi

Harraguchi et al. [23] presented their correlation for two-phase friction pressure drop from their experimental results with R134a, R123, and R22 in a horizontal smooth tube and a horizontal microfin tube in 1993.

Two-phase friction pressure drop is given as follows:

$$X_{tt} = \left(\frac{1-x}{x}\right)^{0.9}\left(\frac{\rho_G}{\rho_L}\right)^{0.5}\left(\frac{\mu_L}{\mu_G}\right)^{0.1} = \left(\frac{1-0.5}{0.5}\right)^{0.9}\left(\frac{35.32}{1129}\right)^{0.5}\left(\frac{1.772.10^{-4}}{1.269.10^{-5}}\right)^{0.1} = 0.2302$$

$$d_e = \sqrt{\frac{4A_c}{\pi}} \cong D = 8.96.10^{-3}\ mm$$

$$Re_{e,v} = \frac{G.x.d_e}{\mu_v} = \frac{220x0.5x0.00896}{1.269.10^{-5}} = 77667.45$$

x	α	dP/dz
[–]	[W/m²K]	[Pa/m]
0.99	8160.4	2448.5
0.95	7930.4	2968.4
0.90	7641.1	3151.4
0.85	7349.0	3173.8
0.80	7053.8	3112.0
0.75	6755.3	2996.4
0.70	6453.0	2843.2
0.65	6146.9	2662.9
0.60	5836.4	2462.4
0.55	5521.1	2247.4
0.50	5200.6	2022.1
0.45	4874.1	1790.4
0.40	4541.0	1555.5
0.35	4200.2	1320.5
0.30	3850.1	1088.6
0.25	3489.1	862.8
0.20	3114.7	646.6
0.15	2722.8	444.1
0.10	2306.4	260.7
0.05	1849.2	104.9
0.01	1413.8	13.1

Table 2.
Heat transfer performance of condensation.

$$f_{e,v} = 0.046\,Re_{e,v}^{-0.2} = (0.046).(77667.45^{-0.2}) = 0.004838$$

$$\Phi_v = 1.1 + 1.3\left\{X_{tt}G/[gd_e\rho_v(\rho_l - \rho_v)]^{0.5}\right\}^{0.35}$$

$$\Phi_v = 1.1 + 1.3\left\{(0.2302).220/[(9.81).(0.00896).(35.32).(1129 - 35.32)]^{0.5}\right\}^{0.35}$$

$$\Phi_v = 2.3378$$

$$\left(\frac{dP}{dz}\right)_f = \Phi_v^2\left(\frac{dP}{dz}\right)_v = \frac{\Phi_v^2 2f_{e,v}(Gx)^2}{(\rho_v d_e)} =$$

$$\left(\frac{dP}{dz}\right)_f = \Phi_v^2\left(\frac{dP}{dz}\right)_v = \frac{\Phi_v^2 2f_{e,v,Go}(Gx)^2}{(\rho_v d_e)} = \frac{(2.3378)^2.2.(0.004838)(220.(0.5))^2}{(35.32).(0.00896) = 2022.105(\frac{Pa}{m})}$$

Condensation process of R1234ze at 35°C inside horizontal microfin tube J60 with quality change from x = 0.99 to x = 0.01 could be estimated with heat transfer performance which are performed in heat transfer and pressure drop in **Table 2**.

6. Conclusions

An overview of the condensation procedure of refrigerant inside microfin tubes has been done with the flow pattern maps and heat transfer performance. The list of some correlations to determine heat transfer and pressure drop during the condensation process has been presented. Besides, an example has been introduced step by step to define heat transfer coefficient and pressure drop of condensation process with refrigerant R1234ze inside horizontal microfin tube J60. Until now, this field is still not getting too much data to get the general correlation for all types of microfin tubes and all kinds of refrigerants. Therefore, these fields are strongly attracted by scientists around the world.

Acknowledgements

We acknowledge the support of time and facilities from Ho Chi Minh City University of Technology (HCMUT), VNU-HCM for this study.

Author details

Thanh Nhan Phan[1,2]

1 Ho Chi Minh City University of Technology (HCMUT), Ho Chi Minh City, Vietnam

2 Vietnam National University Ho Chi Minh City, Ho Chi Minh City, Vietnam

*Address all correspondence to: phannhan@hcmut.edu.vn

References

[1] el Hajal J, Thome JR, Cavallini A. Condensation in horizontal tubes, part 1: Two-phase flow pattern map. International Journal of Heat and Mass Transfer. 2003;**46**(18):3349-3363. DOI: 10.1016/S0017-9310(03)00139-X

[2] Cavallini A, del Col D, Matkovic M, Rossetto L. Frictional pressure drop during vapour-liquid flow in minichannels: Modelling and experimental evaluation. International Journal of Heat and Fluid Flow. 2009; **30**(1):131-139. DOI: 10.1016/j.ijheatfluidflow.2008.09.003

[3] Thome JR, el Hajal J, Cavallini A. Condensation in horizontal tubes, part 2: New heat transfer model based on flow regimes. International Journal of Heat and Mass Transfer. 2003;**46**(18): 3365-3387. DOI: 10.1016/S0017-9310(03)00140-6

[4] Cavallini A, del Col D, Mancin S, Rossetto L. Condensation of pure and near-azeotropic refrigerants in microfin tubes: A new computational procedure. International Journal of Refrigeration. 2009;**32**(1): 162-174. DOI: 10.1016/j.ijrefrig.2008.08.004

[5] Cavallini A et al. Condensation in horizontal smooth tubes: A new heat transfer model for heat exchanger design. Heat Transfer Engineering. 2006;**27**(8):31-38. DOI: 10.1080/01457630600793970

[6] Cavallini A, Bortolin S, del Col D, Matkovic M, Rossetto L. Condensation and Vaporization of Halogenated Refrigerants Inside a Circular Minichannel. International Refrigeration and Air Conditioning Conference. 2008. p. 886. Available online: http://docs.lib.purdue.edu/iracc/886

[7] Cavallini A, Censi G, del Col D, Doretti L, Longo GA, Rossetto L. Condensation of halogenated refrigerants inside smooth tubes. HVAC and R Research. 2002;**8**(4):429-451. DOI: 10.1080/10789669.2002.10391299

[8] Cavallini A, Brown JS, del Col D, Zilio C. In-tube condensation performance of refrigerants considering penalization terms (exergy losses) for heat transfer and pressure drop. International Journal of Heat and Mass Transfer. 2010;**53**(13–14):2885-2896. DOI: 10.1016/j.ijheatmasstransfer.2010.02.007

[9] Colombo LPM, Lucchini A, Molinaroli L, Niro A, Phan TN, Pittoni PG. Flow patterns during flow boiling and convective condensation of R1234ze (E) inside a microfin tube. In: 5-6th Thermal and Fluids Engineering Conference (TFEC), ASTFE Digital Library. 2021. pp. 109-119. DOI: 10.1615/TFEC2021.boi.032115

[10] Muzzio A, Niro A, Garavaglia M. Flow patterns and heat transfer coefficients in flow-boiling and convective condensation of R22 inside a microfin tube of new design. Heat Transfer Conference. 1998;**2**:291-296

[11] Muzzio A, Niro A, Arosio S. Heat transfer and pressure drop during evaporation and condensation of R22 inside 9.52-mm O.D. microfin tubes of different geometries. Journal of Enhanced Heat Transfer. 1998;**5**: 39-52. DOI: 10.1615/JEnhHeatTransf.v5.i1.40

[12] Colombo LPM, Lucchini A, Muzzio A. Flow patterns, heat transfer and pressure drop for evaporation and condensation of R134A in microfin

tubes. International Journal of Refrigeration. 2012;**35**(8):2150-2165. DOI: 10.1016/j.ijrefrig.2012.08.019

[13] Wu Z, Sundén B, Wadekar V, Li W. Heat transfer correlations for single-phase flow, condensation, and boiling in Microfin tubes. Heat Transfer Engineering. 2015;**36**(6):582-595. DOI: 10.1080/01457632.2014.939531

[14] Wu Z, Sundén B. Frictional pressure drop correlations for single-phase flow, condensation, and evaporation in microfin tubes. Journal of Heat Transfer. 2015;**138**(2):022901. DOI: 10.1115/1.4031268

[15] Guo SP et al. Condensation and evaporation heat transfer characteristics in horizontal smooth, herringbone and enhanced surface EHT tubes. International Journal of Heat and Mass Transfer. 2015;**85**:281-291. DOI: 10.1016/j.ijheatmasstransfer.2015.01.115

[16] Breber G, Palen JW, Taborek J. Prediction of horizontal tubeside condensation of pure components using flow regime criteria. Journal of Heat Transfer. 1980;**102**(August):471. DOI: 10.1115/1.3244325

[17] Tandon TN, Varma HK. A new flow regimes l a p for condensation inside horizontal tubes. Journal of Heat Transfer. 1982;**104**(November): 763

[18] Yu J, Koyama S. Condensation Heat Transfer of Pure Refrigerants in Microfin Tubes. International Refrigeration and Air Conditioning Conference. 1998. p. 431. Available online: http://docs.lib.purdue.edu/iracc/431

[19] Kedzierski MA, Goncalves JM. Horizontal convective condensation of alternative refrigerants within a micro-fin tube. Journal of Enhanced Heat Transfer. 1999;**6**(2):161-178. DOI: 10.1615/JEnhHeatTransf.v6.i2-4.90

[20] Han D, Lee KJ. Experimental study on condensation heat transfer enhancement and pressure drop penalty factors in four microfin tubes. International Journal of Heat and Mass Transfer. 2005;**48**(18):3804-3816. DOI: 10.1016/j.ijheatmasstransfer.2005.02.041

[21] Choi J, Kedzierski M, Domanski P. Generalized pressure drop correlation for evaporation and condensation in smooth and micro-fin tubes. Proceedings of the International Institute of Refrigeration (IIR). Thermophysical Properties and Transfer Processes of New Refrigerants, Paderborn, 1, GE. 2001;**1**:986. Available online: https://tsapps.nist.gov/publication/get_pdf.cfm?pub_id=860849

[22] Goto NIM, Inoue N. Condensation and evaporation heat transfer of R410A inside internally grooved horizontal tubes. International Journal of Refrigeration. 2001;**24**(1):528-538. DOI: 10.1016/j.ijrefrig.2009.09.008

[23] Haraguchi TFH, Koyama S. Condensation heat transfer of refrigerants HFC134a, HCFC123 and HCFC22 in a horizontal smooth tube and a horizontal microfin tube. In: Proc. 30th National Heat Transfer Symp. of Japan; Yokohama. 1993. pp. 343-345

[24] Cavallini A, Del Col D, Doretti L, Longo GA, Rossetto L. Heat transfer and pressure drop during condensation of refrigerants inside horizontal enhanced tubes. International Journal of Refrigeration. 2000;**23**(1):4-25. DOI: 10.1016/S0140-7007(99)00032-8

Chapter 9

Internal and External Influences on Hydro-Thermal Behavior of Micro-channel Flow

Naga Ramesh Korasikha, Thopudurthi Karthikeya Sharma, Gaddale Amba Prasad Rao and Kotha Madhu Murthy

Abstract

Microchannel flow is an effective solution for many engineering problems. Application of microchannels is found in various fields such as thermal management of electronics, micro-combustors, biomedical industries, MEMS. In microchannel flow, some internal and external influences such as surface roughness, electric and magnetic fields are very significant and commonly neglected in macro-scale flow. Early research works on microchannels stated that the conventional theories of macro-scale flow were not applicable for microscale flows. Finally, researchers are concluded that the deviation in conventional theories in the case of micro-scale flow is because of neglecting the internal forces, surface roughness, surface wettability, etc., which play a prime role in micro-scale flows. In this chapter, the behavior of microchannel flow under the internal and external influences is discussed. The heat transfer and hydrodynamic characteristics of microchannel flow under the external magnetic field and electric fields are presented. The effect of surface morphology, roughness, electro-osmotic effect, electrophoresis, internal heat generation, and analysis methods is discussed.

Keywords: microchannel flow, electronic cooling, MHD, surface roughness, surface wettability, electro-osmosis, electrophoresis

1. Introduction

With the rapid development of electronic technologies towards miniaturization and high power density, heat generation becomes a major problem in high-powered electronic systems, batteries, fuel cells, etc. High-temperature operation of these electronic devices reduces their performance, reliability, and life span, so thermal management has become the prime concern in modern electronic equipment. Researchers have developed various thermal management methods using compact heatsinks like mini-channel heatsinks, micro-channel heatsinks with forced and natural convection, flow boiling and phase change materials, etc.; according to the heat dissipation requirement, heat transfer modes were implemented. In the natural convection method, fluid was driven by buoyancy force generated by the temperature

IntechOpen

gradient, and this was implemented for low heat dissipation applications. For large heat transfer requirements forced convection and flow boiling was used. Forced convection and flow boiling in microchannels are effective methods to deal with the immense heat dissipation requirement in modern electronics.

From the initial research on microchannel flows, there is uncertainty in implementing conventional theories of hydrodynamics and heat transfer to the micro, and mini channel flows. Some studies on microfluidics stated that the correlations applied for pressure drop in conventional channel flow were not valid for micro and mini channel flow. Finally, this ambiguity came to a concussion that the discrepancies in the outcomes when conventional theories are used for micro and mini channels are because of errors in channel measurements and neglected influences, which are significant microchannels flow.

The experimental study of J. Judy et al. [1] is evident that the uncertainties in the conventional flow theories are significantly influenced by the uncertainties in the diameter measurement, which may cause up to 20% discrepancy in Poiseuille's number. In this analysis, from Poiseuille's number data, it was observed that there is no distinguishable between the microscale and macroscale strokes flow. Bruno Agostini et al. [2] witnessed the 21% divergence in the friction factor due to the 3% error in estimating channel width and height. Apart from measurement errors, other possible causes for uncertainties are viscous dissipation, entrance effect, the influence of electric double layer and surface effects, etc. Some authors developed, advanced thermal characterization techniques for finding the probable sources of uncertainties in the microchannel flow [3]. Some external influences like electric field and magnetic field also show a noticeable influence on the hydrothermal behavior of the microchannel flow.

M.R. Akhtari and Nader Karimi [4] investigated the characteristics of microchannels with varying super-phobic surface roughness with four different microstructured configurations. Outcomes of this work witnessed the decrease of Poiseullie number and Nusselt number with increasing the cavity fraction. They also reported that the triangular pattered structure showed the best performance among the four because in triangular patterned structure, the influence of shear-free interface on the drag reduction is more dominant than the reduction of heat transfer. Mohit Trivedi et al. [5] conducted a numerical study on electro-osmotic coupled pressure-driven flow of simplified Phan Thien Tanner fluids in a diverging micro-channel. They found that the velocity of the flow decreases moderately with the increasing the diverging angle when the other parameters were fixed.

Hang Xu [6] analyzed the mixed convection flow by buoyancy and pressure gradient in inclined micro-channel by considering the electric double layer effect and discussed the various parameters influence on the temperature distribution and velocity. It was noticed that the streaming potential raised with increasing the electrical-double-layer thickness and the zeta potential.

In this chapter, possible internal and external influences on microchannel flow are discussed (**Figure 1**). These influences can be also helpful to improve the heat transfer performance and development of innovative flow and heat transfer devices. Electro-osmosis was used in the development of electro-osmotic pumps, they have considerable advantages of no moving parts and less power requirement. Surf ace roughness of the channel has a considerable effect on the improvement of heat transfer. Surface wettability is also a significant parameter to enhance the boiling heat transfer. This chapter gives an overview of the recent works that have been done on electro-osmosis, surface roughness, surface wettability, external magnetic and electric field in micro-channel flow.

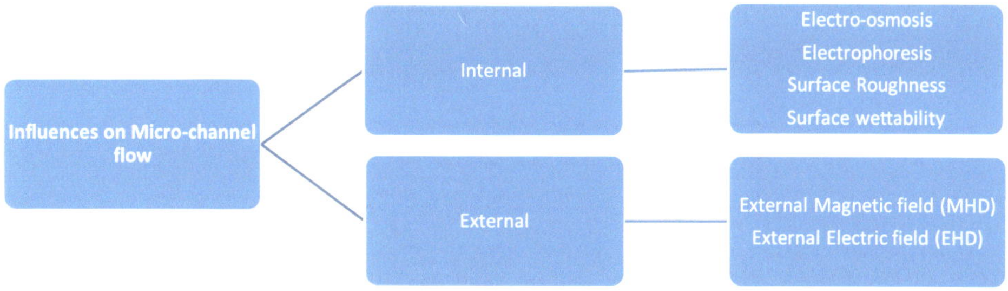

Figure 1.
Influences on micro-channel flow.

2. Internal influences

2.1 Electro-osmosis

Electro-osmosis refers to the movement of ionic fluid concerning the stationary charged surfaces with the development of an electric double layer (EDL) under the imposition of an external electric field [7–10]. Electro-osmotic actuated flow (electro-osmotic flow EOF) captured massive interest in the field of microfluidics because of its superior features of less pressure drop, no moving parts etc. Electro-osmosis is the modern pumping method for fluid transport and has advantages compared to the sure-driven flows [11, 12]. So much research work has been carried out on EOF-based micropumps and electronic cooling systems to analyze the thermo-hydraulic aspects [13]. V K Narla et al. [14] performed the theoretical analysis on EOF of nanofluid in curved microchannel. In this work, authors have studied the influence of Helmholtz-Smoulochowski velocity, inverse EDL thickness, Brinkman number, Joule Number, Curvature parameter, entropy generation and Nano particle concentration. It was found that the increasing inverse EDL thickness generates the flow acceleration in the microchannel upper region. The curved surface of the microchannel acts as the source of entropy generation. Schematic representation of pumping mechanism of electro-osmotic flow is presented in the **Figure 2**.

Michael O. Oni and Basant K. Jha [16] studied the viscous dissipation and Joule heating influence on the EOF in a vertical microchannel with asymmetric heating. Results of this work revealed that viscous dissipation and Joule heating reduces the temperature distribution and the asymmetric heating tends to increase the pressure gradient and induced electric potential. M Peralta et al. [17] was also examined the asymmetric zeta potential influence on the EOF of Maxwell fluid in parallel plate microchannel. They observed the fluctuations in the velocity profiles with increasing the angular Reynolds number and elasticity number because of viscous, elastic, electric and inertial forces in the flow.

An experimental and numerical study performed on the microfluidic chip witnessed the stable EOF on the super wettability surface [18]. M S Saravani and M Kalteh [19] conducted numerical study on combined pressure driven and electroosmotic flow of Newtonian nanofluid in the microchannel using Lattice-Boltzmann method. They have studied the influence of different parameters like slip coefficient, volume fraction of Nano-particles, Smolochowski velocity on the heat transfer. It was observed that the Nusselt number was decreases with increasing the pressure force at fixed electric field and Nusselt number was increases with increasing the electric field at fixed pressure force. Souvik Pabi et al. [20] investigated the hydrothermal and entropy generation characteristics of EOF in a hydrophobic

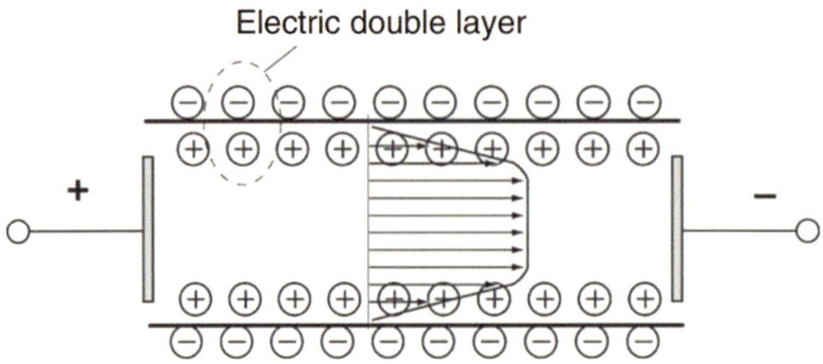

Figure 2.
Pumping Mechanism of electro-osmotic flow [15].

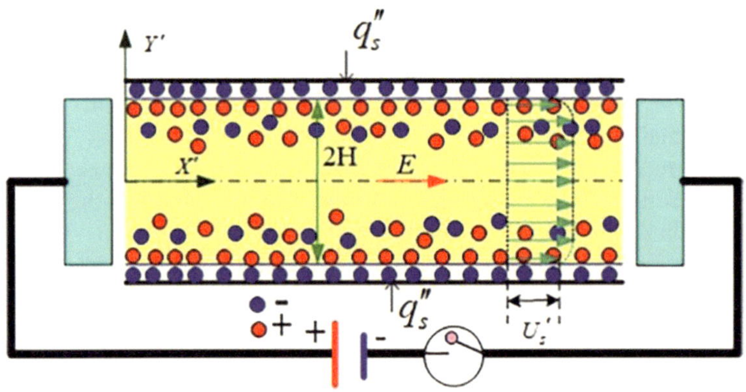

Figure 3.
Schematic image of the physical model developed by Souvik Pabi et al. [20].

micro-channel. Schematic image of physical model developed by Souvik Pabi et al. shown in the **Figure 3**. It was found that the value flow velocity (U) reduces with rise in the viscoelectric coefficient (f). The Nusselt number was found to be enhanced under slip and percentage of enhancements achieved as 8.82% and 7.96% at the Brinkman number of 10^{-5} and 0.1 respectively. Variation of Nu with Peclet number (Pe) at different viscoelectric coefficient (f) and Brinkman number (Br) is presented in the **Figure 4**.

Edgar A. Ramos et al. [21] examined the EOF in rectangular microchannel by considering the temperature dependent absolute viscosity and slip boundary condition at the walls (**Figure 5**). In this study they implemented the asymptotic analysis by using the perturbation methods. The outcomes of this analysis revealed that the influence temperature depended viscosity and slip increases the flow rate in the microchannel. Jacky S.H. Lee et al. [22] studied the behavior of electro-osmotic flow in a cylindrical microchannel with non-uniform zeta potential distribution. Results of this study reveals that, the non-uniform distribution of zeta potential generates the various types of velocity profiles in both downstream and upstream sections. It was noticed that, flow circulation cannot be generated a simple step change, if such a change does not have polarity change of the surface charge. Various heterogeneous patterns with change in zeta potential polarity was observed to be able to produce different types of flow circulations.

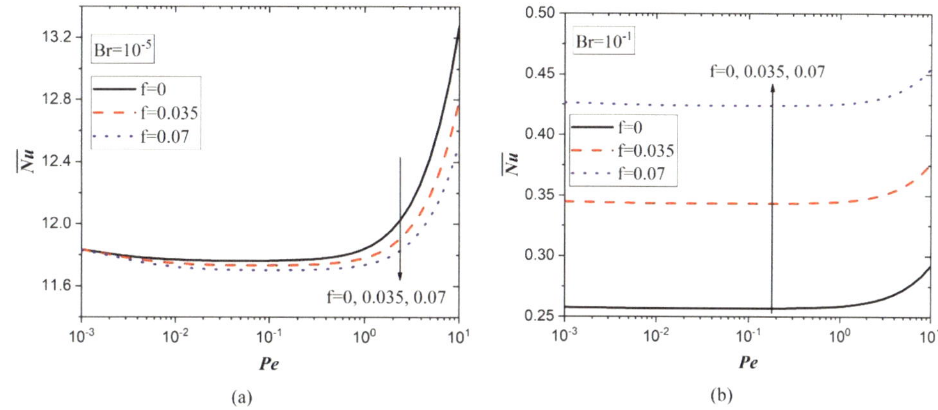

Figure 4.
Variation of Nu with Peclet number (Pe) at different viscoelectric coefficient (f) and Brinkman number (Br) [20].

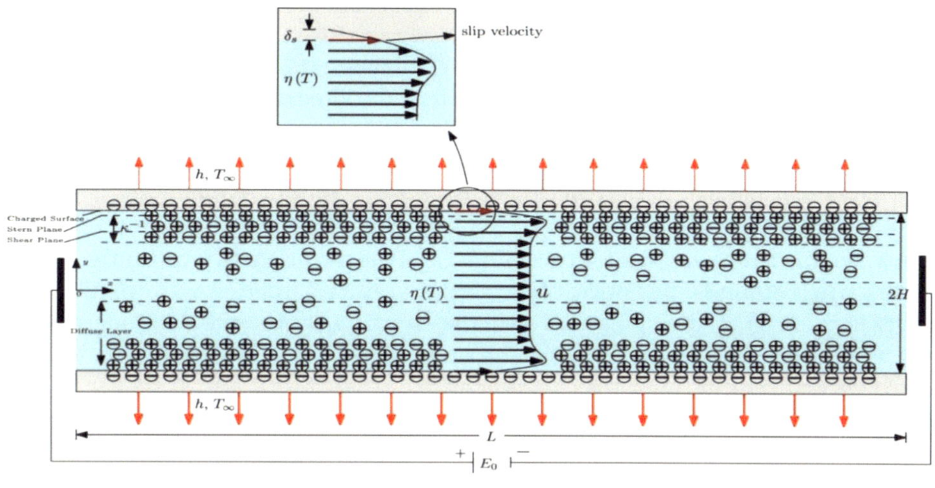

Figure 5.
Schematic image of the physical model used by Edgar A. Ramos et al. [21].

2.2 Electrophoresis

Electrophoresis is the phenomena of movement of charged particles in the colloidal solution under the influence of the imposed electric field. As a result of, applied electric field and viscous resistance imposed by the liquid on the colloidal particles, the stationary particles moves with constant velocity [23]. So many researchers has been working on the electrophoresis in straight and curved micro-channels with constant and variable cross-section by applying the uniform and non-uniform electric field [24–26]. Litao Liang et al. [27] performed the experimental demonstration on uniform straight rectangular microchannel with lateral migration in particle electrophoresis and developed an analytical model to predict the outcomes (**Figure 6**). The results of this work revealed that, the particle stream width at the exit of channel was noticed to be diminishes with the increment in either electric field or particle size and

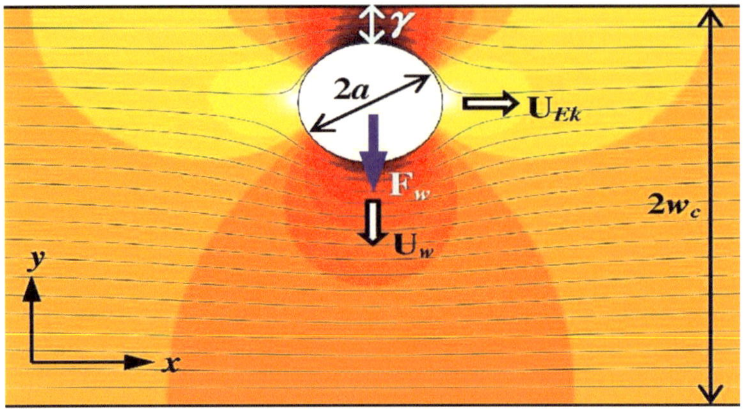

Figure 6.
Velocity analysis of a particle in electrophoretic motion in rectangular micro-channel. The background presents the electric field and lines [27].

the predicted value of width of the steam through analytical model has good agreement with measured values.

P.H. Wiersema et al. [28] presented a correlation between the electrophoresis mobility and the Zeta potential of spherical shaped colloidal particles. They found that the correction for diminished values of dielectric constant in the double layer was negligible. Shizhi Qian et al. [29] performed numerical investigation on electrophoretic mobility generated in the convergent-divergent nanotube containing electrolyte solution with spherical charged particles by applying the electric field. Authors found that, when the nanotube wall was uncharged and the particle was charged the motion of the particle attained the noticeable acceleration and retardation as the particle moves in converging–diverging section in the direction of opposite electrode and achieved the maximum velocity at the throat. However, the mentioned effect is not always correct because when the charge density of wall of tube is of same sign and magnitude as the charge density of particle surface, the flow generated by the electrostatic force and the induced pressure gradient will reverse the direction of motion of particles. The schematic image of converging–diverging nano-tubes presented in **Figure 7**.

2.3 Surface effects

Surface of the fluid flowing channel have noticeable influence on the hydrothermal characteristics of the microchannel flow. Surface morphology, wettability, hydrophilic and hydrophobic behavior of the surface shows significant influence on the heat transfer [30]. Recent studies on surface effects on microchannel flow is presented in this section.

2.3.1 Surface roughness

Earliest investigations on the microchannel flows showed that the hydrothermal behavior of microchannel is differ from macrochannel. However, researchers concluded about this divergence as, some of the factors neglected in the macro scale shows significant effect on microscale and these factors are the reason for

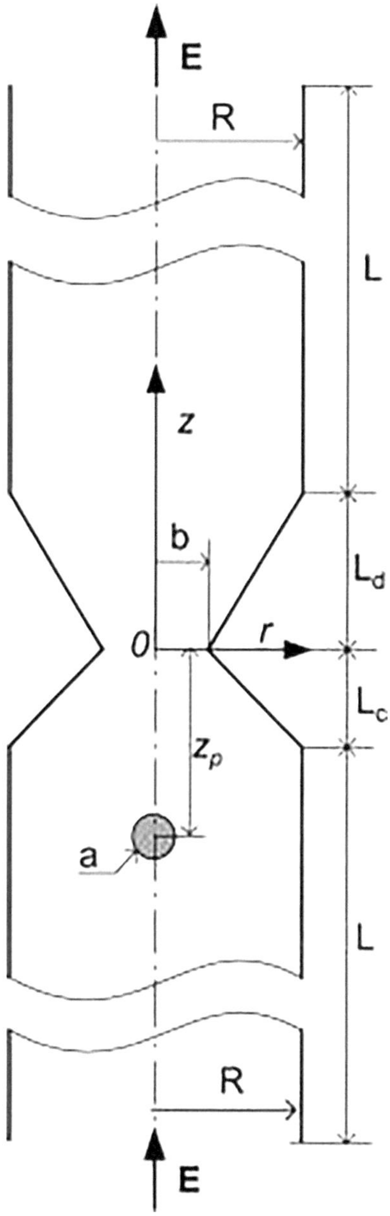

Figure 7.
Schematic image of converging–diverging nanotube filled with an electrolyte containing spherical charged particle [29].

discrepancies [1]. Surface roughness is an important factor which has considerable influence on microchannel flow. An experimental investigation done by Yuan Xing et al. [31] on microchannel surface roughness and witnessed the remarkable influence of roughness on heat transfer and hydraulic characteristics. At larger relative roughness, Nusselt number and the Poiseuille was found to be higher and the friction factor and Nusselt number increases with increasing the Reynolds number

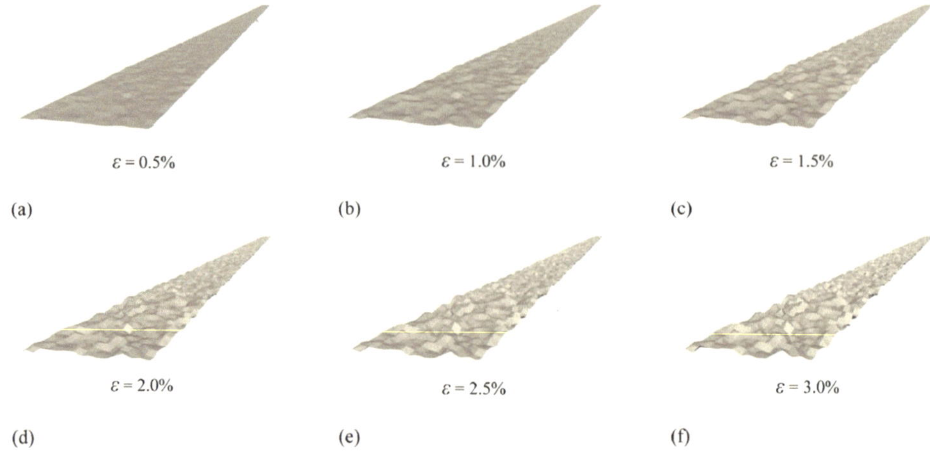

Figure 8.
Isometric view of bottom surface of microchannel with six different cases of relative roughness [32].

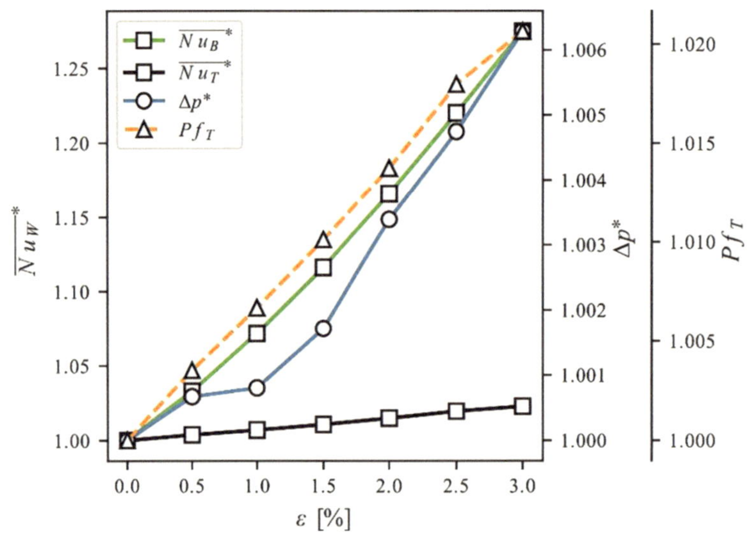

Figure 9.
Normalized average Nusselt numbers ($\overline{Nu}_B^*$ and $\overline{Nu}_T^*$) pressure drop Δp^*, and performance factor P_{fT} variation with maximum relative roughness ε [32].

consistently. Benedikt Sterr et al. [32] performed the stochastic analysis on MCHS with random Roughness (**Figure 8**). It was revealed that the local surface height is inadequate to define the behavior of Nusselt number and the dependency of Nusselt number on various factors was increase with the value of roughness. The variation of normalized average Nusselt numbers, pressure drop and performance factor are depicted in the **Figure 9**.

Rahim Jafari et al. [33] conducted experimental analysis on flow boiling in microchannel to study the influence of surface roughness on micro-scale boiling. They found the 45% enhancement in heat transfer coefficient, when the roughness increased from 2.03 μm to 15.86 μm. Considerable increase of heat transfer during the phase change was obtained with low heatflux value at high roughness.

2.3.2 Surface wettability

Surface wettability is aptness of a fluid to spread over a surface. It is identified as an important factor in microfluidic devices, especially in case two-phase flows in microchannels, where the surface tension becomes presiding force as the channel size reduces [34]. Many researchers are focused on the influence of surface wettability on two-phase flow in microchannels. If water is considered, the characteristics of water droplets on a surface can be explored by surface energies and wettability. When Surface energies are high, the water droplets area attracted towards the surface resulting low contact angle, it is called hydrophilic surface and in the reverse case surface is called hydrophobic surface [35]. Important findings of various researchers on surface wettability is discussed in this section. **Figure 10** indicates the schematic representation of water contact angle on hydrophilic, hydrophobic surfaces.

Konstantinos Vontas et al. [37] studied the flow boiling in and hydrophobic surfaced microchannels under various heat fluxes and mass fluxes and observed the slug flow regime in the both cases. Authors revealed that liquid film and contact line evaporation are the prime heat transfer mechanism for hydrophilic and hydrophobic surfaces respectively. Prakash Rapolu and Sang Young Son [38] conducted an experimental analysis on geometry and wettability influence on pressure drop in 2-Phase microchannel flow and witnessed the rise in pressure drop with reducing the surface wettability (**Figures 11** and **12**). They also revealed that the effect of the capillary resistance on the pressure drop in a hydrophilic channel was comparatively less.

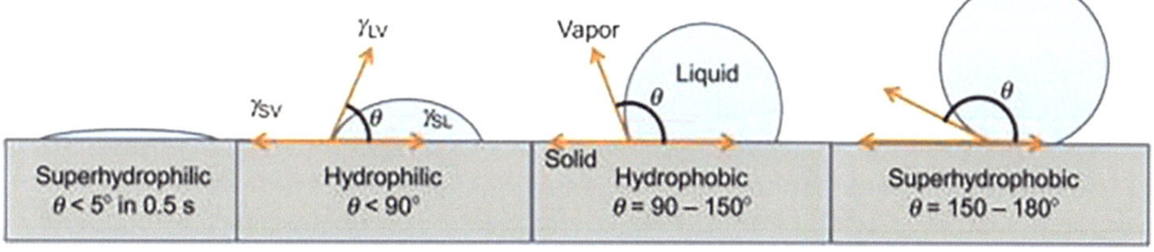

Figure 10.
Schematic representation of water contact angle on hydrophilic, hydrophobic surfaces [36].

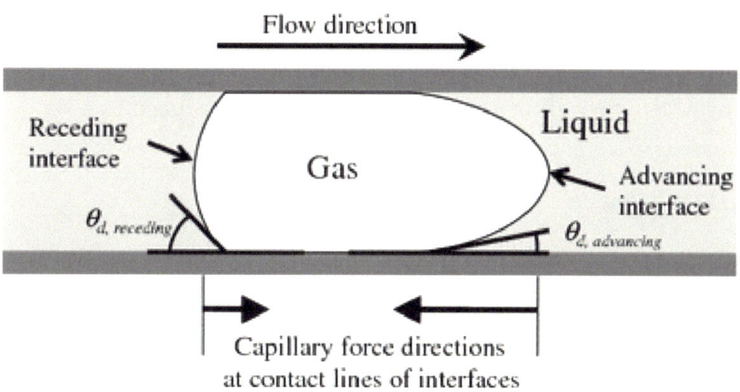

Figure 11.
Schematic image of dynamic contact angles in slug flow [38].

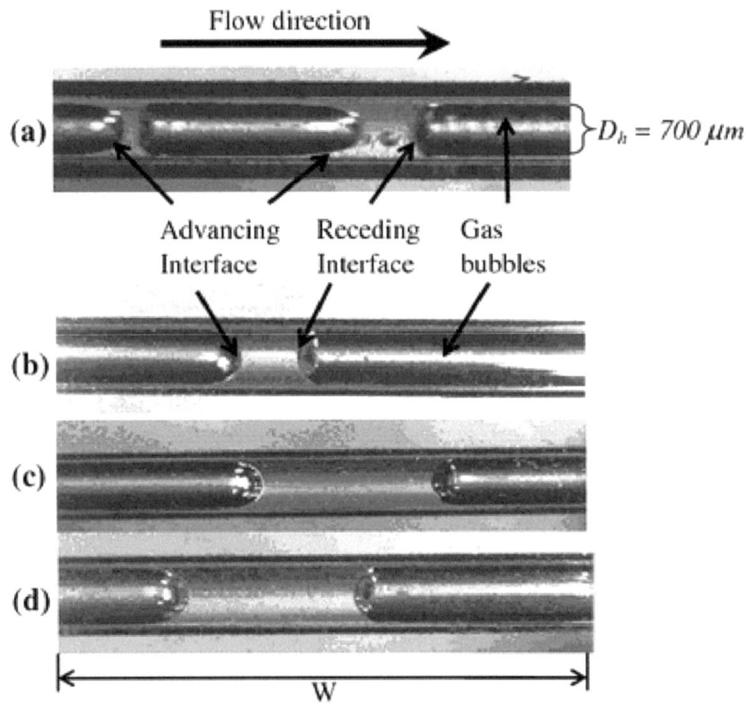

Figure 12.
Slug flow in microchannels at different wettabilities (a) Superhydrophobic (b) Hydrophobic (c) Hydrophilic (d) Superhydrophilic [38].

Ming Yu Zhou et al. [39] examined behavior of monodisperse poly n-isopropylacrylamide microspheres in microchannels with biphilic surfaces. Outcomes of this work disclosed that the mean fluid velocity in hydrophobic surfaced microchannel without microspheres is higher than hydrophilic surfaced microchannel. Wei Deng et al. [40] studied the bubble behavior and heat transfer in a boiling flow over the biphilic surfaces and found that at low heatflux values the bubbles prone to grow first on the hydrophobic surfaces because of low attractive force. Authors also stated that they noticed a different behavior of bubble characteristics at the wall temperature of 150 K and 200 K. Ali Heidarian et al. [41] conducted the hydrodynamic study on the nanofluid flow in the hydrophobic and super hydrophobic surfaced microchannel. Results of this study showed that the pressure drop of nanofluid flow in case of hydrophilic surface was increased by 74% and in case of hydrophobic surface was decreased by 47%. The wall shear stress was found to be reduced by 65%.

3. External influences

3.1 Magneto-hydrodynamics (MHD)

Magneto-hydrodynamics (MHD) is the study of behavior of an electrically conducting fluid under the magnetic field and it has many applications in engineering field including the cooling of liquid metals in nuclear reactors, design of MHD flow meters and MHD pumps etc. [42–44]. In the field of microfluidics, MHD is promising method for pumping applications because of its benefits of simple design, no moving parts and less power consumption [45–48]. Chunhong Yang et al. [49] performed theoretical analysis on the thermal behavior of a incompressible MHD flow in a rectangular microchannel. In this study the cumulative influence of the Joule heating, viscous dissipation and electromagnetic coupled heat was considered. A gradual decrease of normal velocity was noticed with rising the Hartmann number (Ha) without implementing the lateral electric field. When lateral electric field was imposed, increasing-decreasing trend is observed for velocity and temperature but the Nu exhibited the adverse trend (**Figure 13**).

Velocity and temperature but the Nu exhibited the adverse trend.

Mehdi Kiyasatfar and Nader Pourmahmoud [50] studied the hydrothermal characteristics of non-Newtonian, conducting fluid flow in a square microchannel, imposed to the transfers magnetic field. I their study the decrease of velocity gradient was found near the wall with increasing the flow index of Power law which tend to decrease the Nu. Magnetic field showed the large influence on the velocity field, as Ha increases, maximum velocity was decreases and wall velocity gradient was increases. Nu variation with shear rate parameter for the dilatant fluids is presented in **Figure 14**. M.M. Rashidi et al.

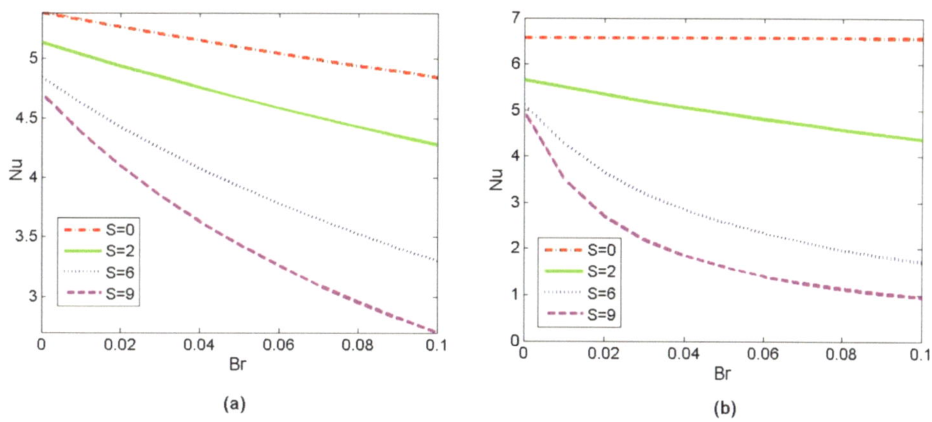

Figure 13.
Lateral electric field influence on Nu with Br (K = 10). (a) Ha = 0.5. (b) Ha = 3 [49].

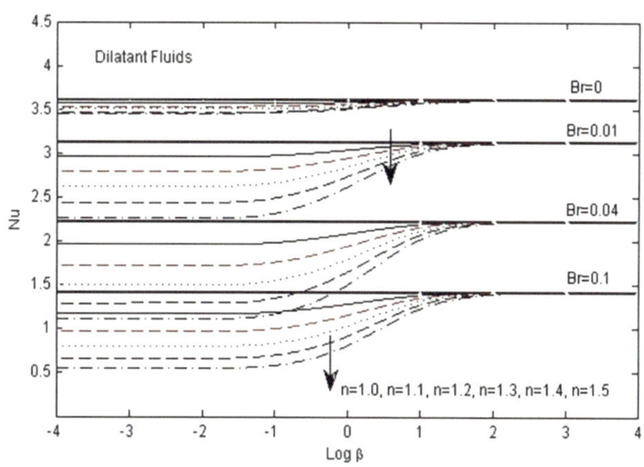

Figure 14.
Nu variation with shear rate parameter for the dilatant fluids (Ha = 0) [50].

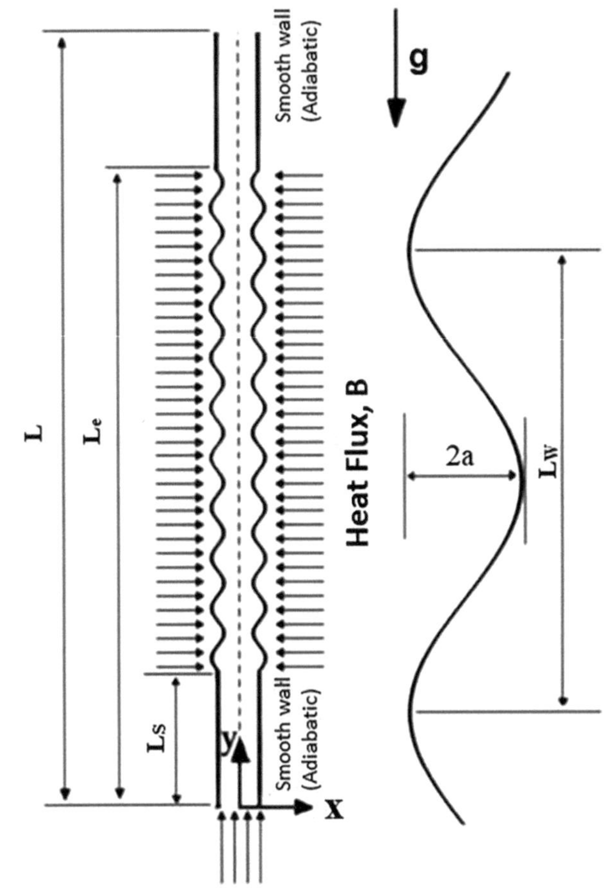

Figure 15.
Schematic image of the physical model used by M.M. Rashidi et al. [51].

[51] examined the mixed convective HT in sinusoidal microchannel with nanofluid flow under the influence of external magnetic field (**Figure 15**). From the results, improvement of average Nu and decrement in Poiseuille's number was observed with increasing the concentration of nanoparticles. An enhancement of Nu and Poiseuille's number was also noticed with increasing Ha at constant Re, Grashof number (Gr) and particles concentration. Basant Kumar Jha et al. [52] performed the magnetohydrodynamic analysis on natural convection flow formed in the micro-gap between two infinitely long parallel vertical plates. By the asymmetric heating of the plates, the natural convection is created between the plates in presence of the Hall current. The increase of fluid velocity was observed with the ratio of wall ambient temperature and rarefaction parameter.

3.2 Electro-hydrodynamics (EHD)

Electro-hydrodynamics (EHD) is the study of characteristics of electrically conducting fluid under the external electric field. The bulk fluid motion created due to the interactions between imposed electric field and the gradients electrical conductivity of the fluid is referred as the Electro-hydrodynamic (EHD) flow [53, 54]. In the field of microfluidics, EHD used for fluid pumping applications. Brian D. Storey [55] conducted a numerical study on instabilities generated in microchannel flow between the two parallel plates due to the gradients of electrical conductivity of fluid. They also investigated the turbulence generated at low Reynolds number due to the instabilities in the flow by considering the electro-osmotic effect. Author found that the existence of shallow walls of the channel can modify the flow stability and flow characteristics remarkably. When walls are shallow with regard to the dimension of gradient of electric conductivity, the fluid motions are considerably suppressed and the wavenumber of instability increased. P. Zangeneh Kazemi et al. [56] investigated the performance of the EHD micro-pump under the influence of spacing of micro pillar electrode. Schematic representation of micropump with symmetric configuration of electrode is presented in **Figure 16**. Considerable higher pressure head was observed in case of micropump with asymmetric design compared to symmetric design of electrode under the equal applied voltage by consuming the lower power. The reduction in the spacing of the micro-pillar electrodes enhanced the performance of pump for the symmetric design, while diminished the performance for the asymmetric design.

Some researchers analyzed the combined effective applied magnetic field and electric field. Guillermo Ibáñez and Sergio Cuevas [57] investigated the electromagnetic interactions in the micro-channel by considering the influence of the Lorentz force generated by injected electric current and transvers magnetic field to diminish the entropy generation. Li Fengqina et al. [58] investigated the electro-magnetohydrodynamic flow in a microchannel with 3D corrugated walls. The flow in the present work was driven by a non-uniform electric field and uniform magnetic field. The effect of corrugations and Hartmann number (Ha) was analyzed by implementing the perturbation method. The results of this study reveal that maximum flow rate was achieved by increasing the Ha (magnetic field strength parameter).

Along with the mentioned influences, some other geometry-related effects like internal structure of the geometry, topology and internal fins etc., were also examined in some research works. Yongqi Lan et al. [59] performed the numerical simulation on effect of relative height and relative offset values of truncated and offset pin-fins on performance and entropy generation of MCHS. Enhancement of heat transfer and flow resistance was noticed in the MCHS with pinfins compared to the smooth MCHS. Increasing pinfin height ratio, increases the heat transfer and the flow resistance. Increasing the offset ratio also improved the heat transfer but extreme increase in offset ration leads to the decrease in heat transfer improvement. Prasenjit Dey et al. [60] conducted the experimental and numerical study on hydrothermal characteristics

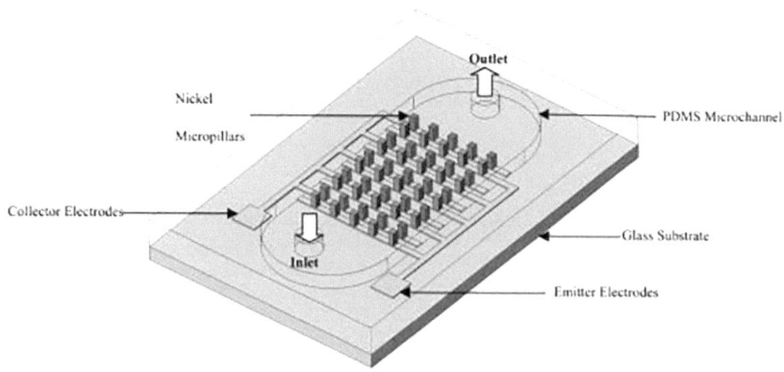

Figure 16.
Schematic representation of micropump with symmetric configuration of electrode [56].

of MCHS with bio-inspired novel fish scale structure. It was found that the fish scale structure can enhance the heat transfer compared to the plan structure. At 0.026 inclination angle of fish scale, friction factor was reduced by maximum 5% and Nusselt number increased by maximum of 14% compared to the plan surface structure. The performance evaluation factor was noticed as 1.75 at the inclination angle of 0.26 and Reynolds number of 1050. Xiaohu Li et al. [61] performed the topology optimization study on MCHS to augment its heat transfer performance by adopting iso-geometric analysis (IGA) approach. It was observed that the IGA approach produced the better outcomes compared to the traditional finite element analysis (FEA) approach.

4. Conclusions

The fluid flow and heat transfer behavior of the micro-channel flow under some internal and external influences is presented in this chapter. Latest works that has been performed on these influences are discussed so far. The prime conclusion from this chapter is,

- In micro-scale studies, the reason for the deviation of conventional theories and correlations was concluded as the neglected effects, which has a considerable influence on microscale flow.

- Electro-osmotic, MHD and EHD pumping methods have enormous significance in the field of microfluidics because of the advantages of no moving parts and low power consumption.

- Surface wettability plays key role hydrothermal performance of two-phase flow in microchannels.

- Surface roughness also has considerable influence on enhancing heat transfer in both single-phase and two-phase flow through micro-channels.

Author details

Naga Ramesh Korasikha[1], Thopudurthi Karthikeya Sharma[1*], Gaddale Amba Prasad Rao[2] and Kotha Madhu Murthy[2]

1 Department of Mechanical Engineering, NIT, Andhra Pradesh, India

2 Department of Mechanical Engineering, NIT, Warangal, India

*Address all correspondence to: tks@nitandhra.ac.in

References

[1] Judy J, Maynes D, Webb BW. Characterization of frictional pressure drop for liquid flows through microchannels. International Journal of Heat and Mass Transfer. 2002;**45**:3477-3489. DOI: 10.1016/S0017-9310(02)00076-5

[2] Agostini B, Watel B, Bontemps A, Thonon B. Liquid flow friction factor and heat transfer coefficient in small channels: An experimental investigation. Experimental Thermal and Fluid Science. 2004;**28**(2-3):97-103. DOI: 10.1016/S0894-1777(03)00027-X

[3] Takács G, Szabó PG, Bognár G. Enhanced thermal characterization method of microscale heatsink structures. Microelectronics and Reliability. 2016;**67**:21-28. DOI: 10.1016/j.microrel.2016.09.019

[4] Akhtari MR, Karimi N. Thermohydraulic analysis of a microchannel with varying superhydrophobic roughness. Applied Thermal Engineering. 2020;**172**:115147. DOI: 10.1016/J.APPLTHERMALENG.2020.115147

[5] Trivedi M, Maurya S, Nirmalkar N. Numerical simulations for electro-osmotic flow of PTT fluids in diverging microchannel. Materials Today: Proceedings. 2022;**57**:1765-1769. DOI: 10.1016/J.MATPR.2021.12.432

[6] Xu H. Fully developed opposing mixed convection in inclined microchannel with electric double layer effects. International Communications in Heat and Mass Transfer. 2022;**131**:105848. DOI: 10.1016/J.ICHEATMASSTRANSFER.2021.105848

[7] Yang C, Li D. Analysis of electrokinetic effects on the liquid flow in rectangular microchannels. Colloids and Surfaces A: Physicochemical and Engineering Aspects. 1998;**143**:339-353. DOI: 10.1016/S0927-7757(98)00259-3

[8] Arulanandam S, Li D. Liquid transport in rectangular microchannels by electroosmotic pumping. Colloids and Surfaces A: Physicochemical and Engineering Aspects. 2000;**161**:89-102. DOI: 10.1016/S0927-7757(99)00328-3

[9] Soong CY, Wang SH. Theoretical analysis of electrokinetic flow and heat transfer in a microchannel under asymmetric boundary conditions. Journal of Colloid and Interface Science. 2003;**265**:202-213. DOI: 10.1016/S0021-9797(03)00513-7

[10] Mukhopadhyay A, Banerjee S, Gupta C. Fully developed hydrodynamic and thermal transport in combined pressure and electrokinetically driven flow in a microchannel with asymmetric boundary conditions. International Journal of Heat and Mass Transfer. 2009;**52**:2145-2154. DOI: 10.1016/J.IJHEATMASSTRANSFER.2008.08.035

[11] Dallakehnejad M, Mirbozorgi SA, Niazmand H. A numerical investigation of magnetic mixing in electroosmotic flows. Journal of Electrostatics. 2019;**100**:103354. DOI: 10.1016/J.ELSTAT.2019.103354

[12] Deng S, Li M, Yang Y, Xiao T. Heat transfer and entropy generation in two layered electroosmotic flow of power-law nanofluids through a microtube. Applied Thermal Engineering. 2021;**196**:117314. DOI: 10.1016/J.APPLTHERMALENG.2021.117314

[13] Li L, Wang X, Pu Q, Liu S. Advancement of electroosmotic pump in microflow analysis: A review. Analytica Chimica Acta. 2019;**1060**:1-16. DOI: 10.1016/J.ACA.2019.02.004

[14] Narla VK, Tripathi D, Bég OA. Electro-osmotic nanofluid flow in a curved microchannel, Chinese. Journal de Physique. 2020;**67**:544-558. DOI: 10.1016/J.CJPH.2020.08.010

[15] Shoji S, Sato H, Zengerle R. Liquid micropumps. Comprehensive Microsystems. 2008;**2**:301-322. DOI: 10.1016/B978-044452190-3.00041-0

[16] Oni MO, Jha BK. Joule heating and viscous dissipation effect on electroosmotic mixed convection flow in a vertical microchannel subjected to asymmetric heat fluxes. Propulsion and Power Research. 2021;**10**:83-94. DOI: 10.1016/J.JPPR.2021.01.001

[17] Peralta M, Bautista O, Méndez F, Bautista E. Pulsatile electroosmotic flow of a Maxwell fluid in a parallel flat plate microchannel with asymmetric zeta potentials. Applied Mathematics and Mechanics. English Ed. 2018;**39**:667-684. DOI: 10.1007/s10483-018-2328-6

[18] Jiang S, Zhang H, Chen L, Li Y, Sang S, Liu X. Numerical simulation and experimental study of the electroosmotic flow in open microfluidic chip based on super-wettability surface. Colloid Interface Science and Communication. 2021;**45**:100516. DOI: 10.1016/J.COLCOM.2021.100516

[19] Sheikhizad Saravani M, Kalteh M. Heat transfer investigation of combined electroosmotic/pressure driven nanofluid flow in a microchannel: Effect of heterogeneous surface potential and slip boundary condition. European Journal of Mechanics - B/Fluids. 2020;**80**:13-25. DOI: 10.1016/J.EUROMECHFLU.2019.11.002

[20] Pabi S, Mehta SK, Pati S. Analysis of thermal transport and entropy generation characteristics for electroosmotic flow through a hydrophobic microchannel considering viscoelectric effect. International Communications in Heat and Mass Transfer. 2021;**127**:105519. DOI: 10.1016/J.ICHEATMASSTRANSFER.2021.105519

[21] Ramos EA, Treviño C, Lizardi JJ, Méndez F. Non-isothermal effects in the slippage condition and absolute viscosity for an electroosmotic flow. European Journal of Mechanics - B/Fluids. 2022;**93**:29-41. DOI: 10.1016/J.EUROMECHFLU.2022.01.001

[22] Lee JSH, Ren CL, Li D. Effects of surface heterogeneity on flow circulation in electroosmotic flow in microchannels. Analytica Chimica Acta. 2005;**530**:273-282. DOI: 10.1016/j.aca.2004.09.026

[23] Ohshima H. Electrophoresis, reference module in chemistry. Molecular Sciences and Chemical Engineering. 2016. DOI: 10.1016/B978-0-12-409547-2.12199-7

[24] Tang YP, Chih MH, Lee E, Hsu JP. Electrophoretic motion of a charge-regulated sphere normal to a plane. Journal of Colloid and Interface Science. 2001;**242**:121-126. DOI: 10.1006/JCIS.2001.7770

[25] Morrison FA. Electrophoresis of a particle of arbitrary shape. Journal of Colloid and Interface Science. 1970;**34**:210-214. DOI: 10.1016/0021-9797(70)90171-2

[26] Davison SM, Sharp KV. Boundary effects on the electrophoretic motion of cylindrical particles: Concentrically and eccentrically-positioned particles

in a capillary. Journal of Colloid and Interface Science. 2006;**303**:288-297. DOI: 10.1016/J.JCIS.2006.07.063

[27] Liang L, Ai Y, Zhu J, Qian S, Xuan X. Wall-induced lateral migration in particle electrophoresis through a rectangular microchannel. Journal of Colloid and Interface Science. 2010;**347**:142-146. DOI: 10.1016/J.JCIS.2010.03.039

[28] Wiersema PH, Loeb AL, Overbeek JTG. Calculation of the electrophoretic mobility of a spherical colloid particle. Journal of Colloid and Interface Science. 1966;**22**:78-99. DOI: 10.1016/0021-9797(66)90069-5

[29] Qian S, Wang A, Afonien JK. Electrophoretic motion of a spherical particle in a converging–diverging nanotube. Journal of Colloid and Interface Science. 2006;**303**:579-592. DOI: 10.1016/J.JCIS.2006.08.003

[30] Zhang P, Lv FY. A review of the recent advances in superhydrophobic surfaces and the emerging energy-related applications. Energy. 2015;**82**:1068-1087. DOI: 10.1016/J.ENERGY.2015.01.061

[31] Yuan X, Tao Z, Li H, Tian Y. Experimental investigation of surface roughness effects on flow behavior and heat transfer characteristics for circular microchannels. Chinese Journal of Aeronautics. 2016;**29**:1575-1581. DOI: 10.1016/J.CJA.2016.10.006

[32] Sterr B, Mahravan E, Kim D. Uncertainty quantification of heat transfer in a microchannel heat sink with random surface roughness. International Journal of Heat and Mass Transfer. 2021;**174**:121307. DOI: 10.1016/J.IJHEATMASSTRANSFER.2021.121307

[33] Jafari R, Okutucu-Özyurt T, Ünver HÖ, Bayer Ö. Experimental investigation of surface roughness effects on the flow boiling of R134a in microchannels. Experimental Thermal and Fluid Science. 2016;**79**:222-230. DOI: 10.1016/J.EXPTHERMFLUSCI.2016.07.016

[34] Zhou K, Coyle C, Li J, Buongiorno J, Li W. Flow boiling in vertical narrow microchannels of different surface wettability characteristics. International Journal of Heat and Mass Transfer. 2017;**109**:103-114. DOI: 10.1016/J.IJHEATMASSTRANSFER.2017.01.111

[35] Crick CR, Parkin IP. Preparation and characterisation of super-hydrophobic surfaces. Chemistry - A European Journal. 2010;**16**:3568-3588. DOI: 10.1002/chem.200903335

[36] Chieng BW, Ibrahim NA, Daud NA, Talib ZA. Functionalization of graphene oxide via gamma-ray irradiation for hydrophobic materials. Synthesis and Applications of Carbon Nanomaterials for Energy Generation and Storage. 2019:177-203. DOI: 10.1016/B978-0-12-815757-2.00008-5

[37] Vontas K, Andredaki M, Georgoulas A, Miché N, Marengo M. The effect of surface wettability on flow boiling characteristics within microchannels. International Journal of Heat and Mass Transfer. 2021;**172**:121133. DOI: 10.1016/J.IJHEATMASSTRANSFER.2021.121133

[38] Rapolu P, Son SY. Characterization of wettability effects on pressure drop of two-phase flow in microchannel. Experiments in Fluids. 2011;**51**:1101-1108. DOI: 10.1007/s00348-011-1129-8

[39] Zhou MY, Xie R, Yu YL, Chen G, Ju XJ, Yang L, et al. Effects of surface wettability and roughness of microchannel on flow behaviors of thermo-responsive microspheres therein during the phase transition.

Journal of Colloid and Interface Science. 2009;**336**:162-170. DOI: 10.1016/J.JCIS.2009.03.048

[40] Deng W, Ahmad S, Liu H, Chen J, Zhao J. Improving boiling heat transfer with hydrophilic/hydrophobic patterned flat surface: A molecular dynamics study. International Journal of Heat and Mass Transfer. 2022;**182**:121974. DOI: 10.1016/J.IJHEATMASSTRANSFER.2021.121974

[41] Heidarian A, Rafee R, Valipour MS. Hydrodynamic analysis of the nanofluids flow in a microchannel with hydrophobic and superhydrophobic surfaces. Journal of the Taiwan Institute of Chemical Engineers. 2021;**124**:266-275. DOI: 10.1016/J.JTICE.2021.04.002

[42] Abdullina KI, Bogovalov SV. 3-D numerical modeling of MHD flows in variable magnetic field. Physics Procedia. 2015;**72**:351-357. DOI: 10.1016/j.phpro.2015.09.109

[43] Gajbhiye NL, Eswaran V. MHD buoyant flow in a cubical enclosure at low to high Hartmann number. International Journal of Thermal Sciences. 2018;**134**:168-178. DOI: 10.1016/j.ijthermalsci.2018.07.028

[44] Xiao X, Kim CN. Numerical analysis of an MHD micro-device with simultaneous mixing and pumping capability. Journal of Industrial and Engineering Chemistry. 2016;**38**:23-36. DOI: 10.1016/j.jiec.2016.04.001

[45] Kang HJ, Choi B. Development of the MHD micropump with mixing function. Sensors and Actuators, A: Physical. 2011;**165**:439-445. DOI: 10.1016/j.sna.2010.11.011

[46] Das C, Wang G, Payne F. Some practical applications of magnetohydrodynamic pumping. Sensors and Actuators, A: Physical. 2013;**201**:43-48. DOI: 10.1016/j.sna.2013.06.023

[47] Al-Habahbeh OM, Al-Saqqa M, Safi M, Abo Khater T. Review of magnetohydrodynamic pump applications. Alexandria Engineering Journal. 2016;**55**:1347-1358. DOI: 10.1016/j.aej.2016.03.001

[48] Xiao X, Kim CN. Magnetohydrodynamic flows in a hairpin duct under a magnetic field applied perpendicular to the plane of flow. Applied Mathematics and Computation. 2014;**240**:1-15. DOI: 10.1016/j.amc.2014.04.049

[49] Yang C, Jian Y, Xie Z, Li F. Heat transfer characteristics of magnetohydrodynamic electroosmotic flow in a rectangular microchannel. European Journal of Mechanics - B/Fluids. 2019;**74**:180-190. DOI: 10.1016/j.euromechflu.2018.11.015

[50] Kiyasatfar M, Pourmahmoud N. Laminar MHD flow and heat transfer of power-law fluids in square microchannels. International Journal of Thermal Sciences. 2016;**99**:26-35. DOI: 10.1016/j.ijthermalsci.2015.07.031

[51] Rashidi MM, Nasiri M, Khezerloo M, Laraqi N. Numerical investigation of magnetic field effect on mixed convection heat transfer of nanofluid in a channel with sinusoidal walls. Journal of Magnetism and Magnetic Materials. 2016;**401**:159-168. DOI: 10.1016/j.jmmm.2015.10.034

[52] Jha BK, Malgwi PB, Aina B. Hall effects on MHD natural convection flow in a vertical microchannel. Alexandria Engineering Journal. 2018;**57**:983-993. DOI: 10.1016/j.aej.2017.01.038

[53] Michael DH, O'neill ME. Electrohydrodynamic instability in

plane layers of fluid. Journal of Fluid Mechanics. 1970;**41**:571-580. DOI: 10.1017/S0022112070000757.

[54] Burgi DS, Chien RL. Optimization in Sample Stacking for High-Performance Capillary Electrophoresis. Analytical Chemistry. 1991;**63**:2042-2047. DOI: 10.1021/ac00018a028

[55] Storey BD. Direct numerical simulation of electrohydrodynamic flow instabilities in microchannels. Physica D: Nonlinear Phenomena. 2005;**211**:151-167. DOI: 10.1016/J.PHYSD.2005.08.010

[56] Zangeneh Kazemi P, Ravi Selvaganapathy P, Ching CY. Effect of micropillar electrode spacing on the performance of electrohydrodynamic micropumps. Journal of Electrostatics. 2010;**68**:376-383. DOI: 10.1016/J.ELSTAT.2010.05.008

[57] Ibáñez G, Cuevas S. Entropy generation minimization of a MHD (magnetohydrodynamic) flow in a microchannel. Energy. 2010;**35**:4149-4155. DOI: 10.1016/j.energy.2010.06.035

[58] Li F, Jian Y, Buren M, Chang L. Effects of three-dimensional surface corrugations on electromagnetohydrodynamic flow through microchannel, Chinese. Journal de Physique. 2019;**60**:345-361. DOI: 10.1016/J.CJPH.2019.05.013

[59] Lan Y, Feng Z, Huang K, Zhang J, Hu Z. Effects of truncated and offset pin-fins on hydrothermal performance and entropy generation in a rectangular microchannel heat sink with variable fluid properties. International Communications in Heat and Mass Transfer. 2021;**124**:105258. DOI: 10.1016/J.ICHEATMASSTRANSFER.2021.105258

[60] Dey P, Hedau G, Saha SK. Experimental and numerical investigations of fluid flow and heat transfer in a bioinspired surface enriched microchannel. International Journal of Thermal Sciences. 2019;**135**:44-60. DOI: 10.1016/J.IJTHERMALSCI.2018.08.042

[61] Li X, Zhang L, Li B. Heat transfer augmentation in microchannel heat sink based on isogeometric topology optimization framework. Applied Mathematical Modelling. 2022;**104**:163-187. DOI: 10.1016/J.APM.2021.11.021

Section 3

Heat Transfer Enhancement and Techniques

Chapter 10

The Combined Method to Improve Heat Transfer Coefficient on Heat Exchanger

Sudarmadji Sudarmadji, Sugeng Hadi Susilo and Asrori Asrori

Abstract

The heat transfer process occurs all the time around us, from simple household appliances to equipment used in large industries. Energy efficiency in large-scale use in industry is necessary because it is related to company profits. One way to save energy use in heat exchangers is to change the thermal properties of the cooling fluid. The addition of particles of the nanometer size (nanofluids) in the working fluid will improve the performance of the heat exchanger, and the main goal is the highest efficiency. In addition, there is another method to increase the heat transfer rate, namely, by vibrating the cooling fluid. This chapter will discuss combining nanofluids and ultrasonic vibrations in heat transfer processes in heat exchangers. The application of these two methods simultaneously gives rise to several advantages to the heat transfer system, will promote higher heat transfer, and at the same time function as cleaning of scale/deposits that often appear on the surface of the heat exchanger.

Keywords: nanofluid, heat exchanger, nanoparticles, heat transfer, overall heat transfer coefficient, ultrasonic vibration

1. Introduction

Energy is an important element in all levels of people's lives. We live in an interdependent world, and access to easy and reliable energy resources is critical for economic growth to maintain our quality of life. World energy consumption continues to increase, especially in developing countries. Global energy demand has tripled in the last 50 years and may even triple in the next 30 years, while current energy consumption and production levels are unsustainable. The energy sources we used to support the entire range of human activities today come from fossil fuels (coal, oil, and natural gas), supplying about 81%. Fossil-fueled power plants convert heat into mechanical energy, which then operates an electric generator. Changing heat energy, of course, involves heat transfer from one point to another.

The heat transfer process certainly involves an essential tool, including a heat exchanger, which is expected to have high performance with smaller dimensions. For large power, the increase in the performance of the heat exchanger will impact financial reductions in a company. Of course, efforts are needed to increase the heat transfer coefficient of a heat exchanger with various methods.

There are several methods to increase the heat transfer of a heat exchanger, namely, the active method and the passive method or a combination of its (compound). Passive methods increase heat transfer in heat exchangers that do not require an external energy supply. For example, one of the passive method is to change the thermal properties of the cooling fluid by adding solid particles with a much higher thermal conductivity than the thermal conductivity of the liquid. While active methods are ways to increase heat transfer in heat exchangers whose systems require external energy, for example, by vibrating the fluid with ultrasonic frequency in the heat exchanger.

The purpose of writing this article is to combine the two methods mentioned above, namely, the use of nanofluid fluid, at the same time vibrated with ultrasonic frequency to get the maximum increase in heat transfer.

2. Nanofluid

The improvement of the heat transfer coefficient can be achieved by modifying the cooling fluid by adding nanometer-sized particles. Hopefully, the application of nanofluid will save energy and reduce the emissions, global warming potential, and greenhouse-gas effect. It enhances the heat transfer for energy-saving purposes that could contribute to a better quality of human life and fulfill sustainable development. Currently, the production of nano-sized particles is made easy because of the nano-technology that was predicted 15 years ago. He predicts that energy is one of the "top 10" problems for humankind worldwide in the next 50 years and can only be solved by nanotechnology [1] and it plays a vital role in heat transfer enhancement [2]. Materials commonly used for the manufacture of nanometer-sized solid particles (nanoparticles) from metal materials (Al, Cu, Zn, Ag, and Au), metal oxides, such as SiO_2, TiO_2, Al_2O_3, ZnO, CuO, and organic particles/organic particles, such as carbon nanotubes, graphene oxide, and diamond [3, 4]. Nanoparticles allow the development of cooling fluids called nanofluids. The fluid is a liquid suspension containing nanoparticles smaller than 100 nm and has better thermal conductivity than the base fluid [5]. Adding nanoparticles into the fluid will change the properties of the fluid as a heat transfer fluid (thermophysical properties).

2.1 Thermophysical nanofluid

The properties of nanofluid for density of nanofluid calculated widely used in research of nanofluid for a wide range of volume concentration can be estimated by using the following equations as [6]:

$$\rho_{nf} = \rho_b(1 - \varnothing) + \rho_{np}\varnothing \tag{1}$$

where, ρ_{nf}, ρ_b, and ρ_{np} are the nanofluids, base fluid, and particle density, respectively.

The nanofluid volume fraction, $\varnothing$, can be calculated as follows:

$$\varnothing = \left(\frac{\frac{m_{np}}{\rho_{np}}}{\frac{m_b}{\rho_b} + \frac{m_{np}}{\rho_{np}}} \right) \times 100\% \tag{2}$$

where m_{np} and m_b are the mass of nanoparticle and base fluid, respectively.

For hybrid nanofluid, the nanofluid volume fraction, calculated by Eq. (3)

$$\varnothing = \frac{m_{np.1}/\rho_{np.1} + m_{np.2}/\rho_{np.2}}{m_{np.1}/\rho_{np.1} + m_{np.2}/\rho_{np.2} + V_b} \tag{3}$$

where V_b is the volume concentrations of base fluid subscript 1 and 2 indicating the first and the second nanoparticles.

The capacity heat of nanofluid is calculated by Xuan and Roetzel [7] as the following equation is widely used.

$$C_{nf} = \frac{\varnothing \rho_{nf} C_{np} + (1-\varnothing)\rho_b C_b}{\rho_{nf}} \tag{4}$$

where C_{np} and C_b are specific heat of the nanofluid and base fluid, respectively.

Nanofluid viscosity was calculated from Brinkman correlations [8], as a shown in Eq. (5)

$$\mu_{nf} = \frac{\mu_b}{(1-\varnothing)^{2,5}} \tag{5}$$

where μ_{nf} and μ_b are the viscosity of the nanofluid and base fluid, respectively.

Various models of the thermal conductivity of nanofluid have been published, the earliest of models developed by Maxwell [9] as shown in Eq. (6)

$$k_{nf} = k_b \frac{k_{nf+}2k_b + 2\left(k_{nf-k_b}\right)\varnothing}{k_{nf+}2k_b - 2\left(k_{nf-k_b}\right)\varnothing} \tag{6}$$

where k_{nf} and k_b are thermal conductivity of the nanofluid and base fluid, respectively.

2.2 Preparation of nanofluid

Nanofluid is made by dispersing nanoparticles into a base fluid. Good dispersion is a prerequisite for the use of nanofluids in various fields. Therefore, surfactants are sometimes used to increase the nanofluid's stability to prevent clogging. In addition, the surface modification of the dispersed particles and the application of strong forces to the clusters of dispersed nanoparticles can improve the stability of nanofluids.

There are two ways to prepare nanofluid: one-step method and two-step method. Each method has advantages and disadvantages. The advantage of the one-step method is that it produces good dispersion, but it is very expensive to manufacture and cannot produce large amounts of nanofluids. While the two-step method produces poor dispersion, the cost of making nanofluids is relatively low and can produce large amounts of nanofluid. Most researchers use this type of method to make nanofluids. Zhu et al. [10] made nanofluids using Al_2O_3 particles with a pure water base fluid using a two-step method. The manufacture of nanofluids using a two-step method is shown in **Figure 1**. The first step is to prepare the base fluid and nanoparticles by weighing them according to the specified concentration. The next step is to mix the nanoparticles into the base fluid by adding a little surfactant and

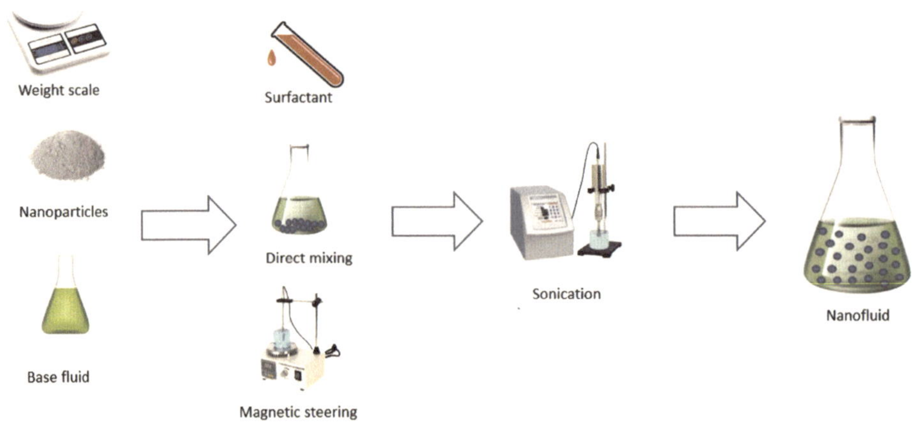

Figure 1.
Preparation of nanofluids using a two-step method [10].

stirring using magnetic steering. After the nanoparticles and base fluid are well mixed, sonification is carried out, namely, vibrating the mixture with ultrasonic frequency to homogeneous for 1 h.

2.3 Heat transfer characteristics

The heat transfer characteristics of nanofluid depend on various factors, including thermophysical properties of base fluid and nanoparticles, nanoparticle concentration, nanoparticle size, presence of surfactants, temperature, etc. Hence, the functional form of the Nusselt number of nanofluid can be expressed by Xuan and Roetzel [7] in Eq. (7)

$$N_{nf} = f\left(Re , Pr, k_{nf}, \rho C_{nf}, \varnothing \right) \tag{7}$$

The Nusselt number of Al_2O_3/water and TiO_3/water nanofluids in turbulent flows proposed by Pak and Cho [6] in Eq. (8)

$$N_{nf} = 0.021\ Re^{0.8}\ Pr^{0,5} \text{ for } 6.54 \le \mathrm{Pr} \le 12.33 \text{ and } 10^4 \le \mathrm{Re} \le 10^5 \tag{8}$$

Xuan and Li [11] studied CuO/water nanofluid in turbulent flow and correlated Eq. (9).

$$N_{nf} = 0.0059\left(1 + 7.628\ \varnothing_p^{0.6886}\ Pe_p^{0.001}\right) Re^{0.9231}\ Pr^{0,4} \tag{9}$$

Several experimental studies have been reported using nanofluid in the heat exchanger. Raei et al. [12] investigated a double-tube heat exchanger with a low volume concentration of particles (0.15%) of Al_2O_3-DI water nanofluids. The nanofluid results in increasing the heat transfer coefficient by 23% without much penalty in pressure drop. Shahrul et al. [13, 14] analyzed nanofluid flow through a shell-and-tube exchanger considering various oxide-based nanofluid at different small volume concentrations (0.01–0.04%) of ZnO, CuO, Fe_3O4, TiO_2, and Al_2O_3

nanoparticles into the water. Heat transfer enhancement was found at about 23–52% compared to the base fluid. In their study, the highest improvement was obtained for ZnO-water nanofluid. Researchers have studied the behavior of nanofluids in automobile radiators due to good heat transfer characteristics. Different nanofluids have been tested to observe their thermal performance and pressure drop to validate their practicality in this application. Heris et al. [15] used base fluid water EG (40/60) to suspend CuO nanoparticles to study the radiator's performance and the maximum enhancement of heat transfer apprehended about 55% compared to the base fluid. Alosious et al. [16] performed an experimental and numerical analysis of Al_2O_3/Water and CuO/water nanofluid, and they found that the Nusselt number, the total heat transfer coefficient, and convective heat transfer coefficient increased by increasing of Reynolds number. They also demonstrated that the thermal conductivity of alumina oxide is more than copper oxide. However, an increase in pressure drop was not significant.

Subhedar et al. [17] recently researched Al_2O_3/Water-EG in the laminar flow regime and found that for just 0.2% concentration volume of nanoparticles, the Nusselt number increased by 30%. Their research, flow rate, and concentration of volume particles were reported as the key factors influencing heat transfer enhancement. Inlet temperature had a lower effect; an increase of about 26% from 70°C to 85° C was recorded. Heat transfer coefficient enhancement is the process of increasing the effectiveness of the heat exchanger. Adding a small amount of nanoparticles to the base fluid increases the thermal conductivity of the nanofluid with ultrasonic vibrations to increase the performance of a heating system, but the viscosity of the nanofluid is also increased slightly. Working conditions at high temperatures in a heat exchanger is an advantage in itself, which causes the viscosity of the fluid to decrease. The addition of small nanoparticles and high working temperature is an ideal combination for improving efficiency of systems.

The main drawback of nanofluids is their poor stability. Long-term stability of nanofluids is a major concern for engineering applications. Nanoparticles tend naturally to aggregate and sediment in the base fluid, due to the interaction between the particles themselves and between the particles and the surrounding liquid [18]. The main cause is the attraction between the particles called the Van der Waals attraction, which causes the particles to form clusters or agglomerate, then settle to the bottom due to gravity. To obtain stable nanofluids, the repulsion of electrical double layers forces must exceed the Van der Waals attraction.

Overcoming this problem by adding surfactants as a dispersant, is an effective stability enhancement method that prevents the agglomeration of nanoparticles within the nanofluids [19]. It is a simple and economical chemical method, which reduces the surface tension of the base fluid and improves the immersion of nanoparticles. Because surfactants consist of the hydrophobic tail portion (e.g., long-chain hydrocarbons) and the hydrophilic polar head group that tends to increase the hydrophilic behavior between the base fluid and the nanoparticles. The disadvantage of using dispersant as a nanofluid stabilizer is its sensitivity to hot temperature because the rise in temperature causes the bonds between the nanoparticles and the surfactant to be damaged. Some common dispersants are sodium dodecylbenzene sulfonate (SDBS), gum arabic, sodium dodecyl sulfate (SDS), polyvinylpyrrolidone (PVP), dodecyltrimethylammonium bromide (DTAB), oleic acid (OA), etc.

The other method to achieve the long-term stability of nanofluids, without the need for surfactants, is surface modification techniques by modifying the nanoparticles' surface via functionalization. Functionalization is the process of adding new functions, features, capabilities, or properties to the material by changing the surface chemistry of the material. This is done by introducing functionalized nanoparticles into the base fluid to obtain a self-stabilized nanofluid. Usually, suitable functional organic groups are selected as they tend to attach to the surface of the atom, enabling the nanoparticles to self-organize and avoid agglomeration [20].

3. Ultrasonic vibration

Ultrasonic waves are sound waves whose frequency is above human hearing (>20 kHz), distinguished according to the power and frequency emitted. When viewed from the magnitude of the frequency, it is divided into three categories, namely, low-frequency ultrasonic waves (20–100 kHz) or ultrasonic waves with high power, high-frequency ultrasonic waves (100 kHz–1 MHz), and very high-frequency ultrasonic waves, namely, above 1 MHz. Ultrasonic with a low frequency of 20 kHz–100 kHz can change the medium in which it passes and is widely used in the chemical industry, which aims to change the chemical properties to generate cavitation bubbles and the effect of surface instability.

When sound waves propagate through a liquid, they produce a series of cycles of pressure compression and tensile (rarefaction). In the compression cycle, at the top of the wave position (max), the liquid gets a compressive force, and in the tensile cycle, where the wave position is at the bottom of the valley (min), the liquid experiences a tensile force and so on. Two important phenomena of ultrasonic waves propagating through liquids are acoustic cavitation and acoustic streaming.

3.1 Acoustic cavitation

When a sound wave consists of a maximum cycle (peak) and a minimum cycle (valley), and propagates in a liquid fluid, compression and expansion cycles occur. The compression cycle produces a positive pressure, while the expansion cycle produces a negative pressure causing the liquid molecules to move away [21]. When the magnitude of the tension exceeds the tensile strength between the liquid molecules in the expansion cycle, the liquid will break down, and a cavity will be created to appear as small vapor-filled voids called cavitation bubbles [22], in other words, the phenomenon. The formation, growth, and contraction of bubbles are known as acoustic cavitation.

Acoustic cavitation is divided into stable cavitation and transient cavitation or unstable cavitation/inertia cavitation. Acoustic cavitation consists of three distinct stages: nucleation, rapid bubble growth (expansion) until it reaches a critical size, and bubble collapse (shown in **Figure 2**.

Stable cavitation due to micro-bubbles in a liquid, resulting from changes in liquid pressure (compression and refraction) ultrasonic waves, changing/oscillating in size and shape that causes the surrounding liquid to move is known as non-inertial cavitation. It produces a strong shearing force around the solid surface, and removes particles. But in unstable cavitation, the micro-bubbles oscillate and in the end, the bubbles burst/collapse due to the acoustic wave strength exceeding the cavitation threshold, producing a shock wave.

When the bubble collapses surrounded by liquid, several physical effects will appear, including shock waves, micro-jets, turbulence, and shear forces. This physical effect has been widely applied in emulsification, extraction, and cleaning [23]. In this process, bubble size changes drastically by several hundred times to reach the

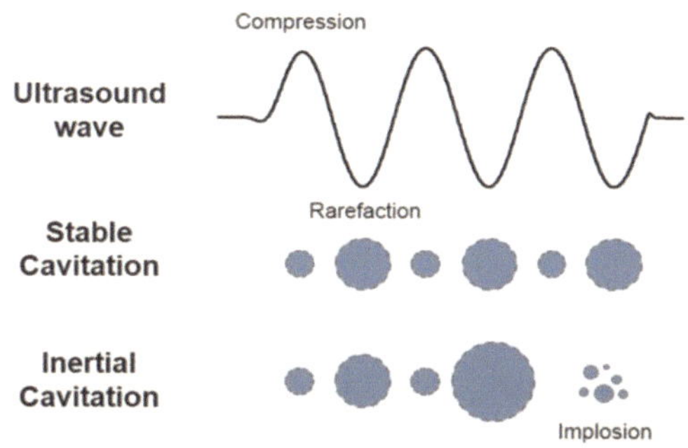

Figure 2.
The formation, growth, and collapse of bubbles due to ultrasonic vibrations in liquid fluids produce stable and unstable cavitation.

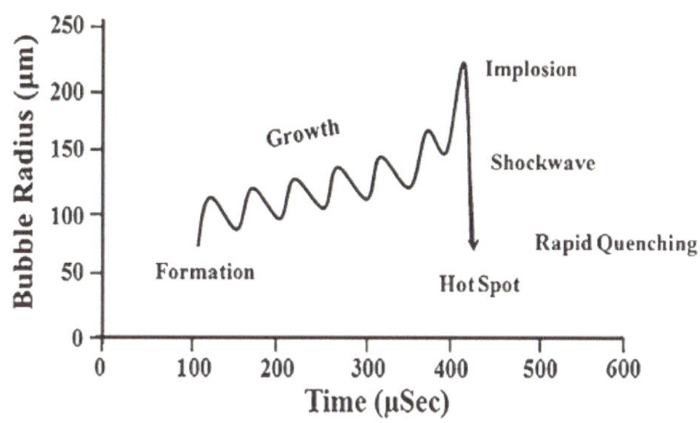

Figure 3.
Bubbles size growth due to ultrasonic waves occurs in just microseconds.

equilibrium size of the bubble radius before an explosion occurs in just a few microseconds. **Figure 3** shows the process of bubble collapse, cavitation releases a large amount of energy and produces local "hotspots" experiencing extreme temperatures and pressures that have a chemical impact because the temperatures and pressures that occur are very high at around 5000°C and 200 atm [24].

3.2 Acoustic streaming

Besides having a chemical impact (sonochemical), ultrasonic waves also impact physical changes (sonophysical). The liquid medium will absorb sound wave energy in the direction of the wave. Physical effects include streaming, micro-jets, and shock waves. Streaming is a term that describes the average mass density at steady conditions or velocity caused by acoustic wave oscillations in a liquid fluid due to a momentum gradient resulting in fluid flow. The sound propagation with large amplitude in the liquid will produce fluid motion. This nonlinear phenomenon is called acoustic streaming. Acoustic streaming caused by ultrasonic vibrations effectively enhances heat transfer and cleans contaminated surfaces.

Unlike bubbles in a liquid, the bursting of bubbles near the solid-liquid interface makes the bubble surface unsymmetrical because the solid surface acts as a flow barrier. The result is a powerful jet coming from the side of the bubble that produces a strong beam toward the surface called micro-jetting, which functions as ultrasonic cleaning or antifouling. This phenomenon is used to clean with the ultrasonic method, while in heat transfer applications, it will interfere with the boundary layer of convection heat transfer, as shown in **Figure 4**.

Unlike cavitation bubble collapse in a liquid, cavitation bubble collapse near a solid surface is not symmetrical due to surface barriers impeding fluid flow. The result is a rush of liquid coming from the side of the bubble away from the surface, creating a strong liquid jet directed toward the surface. The collapsing bubble near the solid-liquid interface disrupts the thermal boundary and velocity boundary layers, reducing thermal resistance and creating micro-turbulence due to shock waves. For this reason, ultrasonic vibrations in the liquid enhance convective heat transfer because of the growth and collapse of bubbles near the surface, which disturbs/compresses the thermal boundary layer. The amount of convection heat transfer is inversely proportional to the thickness of the thermal boundary layer. The thinner the thermal boundary layer, the greater the heat transfer rate, as shown in formula (10).

$$q_{CONV} \approx \frac{1}{\delta_{TBL}} \tag{10}$$

Another phenomenon due to the bursting of cavitation bubbles near the solid surface is widely applied as a cleaning process due to the presence of liquid jets toward the surface. The study of bubble collapses near the solid wall surface was carried out by Li et al. [25] with numerical and experimental methods on the shape and behavior of dynamic bubbles, which state that there are similarities between experimental results and numerical simulation results. The results show that the velocity of the liquid jet is stronger toward the walls of the spherical bubble compared to the non-spherical shape. Other studies by vibrating the working fluid with ultrasonic

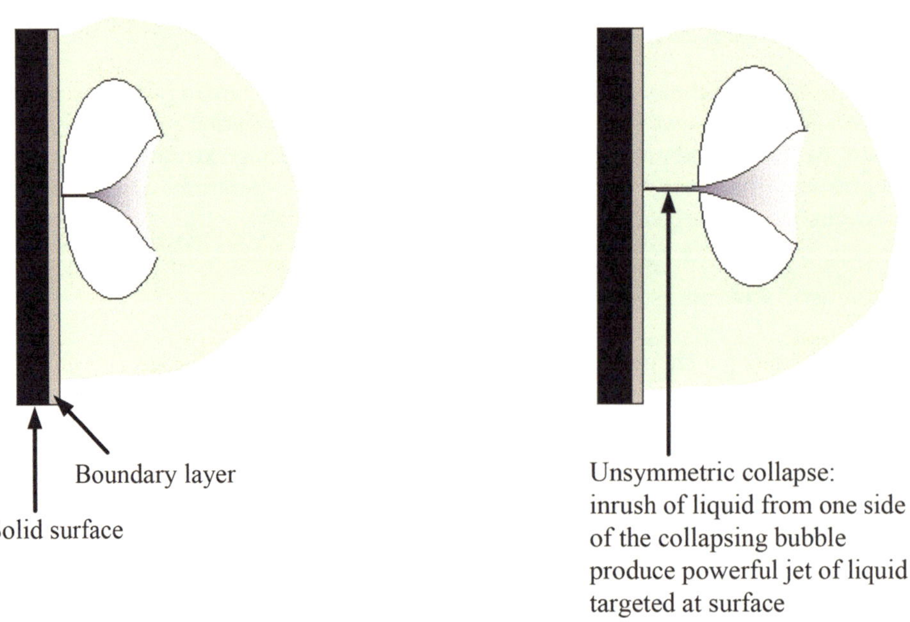

Figure 4.
Bubble collapse near the solid surface.

frequencies have been carried out by several researchers, including Tajik et al. [26], who showed that the increase in heat transfer is proportional to the ultrasonic power and inversely proportional to the distance between the vibration center and the heat source. The highest increase is due to ultrasonic vibration with a frequency of 18 kHz by 390% with power between 56 and 158 W. Legay et al. [27] investigated convection heat transfer using a double pipe heat exchanger using pure water, and an ultrasonic vibrator was installed in the middle of the heat exchanger with a frequency of 35 kHz, the highest increase in heat transfer was in laminar flow of 150%, whereas in turbulent flow it had no effect. Ultrasonic vibrations also decrease the thermal convection heat transfer resistance. A comparison of the performance of shell-and-tube and double-pipe heat exchangers has also been carried out by Legay et al. [28]. In Gondrexon et al. investigated a shell and tube heat exchanger, the increase in the total heat transfer coefficient with ultrasonic vibrations ranged from 123% to 257% depending on the fluid flow rate [29]. The experiment was carried out by Zhou [30] to test convective heat transfer in a horizontal copper pipe under the influence of an acoustic cavitation field. The fluids used are acetone, ethanol, and pure water at constant heat flux. The results show that the highest heat transfer increase of 395% occurs in acetone fluid. The disturbance of the cavitation bubbles and the collision of the cavitation bubble clusters cause thinning of the thermal boundary layer on the pipe surface, leading to increased convection heat transfer.

These properties in terms of heat exchanger performance are very attractive because two advantages, namely, increasing heat transfer and reducing fouling (dirt on the heat exchanger surface) can be achieved by high-power ultrasonic wave propagation.

4. The combined methods are nanofluid and ultrasonic vibration

As discussed earlier, many studies on the technique of increasing heat transfer by adding nanoparticles to pure water (nanofluid) have been carried out by researchers. There are two main problems in using nanofluid as a cooling fluid: high particle deposition and agglomeration rate of nanoparticles. Practically, this deposition can be avoided by vibrating the fluid with ultrasonic frequency. In addition to antifouling and anti-agglomeration effects, ultrasonic waves can also improve heat transfer. The heat transfer characteristics of nanofluids with vibrations imposed on heat transfer surfaces have not been widely studied until now, which provides an important purpose for writing this book.

Delouei et al. [31] studied nanofluids vibrated by ultrasonic waves to determine the effect of increasing heat transfer and pressure drop on forced convection. It uses nanofluid Al_2O_3 particles passed through a fine pipe with an inner diameter of 17 mm, a pipe thickness of 1 mm, and a length of 1 m, which is vibrated using 28 kHz ultrasonic waves with ultrasonic power variations of 75 W and 100 W. The results show that ultrasonic waves can increase heat transfer by 11.37% and reduce the pressure drop by 15.27%. Furthermore, it shows that ultrasonic vibration can increase heat transfer and reduce pressure drop, and the ultrasonic effect is sensitive at a high percentage of particle volume and a low flow rate. Therefore, ultrasonic waves in a heat exchanger that uses nanofluid fluid are very beneficial, besides being antifouling and anti-agglomeration, the third is increasing heat transfer and reducing pressure drop. The use of ultrasonic waves in nanofluids is the answer to the weakness of nanofluids in practical applications, namely, an increase in pressure drop due to nanoparticles, which cause greater energy consumption. Shen et al. [32] tested the effect of ultrasonic vibrations on natural convection heat transfer by using a platinum rod with a diameter of 2 mm and a length of 8 cm immersed in nanofluid Al_2O_3-water with a volume concentration of 0.01% with variations in the temperature of the fluid (nanofluid) 30°C, 40°C, and 50°C. The results show that ultrasonic vibration can increase the convection heat transfer coefficient by 128% at a temperature of 30°C; at a fluid temperature of 50°C, it increases the convection heat transfer coefficient by 87%, and at a temperature of 50°C, it increases the convection heat transfer coefficient by 25%. In addition, it was found that high heat flux will reduce the heat transfer coefficient and the temperature of the test rod.

Research has been carried out by Sudarmadji et al. [33] using Al_2O_3 particles in radiator coolant (40:60 EG/water) on radiators accompanied by ultrasonic vibrations to increase the rate of heat transfer. The effect of adding nanoparticles to the radiator coolant increased heat transfer by 39.6%, while the ultrasonic effect was only 15.7%. The increase in heat transfer due to both methods is 62.7% when compared to pure radiator coolant without vibration. The drawback of this research is the limitation of ultrasonic power, which is relatively small so that the ultrasonic wave energy absorbed by the cooling fluid is relatively low.

5. Conclusions

One of the important parameters in ultrasonic reactions is the size change of bubbles. Generation, growth, and transient collapse of microbubbles could create intense local energy because these changes produce extremely high temperatures and pressure. The bubble implosion near a solid-liquid interface disrupts thermal and velocity boundary layers, reducing thermal resistance and creating microturbulence. This is one of the reasons why acoustic cavitation is often considered the main reason for heat transfer enhancement by ultrasound. It can also be used as a way to promote or control turbulence, which already suggests some possible use in heat exchange devices. The presence of nanoparticles will strengthen the heat transfer coefficient in the heat transfer systems. The advantage of using this combined method, in addition to increasing heat transfer and reducing pressure drop on the heat transfer system, can also function as a descaling agent on the surface of a heat exchanger (cleaning), which is often called antifouling, and also anti-clogging on nanofluids.

Nomenclature

$\varnothing$	the volume fraction of nanofluid
C_{np}	the specific heat of nanofluid
C_b	the specific heat of the base fluid
ρ_b	density of the base fluid
ρ_{nf}	density of nanofluids
ρ_{np}	density of nanoparticles
m_b	mass of base fluid
m_{np}	mass of nanoparticles
μ_b	viscosity of base fluid
μ_{nf}	viscosity of the nanofluid

k_b	thermal conductivity of the base fluid
k_{nf}	thermal conductivity of nanofluid
N_{nf}	Nusselt number of nanofluid
Re	Reynold number
Pr	Prandtl number

Author details

Sudarmadji Sudarmadji*, Sugeng Hadi Susilo and Asrori Asrori
State Polytechnic of Malang, Malang, Indonesia

*Address all correspondence to: sudarmadji@polinema.ac.id;
sugeng.hadi@polinema.ac.id; asrori@polinema.ac.id

References

[1] Smalley RE. Future global energy prosperity. The terawatt challenge. MRS Bulletin. 2005;**30**:412-417

[2] Mahbubul IM, Chong TH, Khaleduzzaman SS, Shahrul IM, Saidur R, Long BD, et al. Effect of ultrasonication duration on colloidal structure and viscosity of alumina-water nanofluid. Industrial and Engineering Chemistry Research. 2014;**2014**(53): 6677-6684

[3] Amrollahi A, Rashidi AM, Emami Meibodi M, Kashefi K. Conduction heat transfer characteristics and dispersion behaviour of carbon nanofluids as a function of different parameters. Journal of Experimental Nanoscience. 2009;**4**: 347-363. DOI: 10.1080/17458080902929929

[4] Murshed SMS, Leong KC, Yang C. Enhanced thermal conductivity of TiO2—water-based nanofluids. International Journal of Thermal Sciences. 2005;**44**:367-373. DOI: 10.1016/j.ijthermalsci. 2004.12.005

[5] Wen D, Ding Y. Experimental investigation into convective heat transfer of nanofluids at the entrance region under laminar flow conditions. International Journal of Heat and Mass Transfer. 2004;**47**:5181-5188

[6] Pak BC, Cho YI. Hydrodynamic and heat transfer study of dispersed fluids with submicron metallic oxide particles. International Journal of Experimental Heat Transfer. 1998;**11**:151-170

[7] Xuan Y, Roetzel W. Conceptions for heat transfer correlation of nanofluids. International Journal of Heat and Mass Transfer. 2000;**43**:3701-3708

[8] Brinkman HC. The viscosity of concentrated suspensions and solution. The Journal of Chemical Physics. 1952; **20**:571-581

[9] Maxwell JC. A Treatise on Electricity and Magnetism. UK Clarendon: Clarendon Press; 1973

[10] Zhu D, Li X, Wang N, et al. Dispersion behavior and thermal conductivity characteristics of Al_2O_3 – H_2O nanofuids. Current Applied Physics. 2009;**9**:131-139

[11] Xuan Y, Li Q. Investigation on convective heat transfer and flow features of nanofluids. Journal of Heat Transfer. 2003;**125**:151-155

[12] Raei B, Sharaki F, Jamialahmadi M, Peyghambarzadeh S. Eperimental study on the heat transfer. Journal of Thermal Analysis and Calorimetry. 2017;**127**: 2561-2575

[13] Shahrul IM, Mahbubul IM, Saidur R, Sabri MFM, Khaleduzzaman SS. Energy and environmental effects of shell and tube heat exchanger by using nanofluid as a coolant. Journal of Chemical Engineering of Japan. 2014;**47**:340-344

[14] Shahrul IM, Mahbubul IM, Saidur R, Khaleduzzaman SS, Sabri MFM, Rahman MM. Effectiveness study of a shell and tube heat exchanger operated with nanofluids at different mass flow rates. Numerical Heat Transfer Part A: Applications. 2014;**65**:699-713

[15] Heris SZ, Shokrgozar M, Poorpharhang S, Shanbedi M, Noie SH. Experimental study of heat transfer of a car radiator with CuO/ethylene glycol-water as a coolant. Journal of Dispersion Science and Technology. 2014;**35**:

677-684. DOI: 10.1080/01932691.2013.805301

[16] Alosious S, Sarath SR, Nair AR, Krishnakumar K. Experimental and numerical study on heat transfer enhancement of flat tube radiator using Al_2O_3 and CuO nanofluids. Heat and Mass Transfer. 2017;**53**:3545-3563. DOI: 10.1007/s00231-017-2061-0

[17] Subhedar DG, Ramani BM, Gupta A. Experimental investigation of heat transfer potential of Al_2O_3/water-mono ethylene glycol nanofluids as a car radiator coolant. Case Study Thermal Engineering. 2018;**11**:26-34. DOI: 10.1016/j.csite.2017.11.009

[18] Dey D, Kumar P, Samantaray S. A review of nanofuid preparation, stability, and thermo-physical properties. Heat Transfer—Asian Research. 2017;**46**(8):1413-1442

[19] Tiara AM, Chakraborty S, Sarkar I, Ashok A, Pal SK, Chakraborty S. Heat transfer enhancement using surfactant based alumina nanofuid jet from a hot steel plate. Experimental Termal and Fluid Science. 2017;**89**:295-303

[20] Neouze M-A, Schubert U. Surface modifcation and functionalization of metal and metal oxide nanoparticles by organic ligands. Monatshefe fur Chemie. 2008;**139**(3):183-195

[21] Vajnhandl S, Majcen Le Marechal A. Ultrasound in textile dyeing and the decolouration/mineralization of textile dyes. Dyes and Pigments. 2005;**65**(2): 89-101. DOI: 10.1016/j.dyepig.2004.06.012

[22] Mason TJ, Lorimer JP. Sonochemistry: Theory, Applications and Uses of Ultrasound in Chemistry. New York: Ellis Horwood Ltd.; 1988

[23] Mason TJ, editor. Advances in Sonochemistry. Vol. 1–6. Amsterdam, The Netherlands: Elsevier; 1990–2001

[24] Suslick KS. Sonochemistry. Science. 1990;**247**(4949):1439-1445. DOI: 10.1126/science.247.4949.1439

[25] Li S, Zhang AM, Han R, Liu YQ. Experimental and numerical study on bubble-sphere interaction near a rigid wall. Physics of Fluids. 2017;**29**(9): 092102. DOI: 10.1063/1.4993800

[26] Tajik B, Abbassi A, Saffar-Avval M, Abdullah A, Mohammad-Abadi H. Heat transfer enhancement by acoustic streaming in a closed cylindrical enclosure filled with water. International Journal of Heat and Mass Transfer. 2013; **60**:230-235

[27] Legay M, Simony B, Boldo P, Gondrexon N, Le Person S, Bontemps A. Improvement of heat transfer by means of ultrasound: Application to a double-tube heat exchanger. Ultrasonics Sonochemistry. 2012;**19**(6):1194-1200

[28] Legay M, Le Person S, Gondrexon N, Boldo P, Bontemps A. Performances of two heat exchangers assisted by ultrasound. Applied Thermal Engineering. 2012;**37**:60-66

[29] Gondrexon N, Rousselet Y, Legay M, Boldo P, Le Person S, Bontemps A. Intensification of heat transfer process: Improvement of shell-and-tube heat exchanger by means of ultrasound. Chemical Engineering and Processing. 2010;**49**(9):936-942

[30] Zhou D, Hu X, Liu D. Local convective heat transfer from a horizontal tube in an acoustic cavitation field. Journal of Thermal Science. 2004; **13**(4):338-343

[31] Amiri Delouei H, Sajjadi M, Izadi RM. The simultaneous effects of nanoparticles and ultrasonic vibration on inlet turbulent flow: An experimental study. Applied Thermal Engineering. 2019;**146**(5):268-277. DOI: 10.1016/j. Applthermaleng.2018.09.113

[32] Shen G, Ma L, Zhang S, Zhang S, An L. effect of ultrasonic waves on heat transfer in Al_2O_3 nanofluid under natural convection and pool boiling. International Journal of Heat and Mass Transfer. 2019;**138**(2019):516-523. DOI: 10.1016/j.ijheatmasstransfer.2019.04.071

[33] Sudarmadji S, Santoso S, Susilo SH. Analysis of the effect of ultrasonic vibration on nanofluid as coolant in engine radiator. Eastern-European Journal of Enterprise Technologies. 2021; **5**(113):6-13. DOI: 10.15587/1729-4061.2021.241694

Chapter 11

Enhancing Surface Heat Transfer Characteristics Using Laser Texturing

Nickan H. Ghahramani, Martin Sharp, Michael Morgan and Mehdi Seddighi

Abstract

The use of a pulsed laser system to manufacture parallel streamwise riblets on the plates of a heat exchanger is reported. There are certain laser system elements that can influence the quality of a micrometre texture geometry; among these, there was a focus on laser incubation effect on obtaining greater depth of the riblets. Surface roughness was always considered to keep the heat transfer efficiency high. The heat exchanging process was measured in two flow regimes: laminar and turbulent. In laminar flow, the surface texture slightly deteriorated the heat transfer rate. However, small improvement in the heat transfer rate was observed in turbulent flow.

Keywords: heat transfer, laser texturing, drag reduction

1. Introduction

Many engineering designs have been inspired by nature. The surface traits of lotus leaf and shark skin are examples of biology being mimicked to inspire designs. Lotus leaves have hierarchical superhydrophobic behaviour which enables the water droplets to roll off and clean the surface by rubbing the dirt particles. The skin of fast swimming sharks is covered with streamwise dermal denticles which have been shown to reduce drag plus having the anti-microorganism fouling feature.

The effect of textures on fluid flow behaviour have been widely studied for drag reduction purposes, more specifically for improving aerodynamic behaviour of solid structures such as aeroplane air foils. The results indicate a considerable drag reduction of approximately 10% compared with a conventional design, successful in creating a more innovative surface design.

Skin drag reduction can also have positive effects on lowering energy consumption in automotive, offshore, and marine industries. There are numerous methods of skin drag reduction such as using wavy surfaces and riblets. Riblets are streamwise parallel patterns whose shape optimization is still in progress.

Skin drag reduction can also improve the efficiency of heat transfer on surfaces dealing with heat transfer. There has not been significant research in this area. Nishida et al. [1] demonstrated the effect of wavy surfaces on heat transfer improvement.

Most studies have been considering the theoretical effects of drag reduction regardless of manufacturing techniques. This project focused on texturing the surface of stainless-steel plates of a heat exchanger by pulsed laser ablation to obtain scalloped riblet design with increased groove depth suggested by Bechert et al. [2].

2. Literature review

Turbulent and Laminar regimes of a flow depend on the relative importance of fluid viscosity (friction) and flow inertia. In turbulent flow the molecules of the fluid move in swirling and cross-stream directions while maintaining the average velocity in the fluid direction.

In turbulent flow around a flat plate, the vortices dominate the area in the viscous layer and the interaction between these vortices and the wall increases the shear stress on the wall causing a rise in the drag factor.

2.1 Drag reduction study

Drag reduction by using textured surfaces has been widely considered in recent studies. Examples in the nature can be used as the best samples for industrial applications. Lotus leaves have natural traits that possess a hierarchical surface structure which leads to a super-hydrophobic behaviour. Another example is Rose petals with super-hydrophobicity and either high or low adhesion [3].

The mucus of fish skin causes a drag reduction and protects the fish from abrasion. The dermal denticles of shark skin are shaped like streamwise riblets that reduces drag up to nearly 10% [2]. The magnitude of drag can be reduced by creating streamlined shapes as the viscous drag is created by the interactions between the molecules of the fluid and a surface parallel to the flow. By moving away from the surface, the flow reaches the mean velocity.

The outer layers of the turbulent boundary layer are disorganised due to the streamwise vortices forming on the surface of the viscous sub-layer. The interaction between the vortices and the surface causes vortices to get ejected from the surface and moved to the outer boundary layers. Reducing this bursting behaviour of such streamwise vortices can cause drag reduction [4].

2.2 Riblets in drag reduction

The first hypothesis on drag reduction was that the surface would redistribute the shear stress with concentration at the protruding parts of the surface. A better explanation is that longitudinal ribs rectify the turbulent flow in streamwise direction and reduce the turbulent momentum transfer close to the surface and hence decrease the shear stress.

Riblets on the skin of a fast-swimming shark reduce the occurrence of vortex ejection and the momentum transfer by impeding the vortex translation in the outer boundary layer. In turbulent flow, fluid drag increases with an increase in wetted surface area due to the shear stress actions. The low velocity fluid in the valleys of the riblets produce very low shear stress across the surface of the riblets and by keeping the vortices above, cross-stream velocity fluctuations between the riblet valleys and above the flat plate increases which is due to the reduction in shear stress and momentum transfer near the surface.

Riblets can reduce the cross-stream translation by protruding into the flow without increasing the drag. As vortices start to form on the surface, they remain above the riblet tips that creates low velocity channel in the riblet valleys. These channels have lower velocity gradient than the flow over a flat plate which reduces the shear stress over the riblet surface. On the other hand, considering the higher shear stress at the riblet tips due to higher velocity gradient, the result of this shear-stress distribution reduces the overall drag.

More recent works investigated the effect of drag reduction on heat transfer using numerical modelling. Zhu et al. investigated the nature-inspired structures in this field [5]. Soleimani and Eckels obtained benefit of riblets to drag reduction in a circular closed channel [6]. However, lack of experimental data to validate the result is still prominent in most works. This experimental research aimed to contribute to the field through an experimental study by employing laser texturing.

2.3 Effect of roughness on heat transfer

Surface roughness increases friction factors and the heat transfer coefficients within a turbulent boundary layer. In flat plate flows, ε/δ decreases along the plate, where δ is the boundary layer thickness and ε being the roughness height [7].

The relative effect of roughness is determined by roughness Reynolds number as follows:

$$Re_k = \frac{u_\tau \varepsilon}{v} \tag{1}$$

Where ε is the surface roughness, and u_τ being the friction velocity. When Re_k is less than 5, the surface is considered hydraulically smooth and for $Re_k > 70$, the flow is in the fully rough regime. For $5 < Re_k < 70$, the flow in transitionally rough regime. In turbulent flows, rough surfaces usually develop larger skin friction coefficients. In heat transfer there is the same behaviour although the skin friction increase is smaller for heat transfer. Surface roughness reduces the Reynold number for the transition from laminar to turbulent flow.

Abuaf et al. conducted an experimental study between air foils with different degrees of surface finish to measure the heat transfer coefficient shows that polishing the surface reduces the average roughness and improves the performance. However, very small differences have been observed between 0.03 and 0.81 micrometre surface roughness.

2.4 Laser texturing

Nanosecond lasers have been widely used in micro-machining and laser texturing. Etsion et al. [8] studied the possibility of reducing friction on piston rings to improve fuel efficiency in diesel engines. Gao et al. [9] investigated the possibility of increasing the smoothness of the surface through adding a second step laser ablation process.

Laser ablation Surface texturing has the advantage of contactless machining with high spatial resolution at fast speed. Among the variety of laser types, nano-second (ns) pulsed laser is widely used due to its affordability and short pulse duration which can create small heat affected zones.

Direct laser ablation (DLA) has been also used for controlling the wettability of metals by either laser texturing at low fluence near the ablation threshold fluence with polarised pulses or laser texturing at high fluence when pulse polarisation is not important. However, direct laser texturing at high fluence has been mostly done through ultra-short pulses (fs and ps). The considerable point is that it is not possible to achieve a hydrophobic surface by adding roughness when the material is already hydrophilic [10]. Menghistu et al. [11] concluded that by keeping the textured surface at atmospheric air, hydrophobic characteristics started to develop. The transformation could not be observed when leaving the textured surface in water.

Appropriate selection of the wavelength and process energy plays an important role in machining quality of nano-second pulsed lasers. The reflectivity of metals decreases in shorter wavelengths [12]. However, the maximum laser power is directly proportioned to the wavelength. The optimum quality of laser surface finishing is achieved when the pulse energy is low; In industrial applications usually a high pulse energy is needed to remove more material. Therefore, excess interaction between the plasma and subsequent pulses should be avoided by either increasing the scanning velocity or reducing the pulse frequency.

3. Methods

Two plate-type heat exchangers were manufactured with identical dimensions, one with smooth plates and the other one with textured plates (only the plates in direct contact with hot fluid) to compare their heat transfer efficiency. The heat exchangers consist of 2 middle plates, a top covering Perspex plate, and an aluminium bed plate. The SS plates are 0.5 mm thick; gaskets are 3 mm thick. Heat exchangers were incorporated into a monitoring rig manufactured by *TecQuipment* model no. *TD-360*. The service module provides two water circuits of cold at the mean ambient temperature of 22°C and hot tank at 60°C.

The water flow can be regulated for both circuits in the range of 0–3.5 lit/min. There are temperature measurement thermocouples near the inlet/outlet hose connectors and show the temperature on the digital displays.

The cold water is from the incoming mains cold water supply which passes through a hand adjusted flow regulator valve and then enters the heat exchanger. The hot water has an electric heating tank which rises the temperature to 60°C. A supply pump then circulates the water between the tank and the heat exchanger through a regulating valve (**Figure 1**).

3.1 Surface imaging

GFM MikroCAD 3D inspection scanner was employed for surface profile measurement using structured light fringe projection profilometry. It can measure and quantify the micro-scale surface structures and geometries based on phase measuring fringe projection by digital micro mirror displays and UV-LED.

3.2 Plate machining and testing

A SPI G3 20W nS laser was used to texture the surface. Laser parameters were power 20 W, frequency 25 kHz, pulse length 200nS. The beam was steered by a Nutfield galvanometer scanning head with a 100 mm focal length f-theta lens with spot size 25 μm, and an engraving field size of 50 mm × 50 mm. This head is placed on an Aerotech 150 mm precision ballscrew slide for z-axis focus control.

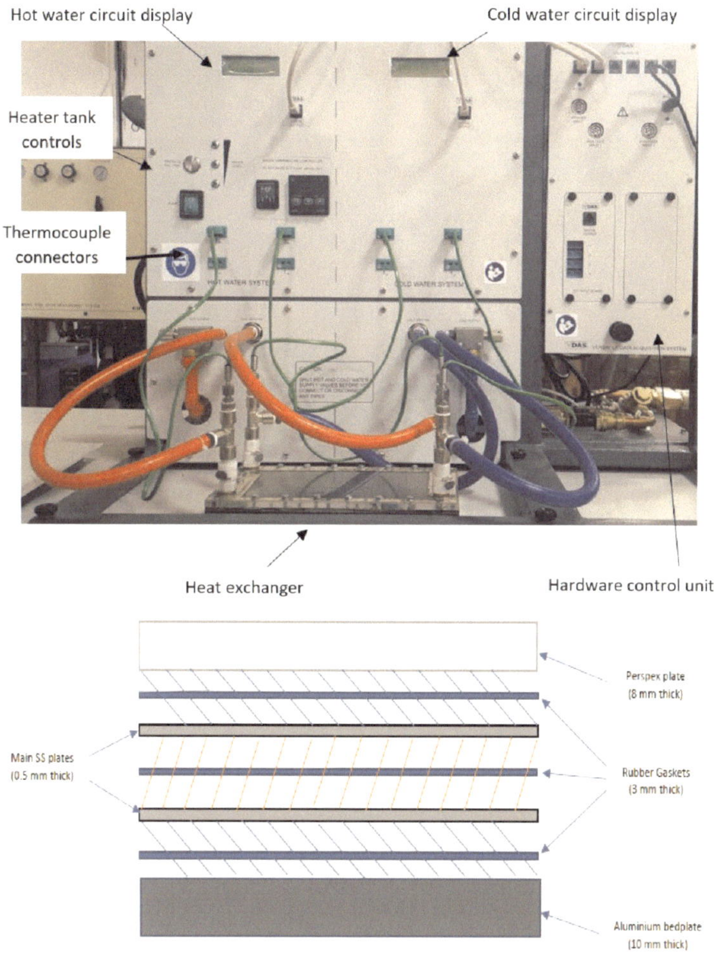

Figure 1.
Heat exchanger console and exploded side-view.

The heat exchanger plates to be engraved were mounted on a fixture mounted upon an Aerotech precision ballscrew 800 mm × 600 mm X-Y table (**Figure 2**). The levelling of the plate was checked using a dial turn indicator (DTI) attached to the z axis, while the heat transfer plate was traversed in the X and Y axes.

The riblet texture was laser engraved by using the galvanometer head to scan the focused laser beam along the parallel lines within the 50 mm × 50 mm field of the head. Upon completion of one patch of the riblets, the X-Y table was used to reposition the heat exchanger plate and the engraving process was repeated, **Figure 3**. By this means the necessary area of the plate was fully engraved with the riblet texture.

4. Heat exchanger parameters calculations

Hydraulic diameter is the ratio of cross-sectional to the wetted perimeter of the channel. For a rectangular channel where the height (b) is much smaller than the width (W) is (**Tables 1** and **2**):

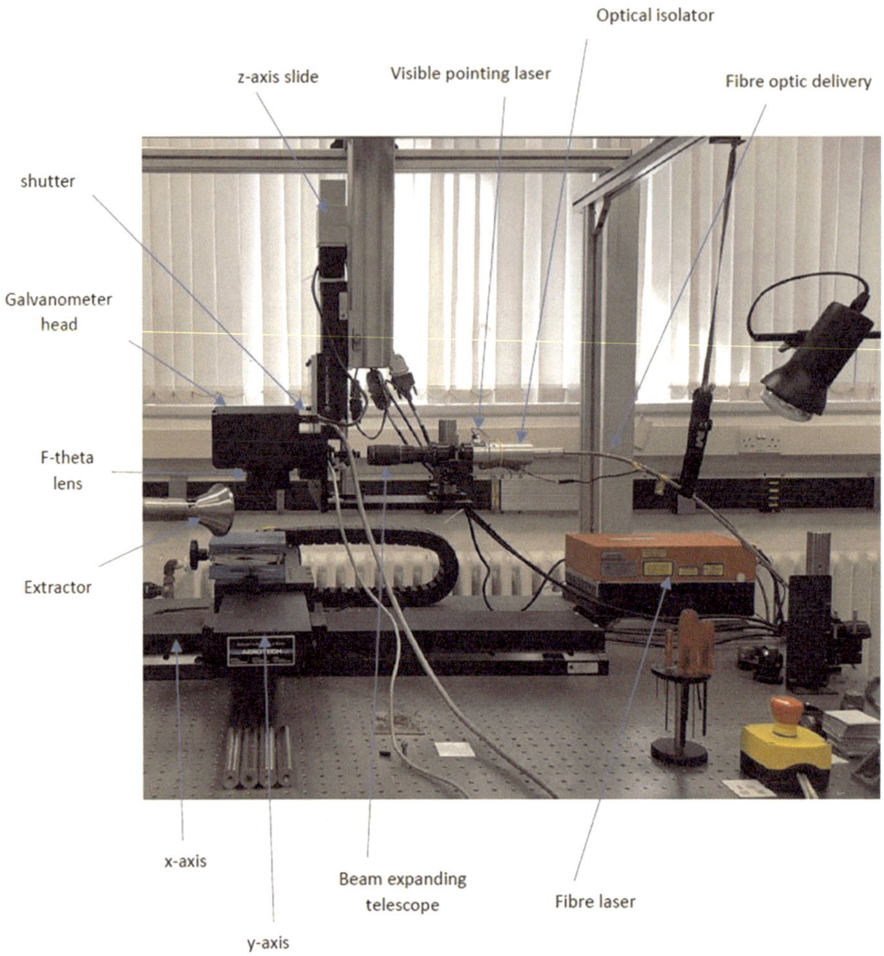

Figure 2.
Heat transfer plate mounted on table and undergoing laser machining.

$$D_H = 2b = 2 \times 0.003 = 0.006 \text{ m} \tag{2}$$

Mean temperature efficiency is defined as:

$$\bar{\eta} = \frac{\eta_H + \eta_C}{2} \tag{3}$$

Where and T_{H1} = hot in; T_{H2} = hot out; T_{C1} = cold in and T_{C2} = cold out; Accordingly $\eta_H = \frac{T_{H1} - T_{H2}}{T_{H1} - T_{C1}} \times 100$ and $\eta_C = \frac{T_{C2} - T_{C1}}{T_{H1} - T_{C1}} \times 100$.

Heat transfer emitted rate is defined as:

$$Q_e = m_h C_{ph} \Delta T_h \tag{4}$$

Where m is the mass flow rate in $\frac{Kg}{s}$; C_p is the specific heat capacity in $\frac{J}{Kg}$ °K and ΔT is the temperature difference between the inlet and outlet fluid in K.

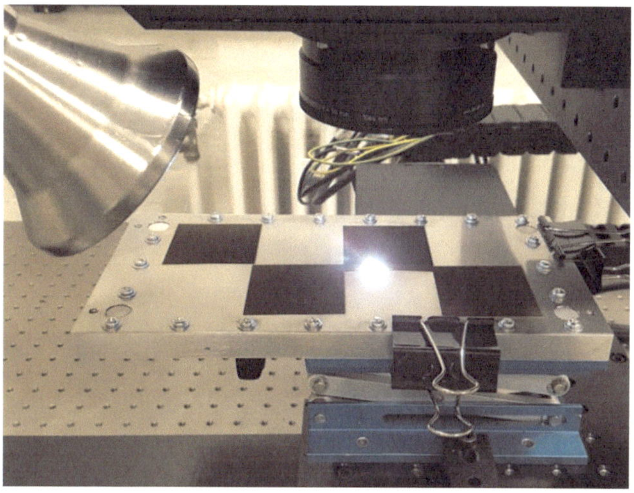

Figure 3.
Heat transfer plate mounted on table and undergoing laser machining.

Property	Unit	Hot water at 60°C	Cold water at 22°C
Heat capacity Cp	J/kg K	4190	4180
Thermal conductivity K	W/mK	0.65	0.598
Density ρ	kg/m^3	983	998
Dynamic viscosity μ	Pa s	4.6×10^{-4}	0.0010518
Flow area A = NWb	m^2	0.0003	0.0006

Table 1.
Water properties.

Property	Unit	Laminar	Turbulent
Mass flow rate m	kg/s	0.0583	0.025
Flow velocity $v = \frac{m}{A\rho}$	m/s	0.198	0.084
Reynolds number $Re = \frac{\rho v D_H}{\mu}$	—	2580	1086

Table 2.
Flow properties.

Logarithmic Mean Temperature Difference (LMTD) is the logarithmic average of the temperature difference between the hot and cold fluids and is a measure of the heat driving force which causes the heat transfer. LMTD is used when the temperature difference varies within the heat exchanger and reflects the difference more accurately.

$$LMT = \frac{(T_{H1} - T_{C2}) - (T_{H2} - T_{C1})}{\ln\left(\frac{T_{H1} - T_{C2}}{T_{H2} - T_{C1}}\right)} \tag{5}$$

Heat transfer coefficient measures the heat exchanging rate of the wall and boundary layers. The higher this coefficient, the more efficient the system is.

$$U = \frac{Q_e}{A \times LMTD} \quad (6)$$

Where Q_e is the energy emitted and $A =$ the heat exchange area

4.1 Plates' surface geometry calculations

Turbulence level scales on the shear and the shear strength is represented by a velocity scale classed shear velocity which characterises the shear at the boundary.

$$u_\tau = \sqrt{\frac{\tau}{\rho}} \quad (7)$$

Where τ is shear stress. Friction coefficient derived from Moody graph [13] equals to 0.03 with Reynolds number of 2580 for the hot chamber. Friction coefficient is defined as:

$$C_f = \frac{\tau}{\frac{1}{2}\rho u^2} \quad (8)$$

Where u is the flow velocity of the hot chamber in m/s. Therefore, the wall shear stress can be calculated as $\tau = 0.03 \times 0.5 \times 983 \times 0.198^2 = 0.578$. Using Eq. (7), friction velocity (shear velocity) would be $u_\tau = 0.024\ m/s$. The shape of velocity profile within a turbulent boundary later has specific characteristics close to the bed where viscosity controls the vertical transport of momentum. The region is called the "viscous sublayer" where turbulence is suppressed by viscosity. The profile shape depends on friction velocity and bed texture described by the roughness. The thickness of viscous sublayer is:

$$\delta_s = \frac{5\upsilon}{u_\tau} \quad (9)$$

Where υ is the kinematic viscosity in $\frac{m^2}{s}$ and u_τ is the shear velocity in m/s. When the surface texture is smaller than the viscous sub-layer, the flow above does not feel the texture [3]. Having kinematic viscosity of water at $60°C\ \upsilon = 4.736 \times 10^{-7} \left(\frac{m^2}{s}\right) \rightarrow$ $\delta_s \cong 90$ μm. Therefore, the texture depth should be minimum 90 μm to have an effect on the flow. However, for optimum riblet height should be:
$10 = \delta \frac{u_\tau}{\upsilon}$ [2] $\rightarrow \delta = 150$ μm.

5. Laser ablation

Figure 4 shows the textured geometry. The riblets are stretched in a regular pattern along the plate with average height of 90 μm which is close to the viscous sublayer thickness and is expected to influence the flow (**Table 3**).

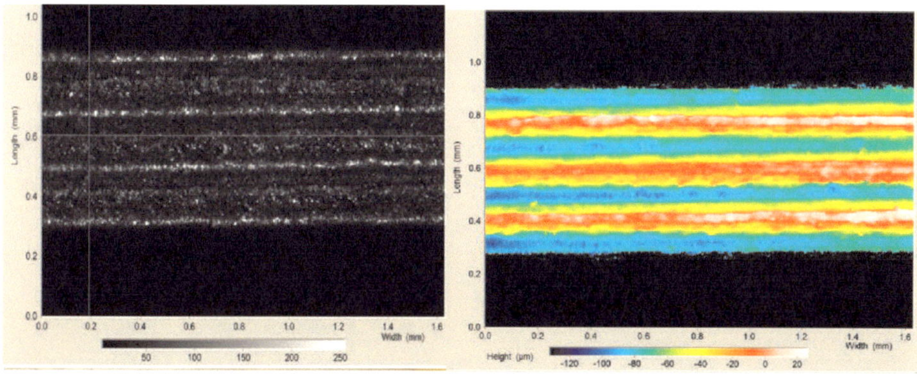

Figure 4.
Schematic display of the riblets with colour variation indicating depth of the geometry.

	Ra (µm)	Rt (µm)	Rz (µm)
Smooth plate	0.16	2.18	1.42
Textured plate (streamwise direction)	7.5	54	38
Textured plate (cross-stream) direction	24.5	103	97

Table 3.
Roughness of the plates before and after texturing.

6. Results and discussion

Two heat exchangers were tested in two separate days to get the accurate results due to the possible environmental effects. The heat exchangers were tested and compared at two flow rates of 1.5 l/m and 3.5 l/m which create Reynolds numbers of 1086 and 2580 respectively. Previous studies [14, 15] suggest that Reynolds number of greater than 2000 in plate type heat exchangers create turbulent flow.

When the working temperatures of the heat exchanger became static inlet and outlet temperature data of hot and cold fluid were collected for 300 seconds at intervals of 5 seconds (**Tables 4–6**).

Surface texture deteriorated the mean temperature efficiency and heat transfer coefficient for laminar flow. This is due to the increase in roughness of the plates which results in larger skin coefficient friction. A slight improvement in the mean temperature efficiency and heat transfer coefficient was observed in turbulent flow of

Flow regime	Plates	ΔT Hot	ΔT Cold	T_{C1}	T_{C2}	T_{H1}	T_{H2}
Laminar	Smooth	7.1	2.7	23.3	26.0	60.9	53.8
	Textured	6.9	2.8	23.3	26.1	61.0	54.1
Turbulent	Smooth	3.9	4.0	22.0	26.0	60.2	56.3
	Textured	3.8	3.9	23.2	27.1	60.1	56.3

Table 4.
Data collection of the heat exchanger in/out flow temperature.

Flow regime	Plates	η_H	η_C	$\bar{\eta}$
Laminar	Smooth	18.9	7.2	13.03
	Textured	18.3	7.4	12.86
Turbulent	Smooth	10.2	10.5	10.34
	Textured	10.3	10.6	10.43

Table 5.
Mean temperature efficiency of the heat exchangers.

Flow regime	Plates	area	m_h	C_p	Q_e	LMTD	U
Laminar	Smooth	0.04	0.025	4190	743	32.65	569
	Textured	0.04	0.025	4190	722	32.81	550
Turbulent	Smooth	0.04	0.0583	4190	952	34.25	695
	Textured	0.04	0.0583	4190	928	33.05	702

Table 6.
Heat transfer coefficient of the heat exchangers.

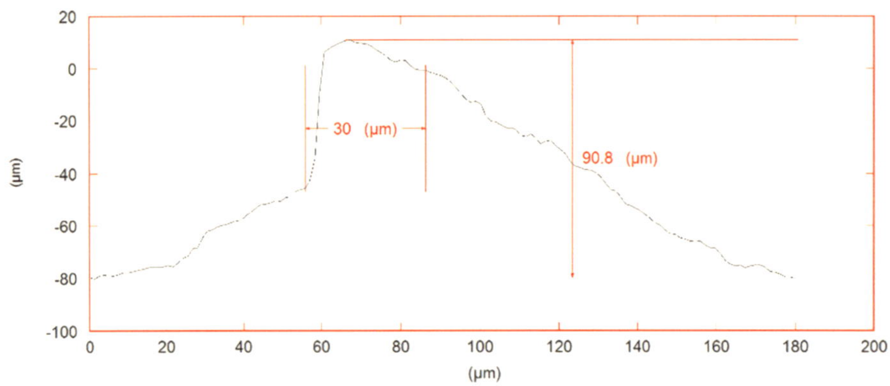

Figure 5.
Profile depth graph of a random riblet along the path shown in black, representing a riblet pitch of 180 µm – laser machining has caused a deeper slope on one side of the riblets and a smoother curve on the other side.

the textured heat exchanger. This improvement is small, and no significant change could be concluded [16].

Machining the plates to create riblets with optimum height of 150 μm was not achievable using this laser set up because of the creation of high surface roughness which is not desirable for heat transfer. Moreover, deeper laser ablation could cause excess deflection of the plates. Asymmetrical geometry of riblets as a result of laser machining, could not be avoided either (**Figure 5**).

7. Conclusions

Riblets are narrow microstructures, which require very precise manufacturing methods. A 20 W nanosecond pulsed fibre laser was used towards this aim.

The texture increased the wetted surface area of the plates; nevertheless, the increase in drag reduction was not enough to predominate over the friction factor rise due to the roughness created. Manufacturing the optimum riblet height of about 150 μm could not be achieved at a reasonable timescale within this project using a 20w pulsed laser system.

The textured plates were oxidised during the laser ablation and may explain the results. Laser ablation also caused distortion of the metal. Suitable jigs and fixtures are necessary to secure the material while being machined.

The project has shown that pulsed laser machining is highly capable of producing precise geometric designs, in particular ribs of micrometre scale. While the results are not significant, the project has demonstrated enough evidence to continue to pursue this method, refining the geometry and surface condition to see if significantly higher heat transfer efficiencies can be achieved.

The overall performance of the textured heat exchanger in turbulent flow, considering the transition from laminar deterioration to turbulent slight enhancement given the undesired oxidation of the plate, was a positive indicator for further works in this field.

Acknowledgements

This work was supported by the Engineering and Physical Sciences Research Council EP/S037292/1.

Author details

Nickan H. Ghahramani*, Martin Sharp*, Michael Morgan and Mehdi Seddighi*
Liverpool John Moores University, Liverpool, UK

*Address all correspondence to: hgh.info@yahoo.com; m.sharp@ljmu.ac.uk; m.seddighi@ljmu.ac.uk

References

[1] Nishida A, Nakatsuji K, Hagiwara Y. Direct numerical simulation of the effects of slip velocity of wavy walls on turbulent heat transfer and drag. In: Proceedings of the 9th International Symposium on Turbulence and Shear Flow Phenomena. Danbury, USA. 2015. pp. 1-6

[2] Bechert D, Bruse M, Hage W, Van Der Hoeven J, Hoppe G. Experiments on drag-reducing surfaces and their optimization with an adjustable geometry. Journal of Fluid Mechanics. 1997;**338**:59-87

[3] Dumanis M. First course in turbulence: Dean Young. Boston Review. 1999;**24**(5):66

[4] Kline S, Reynolds W, Schraub F, Runstadler P. The structure of turbulent boundary layers. Journal of Fluid Mechanics. 1967;**30**(4):741-773

[5] Zhu Z, Li J, Peng H, Liu D. Nature-inspired structures applied in heat transfer enhancement and drag reduction. Micromachines. 2021;**12**:6

[6] Soleimani S, Eckels S. A review of drag reduction and heat transfer enhancement by riblet surfaces in closed and open channel flow. International Journal of Thermofluids. 2021;**9**:10053

[7] Abuaf N, Bunker RS, Lee CP. Effects of surface roughness on heat transfer and aerodynamic performance of turbine airfoils. In: ASME 1997 International Gas Turbine and Aeroengine Congress and Exhibition. New York, UK: American Society of Mechanical Engineers; 1997

[8] Etsion, Sher. Improving fuel efficiency with laser surface textured piston rings. Tribology International. 2009;**42**(4):542-547

[9] Gao Y, Wu B, Zhou Y, Tao S. A two-step nanosecond laser surface texturing process with smooth surface finish. Applied Surface Science. 2011;**257**(23): 9960-9967

[10] Gregorčič P, Šetina-Batič B, Hočevar M. Controlling the stainless steel surface wettability by nanosecond direct laser texturing at high fluences. Applied Physics A. 2017;**123**(12):1-8

[11] Menghistu K, Sharp M. An Investigation into the Properties of Laser Machined Antifouling Microstructure. Liverpool, UK: Liverpool John Moores University; 2018

[12] Knowles M, Rutterford R, Karnakis H, Ferguson G. Micro-machining of metals, ceramics and polymers using nanosecond lasers. The International Journal of Advanced Manufacturing Technology. 2007;**33**(1):95-102

[13] Munson BR, Young DF, Okiishi TH, Huebsch WW. Fundamentals of Fluid Mechanics. Hoboken, US: John Wiley & Sons; 2014

[14] Tambe Shahanwaj K, Pandhare Nitin T, Bardeskar Santosh J, Khandekar SB. Experimental investigation of performance of plate heat exchanger for water as working fluid. International Journal of Research in Engineering and Technology (IJRET). 2015;**2015**:372-380

[15] Ayas M. Model deskového výměníku tepla. Master's thesis, Czech Technical University in Prague. Computing and Information Centre. 2015

[16] Ghahramaninia H, Sharp M, Morgan M. Heat transfer of a laser textured surface. In: Advances in Manufacturing Technology XXXIII. Amsterdam, NL: IOS Press; 2019. pp. 340-345

Chapter 12

Digital Twin of Heat Exchanger

Miha Bobič

Abstract

How to create proper digital twin of plate heat exchange. The chapter will investigate alternatives for modeling and propose the best choices in modeling to trade-off between accuracy of the calculation vs. speed. Lump sum vs. distributed model, how to make plate model discrete, temperature distribution models, and details in heat exchange wall modeling will be considered. Last part would focus on experimental evaluation of the simulation using IR temperature vision measurements. Conclusions would take concrete examples of dynamic responses of heat exchange units, comparing numeric simulation with experimental results.

Keywords: dynamic behavior, heat exchanger, models with distributed parameters, digital twin

1. Introduction

Heat exchangers are used for heat and cool transfer in different process industries, heating/cooling, and other applications. Plate heat exchangers are for majority of applications where pressure is not so high, the most cost-efficient solution. Brazed are used for smaller sizes, and more clean media, while gasketed are meant more for applications where occasional cleaning is needed.

The simplest control loop depicted in **Figure 1** is to control the outlet temperature on secondary side of the heat exchanger by controlling the flow on primary at given conditions of inlet temperatures on primary and secondary sides.

The system consists of a heat exchanger as the central component of the control system. The control can be performed using an electronic controller (usually due to the speed of response of the proportional-integral controller system) and the valve with a drive. In the return control loop, the measuring sensor for the temperature represents the guiding control variable.

The speed of process, seen from dynamics point of view, is a major challenge, as the time constants of the measuring system for the temperature of the heat carrier or hot water need to be small due to a relatively small control volume. Time constants may be less than few seconds.

2. Literature overview

In the field of heat exchangers, we can highlight the contribution [1], which is an overview of the current literature on the state of heat exchangers. Aslam Bhutta et al. [2] and Zhang et al. [3] discuss the review of contributions dealing with the topic of

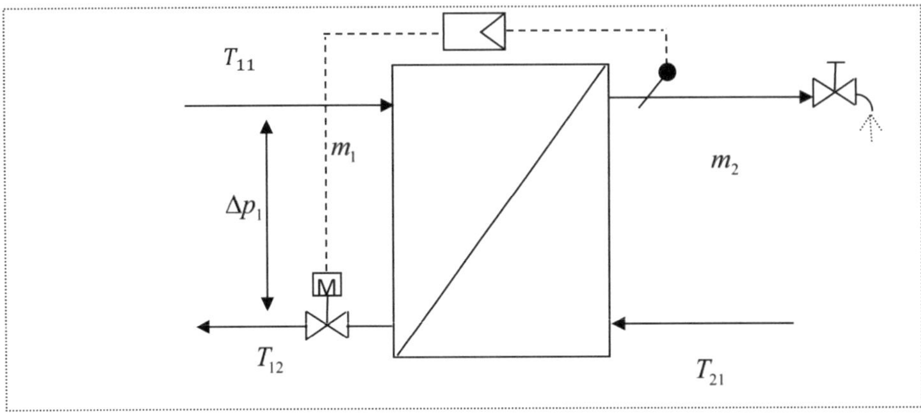

Figure 1.
Control loop schematics.

numerical simulation of heat transmissions. In Ref. [4] the case of a numerical simulation of a plate heat exchanger is considered. Freund and Kabelac [5] also deals with modeling of dynamic responses of plate heat transmissions [6]. The topic of evaporation in heat exchangers is discussed. The paper [7] deals with the experimental determination of the operational factors of heat exchangers in district heating for heating and hot sanitary water. Sources [8, 9] present an experimental method of visualizing the flow in a flat heat exchanger with a microplate structure [10]. The authors are also involved in visualizing the flow and determining local heat transfer. The source [11] analyzes the local heat transfer conditions. While the source [12] addresses the impact of the final plates on the operation of the heat exchanger. The paper [13] focuses on a study of the sensitivity of the geometric properties of a heat exchanger to its efficiency. The paper [14] deals with the technical requirements and cost analysis of heat transmissions in low-temperature systems. A similar theme of low-temperature systems is also in the paper [14, 15], with the emphasis in publication [15] on the impact of low-temperature systems on heating and the preparation of hot water without accumulation and in the source [14], on the main changes in the structure of the exchangers and thus the dynamic properties of heat transfer for such systems. A comparison between the plate and the tube heat exchanger is made in Ref. [16]. The source [17] discusses experimental and simulation methods in the field of heat exchangers. In Ref. [5], a study is carried out on how to determine local heat transfer coefficients in heat exchangers [18]. It also addresses the sensitivity of the parameter analysis to the numerical model of the heat exchanger. The link between the flow rate and the heat transfer under the micro heat exchanger conditions is given in Ref. [19]. Laszczyk [20] provides a method of a simplified heat exchanger model. Improved heat transfer in heat exchangers deals with [21]. Heat exchangers shell and tube are treated in Refs. [22]. Sharifi et al. [23] present a dynamic simulation of the heat exchanger. General dimensioning and dynamic analysis can be found in Refs. [24, 25]. The more generic described heat transfer dynamics depend, of course, on the configuration and sample on the heat exchanger plate, the various examples described in Refs. [26–28]. Many studies use CFD to analyze current patterns, distribution of pressure field, and temperature field, such as Ref. [4]. Sharifi et al. [23] provide a general model of heat transfer. Semiempirical equations are important for determining the coefficient of heat transfer, such as in Refs. [29–31]. Carlomagno and Cardone [17] describe the method of noninvasive measurement of heat exchanger parameters. Aprea and Renno [32] describe the design of tests for the portable evaporator function, that is, similar functionality to a heat exchanger. The use of MATLAB Simulink is shown in Ref. [33]. Integrated thermo-fluid dynamics and modeling reliability are shown in Ref. [34]. The literature also lists several approaches to modeling heat exchangers for different purposes, for example in Ref. [35], a model mode that is object-oriented using the Modelica programming language and provides a model based on the control volume method. Using Laplace's transformation, dynamic analysis of plate heat transfer is given in Ref. [36]. While in the Ref. [37] heat exchanger is described with the hyperbolic first order differential equations. Furthermore, in Ref. [38], a simplified mathematical model of a heat exchanger based on the final volume method is also used in the interpolation scheme. Similar is used in Ref. [39]. Further more modelling of the heat exchangers in the system and elements are the system are described in Refs. [40–44].

3. Basics of modeling

Seen from dynamic behavior point of view heat exchanger is a system with distributed parameters, meaning that the temperature continually changes throughout its length, thus traditional approach of direct Laplace transformation becomes too complex to solve, as we end up in double Laplace transformation, causing numeric problems with solution.

If we want to reduce the model to lump sum, then discretization is needed. Of course, the question is how many sections we should use to get good approximation with minimum numeric calculation load. The answer is in what kind of constitutional equation we use for temperature distribution in every cell. One has three options, one is if cells represent perfect mixing cell, and the assumption is that temperature immediately drops to exit temperature, and the second option is the opposite where the assumption is that mixing is imperfect, so the temperature is kept on inlet value until it leaves the cell. One can intuitively feel that those options are not realistic. The third option is to use distribution, it can be linear or to get closer to reality logarithmic. Experiments show that linear distribution is the best choice between number of cells, calculation effort, and minimized calculation error.

The second important question is how to model the thin metal wall between two media streams. One can model it as a conductive element of just ignoring the conduction. Given the ratio between heat convection resistance factor and heat conduction in a thin metal wall, it can be very well assumed that this is low resistance to heat compared to heat transfer for media to wall.

It is also important that proper material and turbulence models are used, thus in next chapter, we will investigate properties of water.

3.1 Basic properties

The material properties depend, of course, on different parameters. In the example shown, the impact of pressure and temperature is greatest, but we are particularly interested in variables:

- Density,
- Viscosity, and
- Specific heat.

The density gives us the dependence of the mass change per volume unit. In general, we can write down the dependence:

$$d\rho = \left(\frac{\partial\rho}{\partial T}\right)_p dT + \left(\frac{\partial\rho}{\partial p}\right)_T dT \tag{1}$$

When the coefficient of temperature extension is applied, the following shall be obtained:

$$\frac{d\rho}{\rho} = -\beta dT + \frac{1}{\varepsilon}dp \tag{2}$$

The eq. (2) specifies that the density depends on the temperature and the suitability of the substance. In this case, the heat carrier will be used as a heat carrier for district heating, and for this purpose, the constitution equation is by source [45]:

$$\rho = \left(1000 + 0{,}067\left(1 - e^{-0{,}016T}\right) - 0{,}001T\right) za\ 0^{\circ}C < T < 50^{\circ}C \tag{3}$$

$$\rho = \left(1.067 - e^{-0{,}023(T-50)} - 0{,}001T\right) za\ 50^{\circ}C < T < 200^{\circ}C \tag{4}$$

Viscosity can be obtained from the viscous tensor as:

$$\tau = \eta \frac{dv_x}{dy} \tag{5}$$

The viscosity is therefore greater, the smaller the shear deformity. For water, the kinematic viscosity equation can be summarized as [45]:

$$\nu = 1{,}75e^{-0{,}029T}10^{-6} \quad \text{for } 0^{\circ}\text{C} < \text{T} < 20^{\circ}\text{C} \tag{6}$$

$$\nu = 0{,}98e^{-0{,}02(T-20)}10^{-6} \quad \text{for } 20^{\circ}\text{C} < \text{T} < 50^{\circ}\text{C} \tag{7}$$

$$\nu = 0{,}55\left(\frac{50}{T}\right)^{0{,}9} 10^{-6} \quad \text{for } 50^{\circ}\text{C} < \text{T} < 200^{\circ}\text{C} \tag{8}$$

Specific heat is a greatness that determines how much energy we need to produce to heat the substance by 1 K.

$$c_p(T) = \frac{1}{m}\frac{dh}{dT} \tag{9}$$

or it can be written as an approximation equation as:

$$c_p = -7 \cdot 10^{-7}T^3 + 8 \cdot 10^{-5}T^2 - 0{,}0031T + 4{,}2161 \text{ for } 0^{\circ}\text{C} < \text{T} < 50^{\circ}\text{C} \tag{10}$$

$$\begin{aligned} c_p = {} & -6 \cdot 10^{-11}T^5 + 2 \cdot 10^{-8}T^4 - 3 \cdot 10^{-6}T^3 + 0{,}0002T^2 - 0, \\ & 0069T + 4{,}269 \text{ For } 51^{\circ}\text{C} < \text{T} < 100^{\circ}\text{C} \end{aligned} \tag{11}$$

4. Heat exchanger

Heat exchanger is the heart of the thermal substation, as we control heat transfer with changing thermal and hydraulic parameters. Despite the different types of heat transferors, in the practice of district heating, plate heat exchangers have established themselves, namely, for smaller dimensions and relatively good media quality (clean

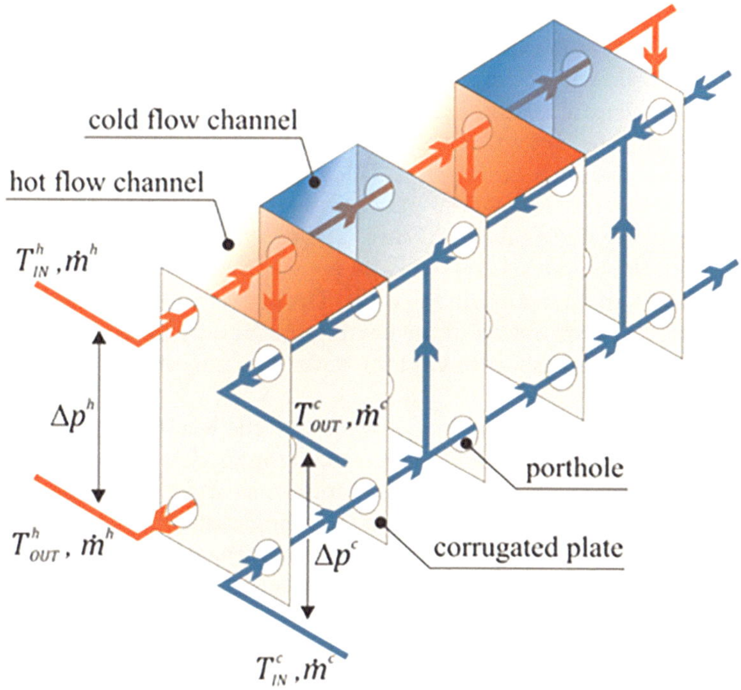

Figure 2.
Schematics of heat exchanger.

and noncorrosive), dedicated laptops, and all other applications are covered by gasketed heat exchangers.

In general, all plate laptops have a sample on plates shaped like a fish bone, the only other ones on the market are Danfoss's microplate heat exchangers, which have a point-shaped sample. The sample on the heat transfer plates is intended to increase the heat transfer area and increase turbulence in order to increase the heat transfer coefficient.

Modern heat exchangers also have an input distribution section and an output collection section to increase the efficiency of the panels, and the task of these two sections is to redistribute the medium to flow evenly throughout the plate and thus increase the efficiency of the surface.

There are quite a few dynamic models of heat transfers in the literature [1]. The simplest assumes a first-order system with a possible delay, but such a heat exchanger model is inappropriate because the validity of the mathematical model is too limited. More interesting are the models where the dual Laplace transformation is used, but they need to be linearized, which in turn is in too limited force range for our case. Therefore, the heat exchanger will be considered as a system with distributed parameters which means that the heat exchanger will be considered as several consecutive systems with no distributed parameters. The directions of the currents in the heat exchanger are schematically shown in **Figure 2**.

4.1 General assumptions and limitations of the mathematical model

When the heating medium enters the heat exchanger, the more distant plates (for example, 100 to 200), represent a greater hydraulic resistance against the water flow.

Therefore, for the first time, water flows over nearby plates faster and the second time there is a certain delay due to the flow of water. As this will be negligible for our case (about 20 panels), we will not take this effect into account.

In general, the heating medium has a higher temperature than the surrounding area, so the heat transfer takes place not only from the warmer medium to the cooler medium but also in the surrounding area. Due to the relatively good thermal insulation, we will neglect these losers. The translation of heat over steel plates exists, but due to the thickness of these (about 0,2 mm) and the relatively high coefficient of heat translation, this will be ignored. We will also pre-bet that the heat exchanger is clean without the suspended limestone on the walls. The most unrealistic assumption will be the uniformity of the distribution of the heating medium by plate. Assuming that the mixing section is relatively short toward the entire surface, we can also take this assumption into account.

In the heat exchanger model, we considered that the heat exchanger is a nonlinear system with distributed parameters. Therefore, in the model, the heat exchanger will be divided into several smaller subunits, in which linearity will be assumed. The heat exchanger has different dynamic responses according to the input parameters, whether it is the temperature or the mass flow rate of the heat carrier. It is important to determine the dynamic properties of the heat exchanger, however, to respond to the temperature change.

4.2 Static characteristics

The static characteristic of the heat exchanger is the ratio of mass flow rate to the heat ousted by the energy equation:

$$P = \dot{m} c_p \Delta T \tag{12}$$

If only this is considered and when the efficiency of the heat exchanger is introduced, which determines the ratio between actual thermal power and maximum thermal power:

$$\varepsilon = \frac{P_{dej}}{P_{maks}} \tag{13}$$

and for primary side

$$\varepsilon_h = \frac{T_{11} - T_{12}}{T_{11} - T_{21}} \tag{14}$$

4.3 Dynamic characteristics

The mathematical model of the heat exchanger is basically the problem of switching heat from a hot medium over the wall to a cold medium. This can be described using the energy equation for a cold and hot medium:

$$\rho_i c_{pi} \frac{\partial^2 T_i}{\partial x \partial t} = \nabla \cdot (\lambda_i \nabla T_i) - \underline{w_i} \rho_i c_{pi} \nabla T_i + r_i(-\Delta I_i) + \upsilon_{Ti} \left(\rho_i T_i \frac{\partial p_i}{\partial t} + \underline{w_i} T_i \nabla p_i \right) \tag{15}$$

Just as in the case of a temperature sensor, there are no chemical reactions and generated heat due to the mechanical part, which is why the last two articles of the eq. (15) are off. However, since the temperature gradient and the translation of heat over the wall are ignored, the equation can be simplified by consideration in:

$$m_i c_{pi} \frac{dT_i}{dt} = \dot{m}_i c_{pi} \frac{dT_i}{dx} \pm \alpha_i A_i (T_{hi} - T_{ci}) \tag{16}$$

Therefore, assuming one dimension model with complete mixing in the cell we obtain a model based on the energy balance of the heat exchanger:

- heat exchanger hot side:

$$m_h c_{ph} \frac{dT_h}{dt} = \dot{m}_h c_{ph} (T_{hi} - T_{ho}) - \alpha_h A_h (T_h - T_w) \tag{17}$$

- cool side of the heat exchanger:

$$m_c c_{pc} \frac{dT_c}{dt} = \dot{m}_c c_{pc} (T_{ci} - T_{co}) + \alpha_c A_c (T_w - T_c) \tag{18}$$

- and the wall of heat exchanger:

$$m_w c_{pw} \frac{dT_w}{dt} = \alpha_h A_h (T_h - T_w) - \alpha_c A_c (T_w - T_c) \tag{19}$$

However, since the heat exchanger system is a system with distributed parameters, the analytical solutions of the above model are impossible. In the trace of this, the heat exchanger will be divided into several units (**Figure 3**). For each unit, the balance sheet equations can be recorded:

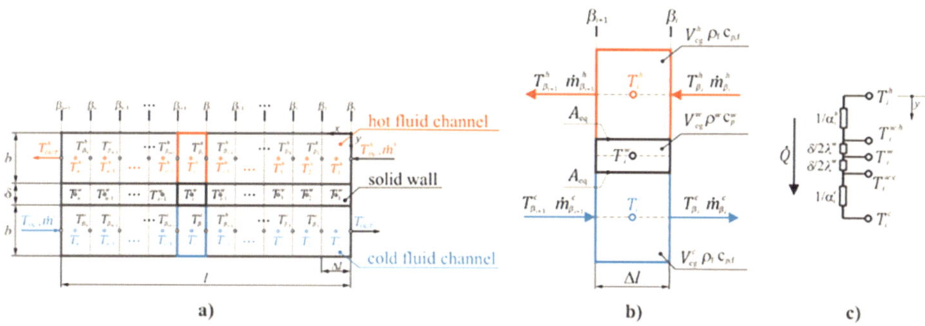

Figure 3.
(a) Heat exchanger section, (b) one cell, (c) resistance scheme.

- heat exchanger hot side:

$$m_h c_{ph} \frac{dT_{hi+1}}{dt} = \dot{m}_h c_{ph} (T_{hi} - T_{hi+1}) - \alpha_h A_h \left(\frac{T_{hi} + T_{hi+1}}{2} - T_{wi} \right) \tag{20}$$

- cool side of the heat exchanger:

$$m_c c_{pc} \frac{dT_{ci-1}}{dt} = \dot{m}_c c_{pc} (T_{ci-1} - T_{ci}) + \alpha_c A_c \left(T_{wi} - \frac{T_{ci-1} + T_{ci}}{2} \right) \tag{21}$$

- and the wall of heat exchanger:

$$m_w c_{pw} \frac{dT_{wi}}{dt} = \alpha_h A_h \left(\frac{T_{hi} + T_{hi+1}}{2} - T_{wi} \right) - \alpha_c A_c \left(T_w - \frac{T_{ci-1} + T_{ci}}{2} \right) \tag{22}$$

An important assumption is what the temperature distribution in each cell will be. In general, it can be assumed to jump (which is impossible in the heat exchanger, but if the sections are infinitesimally narrow, the assumption is good enough for the apportionment, while greatly simplifying the calculation), which means that the medium has an exit temperature from the cell immediately upon entering the cell. Such a model is very easy to use but the assumption is valid only if mixing is practically ideal. A more realistic model is where the temperature changes linearly along the cell. Thus, the model is numerically still sustainable, but to us, results that are close to real are why this model is used in the above equations. The third model considers the real logarithmic distribution of the temperature in the cell (**Figure 4**).

By rearranging eqs. (20)–(22) and inserting variables:

$$\tau_{h/c_i} = \frac{m_{h/c_i}}{\dot{m}_{h/c_i}} \tag{23}$$

$$\delta_{w_i} = \frac{m_{w_i} c_{p_{h/e_i}}}{A_i} \tag{24}$$

$$\gamma_{h/c_i} = \frac{A_i \alpha_{h/c_i}}{m_{h/c_i} c_{p_{h/c_i}}} \tag{25}$$

One can get a matrix for the i-th cell:

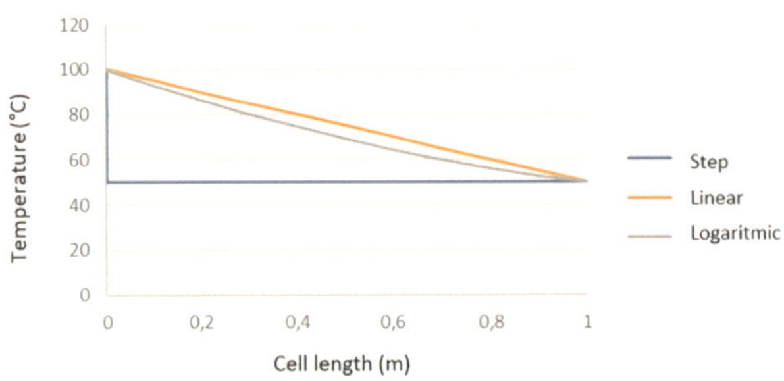

Figure 4.
Temperature distribution in the cell.

$$\begin{bmatrix} \dots \\ \dot{T}_{h_{i-1}} \\ \dot{T}_{h_i} \\ \dots \\ \dot{T}_{w_i} \\ \dots \\ \dot{T}_{c_i} \\ \dot{T}_{c_{i-1}} \\ \dots \end{bmatrix} = \begin{bmatrix} \dots & \dots & \dots & \dots & \dots & \dots & \dots & \dots & \dots & \dots & \dots & \dots & \dots & \dots \\ \dots & \dots & \dots & \dots & \dots & \dots & \dots & \dots & \dots & \dots & \dots & \dots & \dots & \dots \\ \dots & 0 & \frac{1}{\tau_{h_i}} & -\frac{1+\gamma_{h_i}}{\tau_{h_i}} & 0 & \dots & 0 & \frac{\gamma_{h_i}}{\tau_{h_i}} & 0 & \dots & \dots & \dots & \dots & 0 \\ \dots & \dots & \dots & \dots & \dots & \dots & \dots & \dots & \dots & \dots & \dots & \dots & \dots & \dots \\ \dots & 0 & \frac{\gamma_{h_i}}{\delta_{w_i}} & 0 & \dots & 0 & -\frac{\gamma_{h_i}+\gamma_{c_i}}{\delta_{w_i}} & 0 & \dots & 0 & \frac{\gamma_{c_i}}{\delta_{w_i}} & 0 & \dots & 0 \\ \dots & \dots & \dots & \dots & \dots & \dots & \dots & \dots & \dots & \dots & \dots & \dots & \dots & \dots \\ \dots & \dots & 0 & -\frac{1}{\tau_{c_i}} & \frac{1+\gamma_{c_i}}{\tau_{c_i}} & 0 & \dots & 0 & -\frac{\gamma_{c_i}}{\tau_{c_i}} & 0 & \dots & \dots & \dots & \dots \\ \dots & \dots & \dots & \dots & \dots & \dots & \dots & \dots & \dots & \dots & \dots & \dots & \dots & \dots \\ \dots & \dots & \dots & \dots & \dots & \dots & \dots & \dots & \dots & \dots & \dots & \dots & \dots & \dots \end{bmatrix} \cdot \begin{bmatrix} \dots \\ T_{h_{i-1}} \\ T_{h_i} \\ \dots \\ T_{w_i} \\ \dots \\ T_{c_i} \\ T_{c_{i-1}} \\ \dots \end{bmatrix} + \begin{bmatrix} \frac{1}{\tau_{h_1}} & 0 \\ 0 & 0 \\ 0 & 0 \\ 0 & 0 \\ 0 & 0 \\ 0 & \frac{1}{\tau_{c_1}} \\ 0 & 0 \\ 0 & 0 \\ 0 & 0 \end{bmatrix} \cdot [T_{h_{in}} \; T_{c_{in}}] \tag{26}$$

or:

$$\underline{\dot{T}}_{H/w/c} = \underline{A}\left(\tau_{H/c}, \gamma_{H/c}, \delta_w\right) \cdot \underline{T}_{H/w/c} + \underline{B}\left(\tau_{H/c_1}\right) \cdot \underline{T}_{H/c_{in}} \tag{27}$$

4.4 Laplace transformation

The above-mentioned equation system can be mapped to:

$$T_{hi+1} = \frac{\left(\frac{1}{\tau_h} - \frac{1}{\delta_h}\right)}{s + \left(\frac{1}{\tau_h} + \frac{1}{\delta_h}\right)} T_{hi} + \frac{\frac{2}{\delta_h}}{s + \left(\frac{1}{\tau_h} + \frac{1}{\delta_h}\right)} T_{wi} = G_{h1} T_{hi} + G_{h2} T_{wi} \tag{28}$$

$$T_{ci-1} = \frac{\left(\frac{1}{\tau_c} - \frac{1}{\delta_c}\right)}{s + \left(\frac{1}{\tau_c} + \frac{1}{\delta_c}\right)} T_{ci} + \frac{\frac{2}{\delta_c}}{s + \left(\frac{1}{\tau_c} + \frac{1}{\delta_c}\right)} T_{wi} = G_{c1} T_{ci} + G_{c2} T_{wi} \tag{29}$$

$$T_{wi} = \sum_{j=1}^{n} \frac{\gamma_h}{s + 2(\gamma_h + \gamma_c)} T_{hj} + \sum_{j=1}^{n} \frac{\gamma_c}{s + 2(\gamma_h + \gamma_c)} T_{cj} \tag{30}$$

to obtain a simplified record:

$$T_{cn} = \prod_{j=1}^{n} G_h T_{hj} + \prod_{j=1}^{n} G_c T_{cj} \tag{31}$$

From what we see, we get an nth-order system that depends on changes in input temperatures as well as material, geometric, and flow parameters.

4.5 Constitution equations

The material properties are generally described in the literature [45] and will not be further stated, but the determination of the heat transfer coefficient is more important. When using no dimensional numbers and links between them:

$$Nu = CPr^{m}\,Re^{n} \tag{32}$$

And for our case, Reynolds' number of

$$Re = \frac{\dot{m}}{o\nu} \tag{33}$$

Nusselt number:

$$Nu = \frac{\alpha A}{o\lambda} \tag{34}$$

The coefficients depend on the shape of the channel and the Reynolds number. For our example, the coefficients are:

$$\mathrm{C} = 0{,}2;\ \mathrm{m} = 0{,}67;\ \mathrm{n} = 0{,}4 \tag{35}$$

Valid for Reynolds numbers:

$$Re > 100 \tag{36}$$

5. Experimental model

Although most of the models of heat exchangers are on the market with a fish bone sample, it was still selected to test the heat exchanger with a microplate sample. The reason is that heat exchangers with a microplate are more efficient because the flow between the plates is more subdued and the distribution by plate is generally better, which gives us about a 20–30% lower pressure drop or a much better heat transfer. The comparison between the plates is shown in **Figure 5**.

The test was performed on a heat exchanger with three plates in different configurations. Four cases were tasted with different flow configurations and speeds as per **Table 1**. Although the test was carried out on only three channels, the multi plates' behavior is similar. The geometry of the heat exchanger is shown in **Figure 6**.

For this purpose, a device was designed to allow for a rapid switch between hot and cold water. The heat exchanger was flushed with cold water and, at the moment of disturbance, a stream of hot water was flushed in. The switching was made with magnetic valves. Despite the fast-switching magnetic valves, the disturbance was not stepped but was approximately similar to the second-order response. To record this phenomenon, the entry signal was recorded, simulated and compared to the real state. Comparisons are shown in **Figures 7–10**.

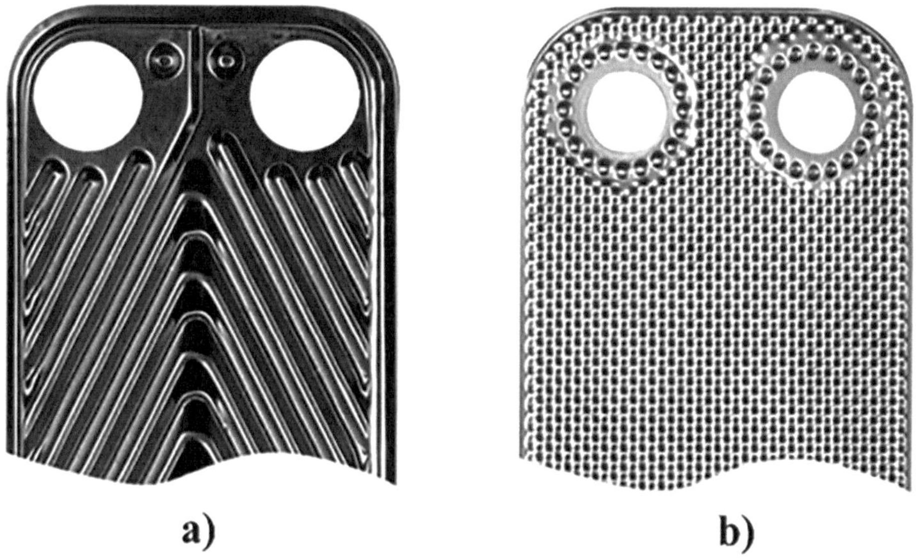

Figure 5.
Comparison between (b) microplate and (a) fishbone heat exchanger.

	Case A	Case B	Case C	Case D
Flow	Hot out, cold in	Cold out, hot in	Hot out, cold in	Hot out, cold in
Hot water	0,113	0,113	0,113	0,057
Cold water	0,113	0,113	0,057	0,057
Unit	l/s	l/s	l/s	l/s

Table 1.
Test design.

In **Figure 7**–**10**, input disturbances are simulated, while in **Figure 11**, deviations are depicted.

The discrepancies between the approximated and the measured disturbance are shown in **Figure 11**.

A FLIR A600 IR thermal camera was used to measure the dynamic temperature change. Flow measurements were carried out through weighing. Since the thermal camera can take a large number of images per second of 4000 images per second, it was assumed that the images are static at 0.1 s, which virtually eliminates the dynamic error of measurement. The temperature measurement error is +/−0,4°C.

In addition to the temperatures on the outside of the plate, the thermal camera also visualized the temperature field. The change in the temperature field is shown in **Figures 12–14** as a comparison between type A and other types.

The results show us that in all cases the distribution of temperature is uneven, which is probably due to the flow conditions on the panel itself. Despite the use of microchannel technology, the apparent distribution of water across the heat exchanger plate is still uneven, or not ideal for heat transfer.

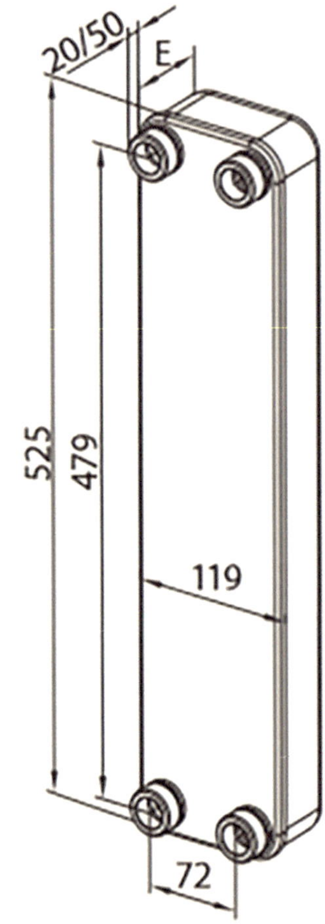

Figure 6.
Heat exchanger geometry.

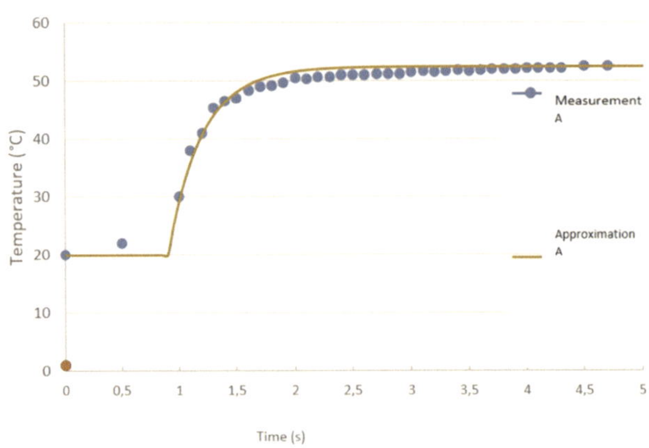

Figure 7.
Disturbance A.

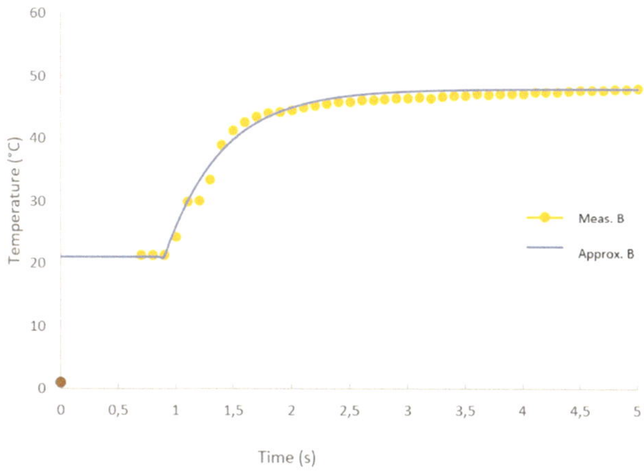

Figure 8.
Disturbance B.

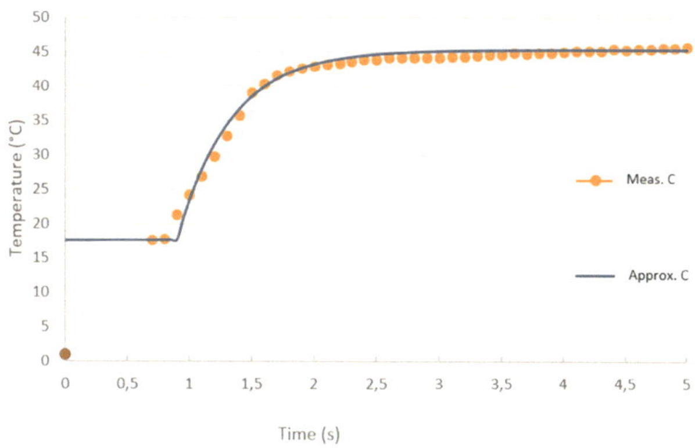

Figure 9.
Disturbance C.

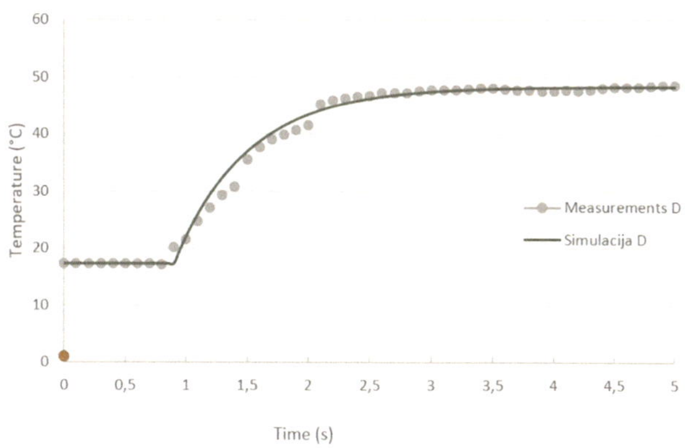

Figure 10.
Disturbance D.

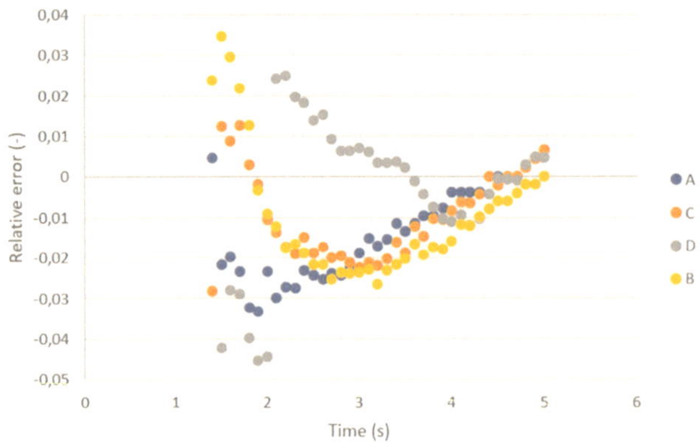

Figure 11.
Disturbance error between approximation and real one.

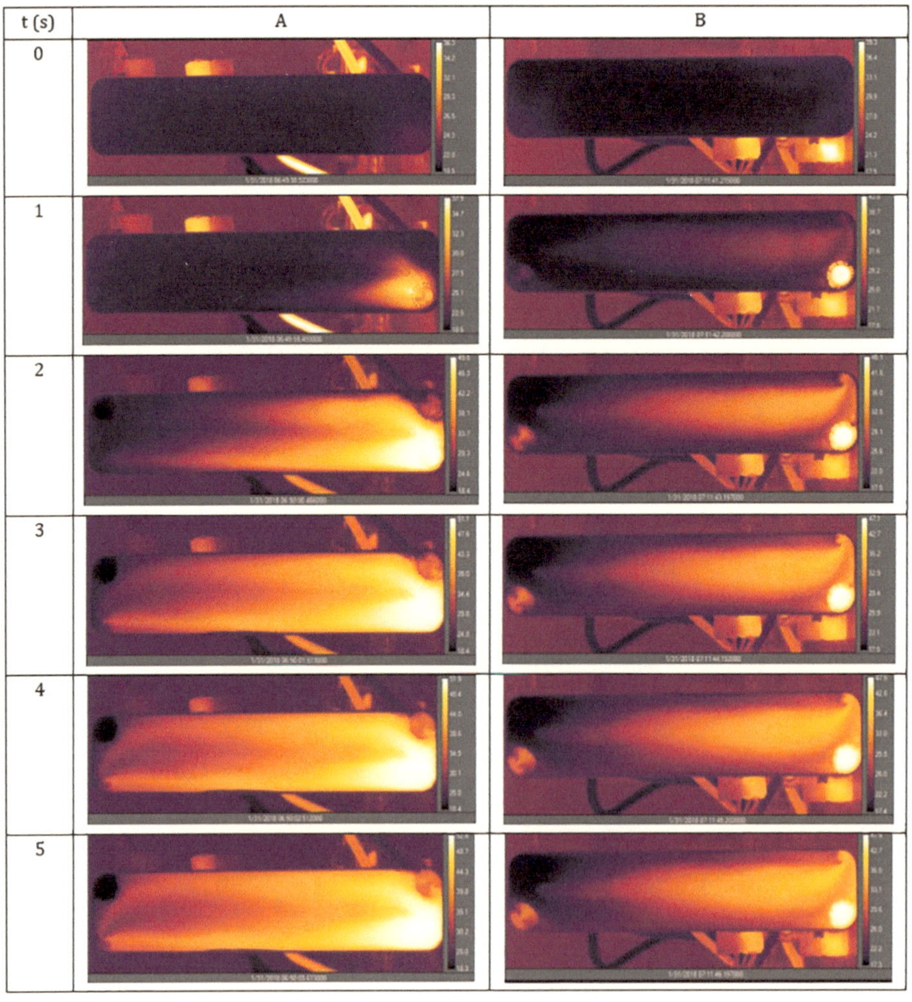

Figure 12.
Case A and B.

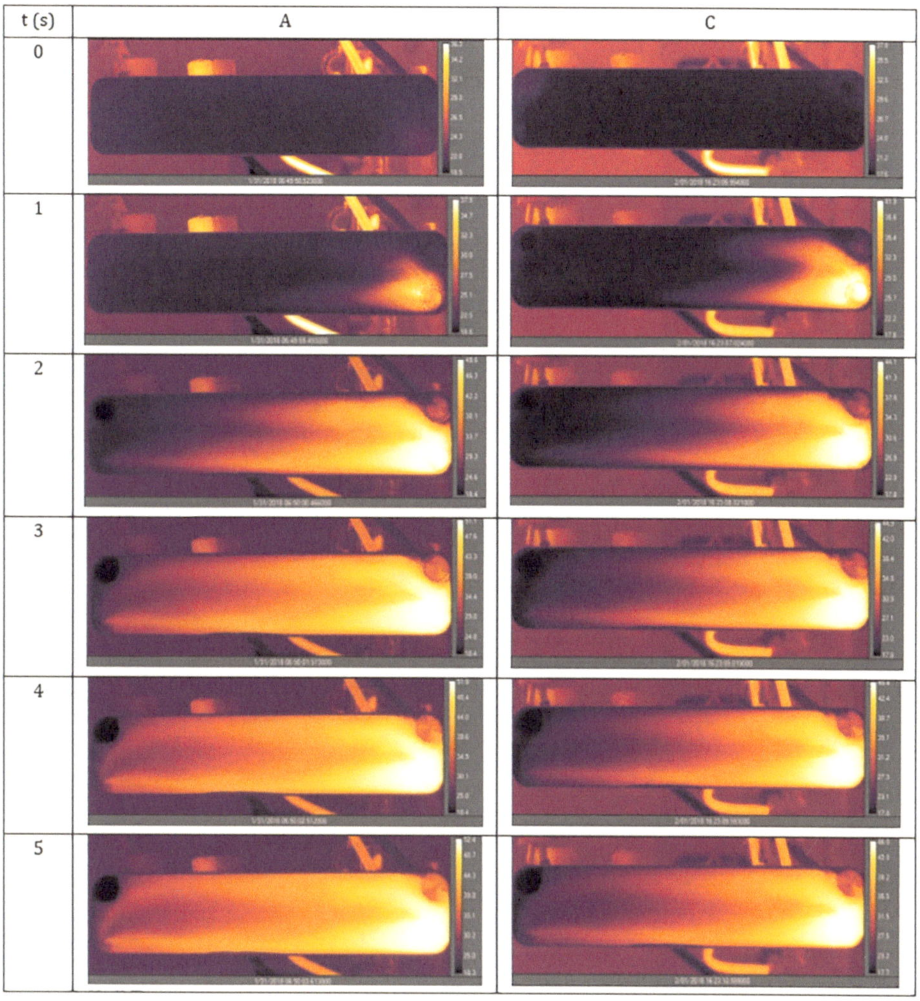

Figure 13.
Case A and C.

The main observations are:

- In the case of hot water flow in the middle, the spread of hot temperature disturbance is more limited to the center of the heat exchanger and therefore better, the reason for this is the higher heat take- and the "filtered" view of the hot plate,

- The C and D responses are quite similar. The reason is simple and is based on 2. The law of thermodynamics, says that the heat currents are the primary and secondary sides in equilibrium and therefore, with the temperature regime unchanged, the main influential parameter may be a flow rate that is smaller relative to A, and therefore the response is slower,

- A similar reason for the slower response in case B is that we have more cold water than hot water, and therefore the overheating of cold water is slower.

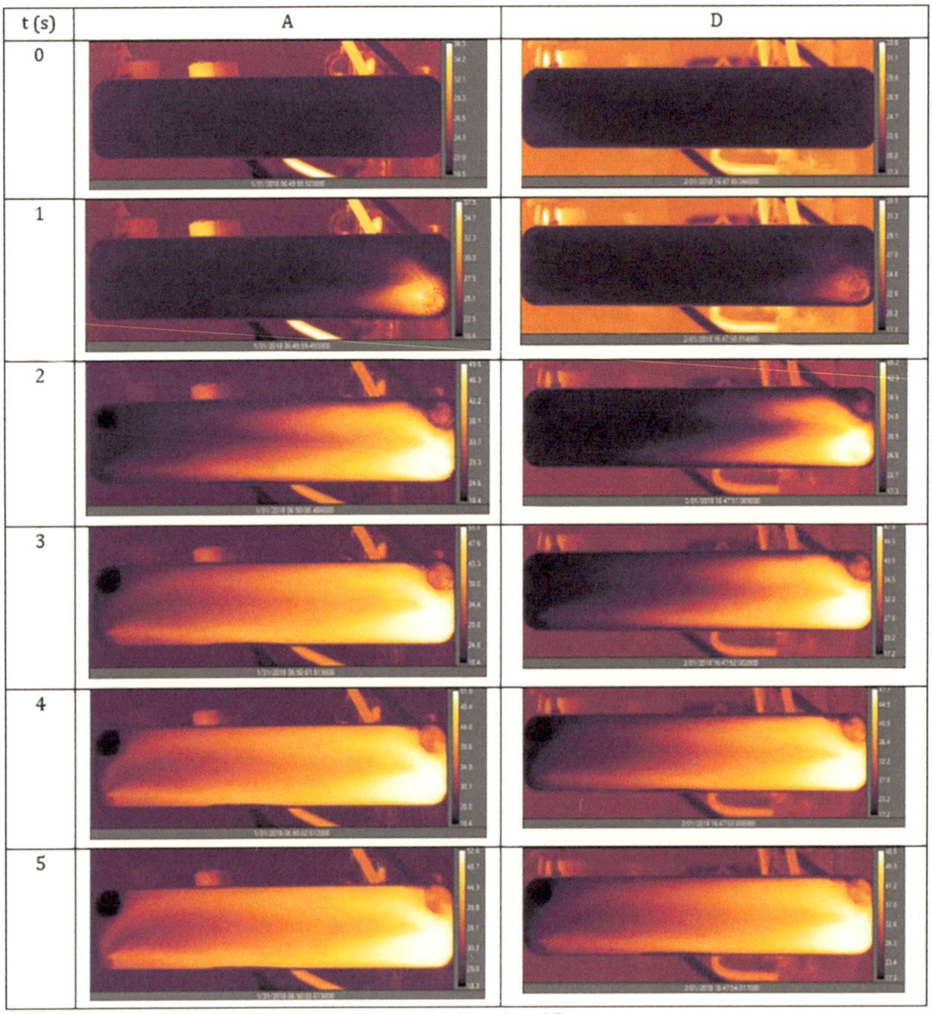

Figure 14.
Case A and D.

The results of the dynamic time response tests of the heat exchanger are shown in **Figure 15**.

One can conclude from **Figures 12** and **13**, that case B is not relevant for temperature front propagation. Comparing then cases A, C, and D one can see that speed of water is main contributor to temperature front propagation. As in A-D cases hot is on outside border, then there is very little difference between A and C, while D with half speed, is noticeable slower than case A.

6. Numeric experiment

The solution to the eq. (27) considering the geometry of the heat exchanger in **Figure 16** (Heat exchanger was divided into 10 segments — **Table 2** indicates a shift from the starting position) and the specifics of the experiment (input disturbance does not step)) is shown in **Figure 17**–**20**. The equation system was solved with the MATLAB Simulink software package using the RK4 numerical method.

Segment	0	1	2	3	4	5	6	7	8	9	10
Distance mm	0	47,9	95,8	143,7	191,6	239,5	287,4	335,3	383,2	431,1	479

Table 2.
Segments position.

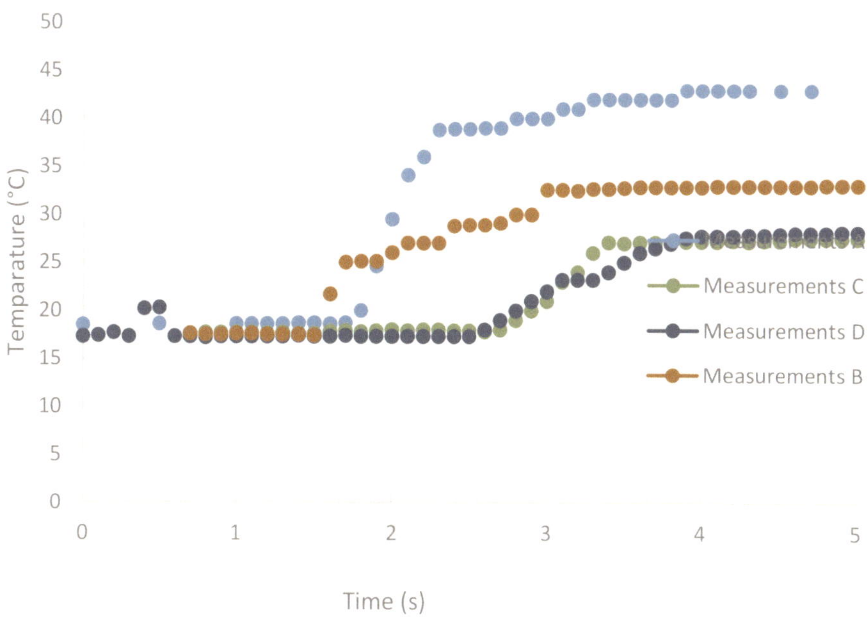

Figure 15.
Experimental results.

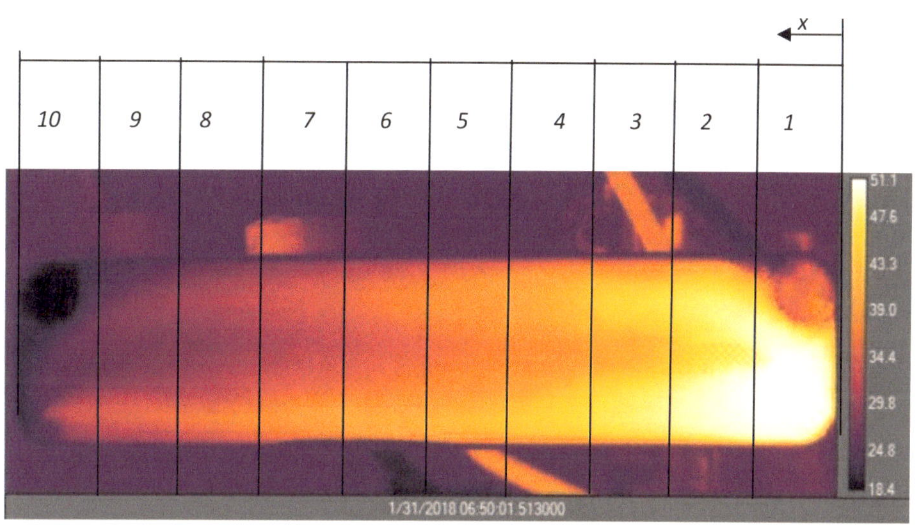

Figure 16.
Heat exchanger sections.

Figure 17–20 show simulation results split into sections. One can see the development of temperature as a function of time. The last 10th curve represents result of heat exchanger outlet.

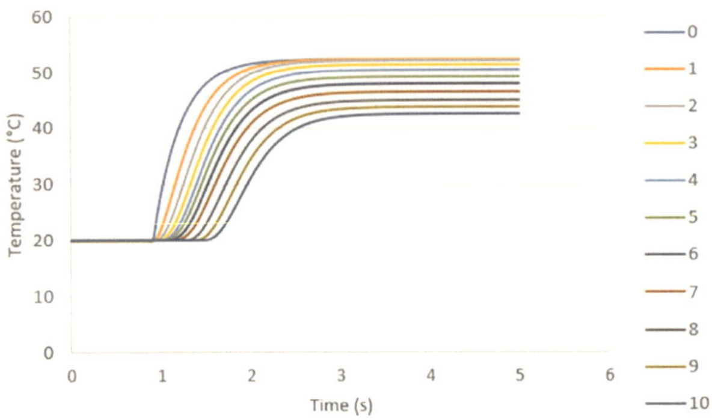

Figure 17.
Simulation A.

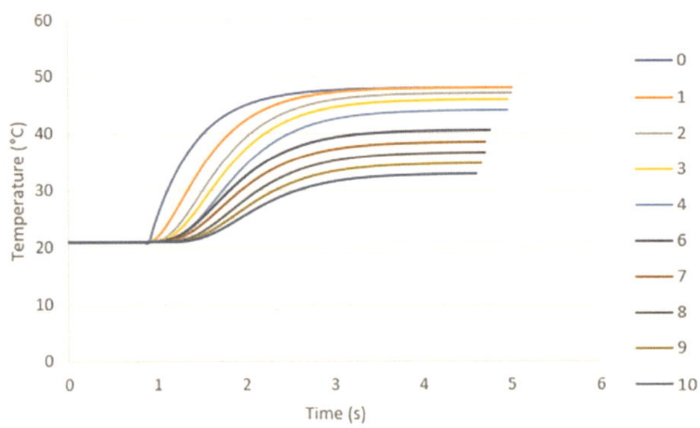

Figure 18.
Simulation B.

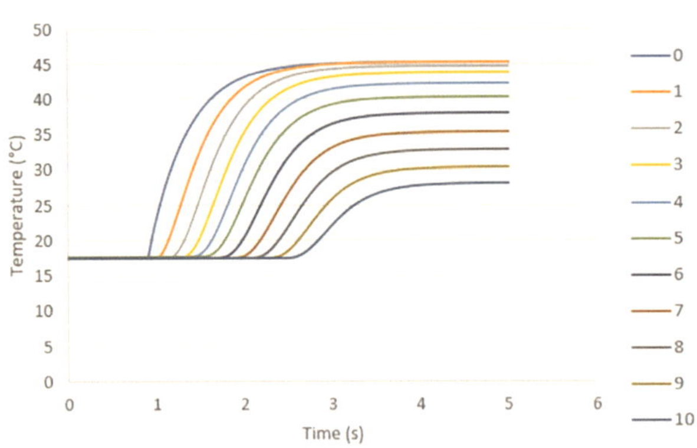

Figure 19.
Simulation C.

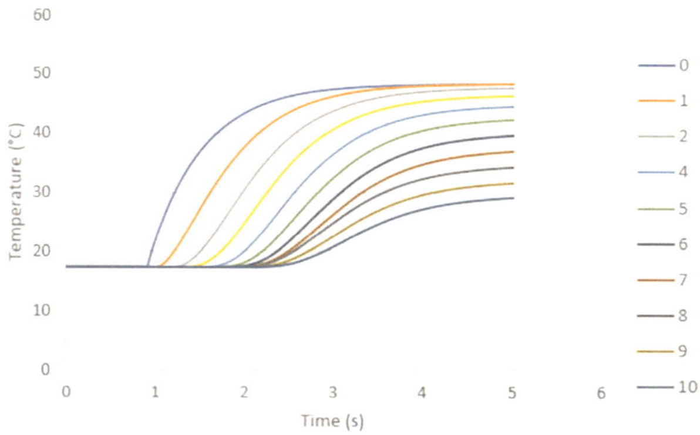

Figure 20.
Simulation D.

7. Comparison between numerical calculation and tests results obtained and discussion

A comparison between the measured and simulated results is shown in **Figure 21**–**24**.

The deviation between the measured and simulated results is relatively small and is less than 10%. The only difference is case D in the initial state because the heat exchanger was flushed with cold water and was not at ambient air temperature, as assumed in the simulation. Therefore, the initial situation is different. The deviations are shown in **Figure 25**.

8. Conclusion

Obtaining digital twin from plate heat exchanger is relatively easy when behavior of the heat transfer coefficients is known. It is necessary to sectionize heat exchanger to get lump sum model, which is easy to resolve. A number of sections determine the accuracy, but also the distribution within the cell. The model can be further simplified and improved.

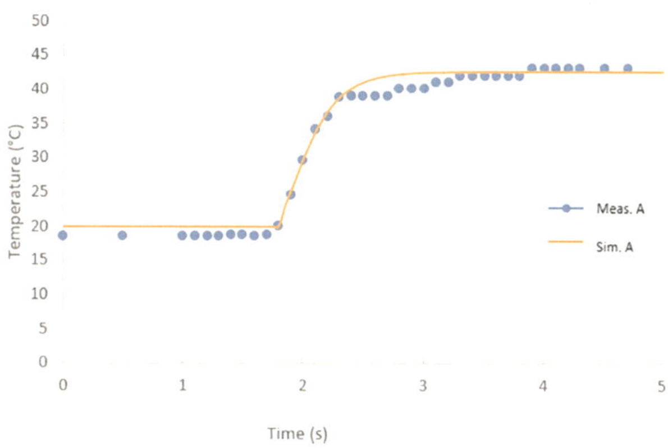

Figure 21.
Comparison A.

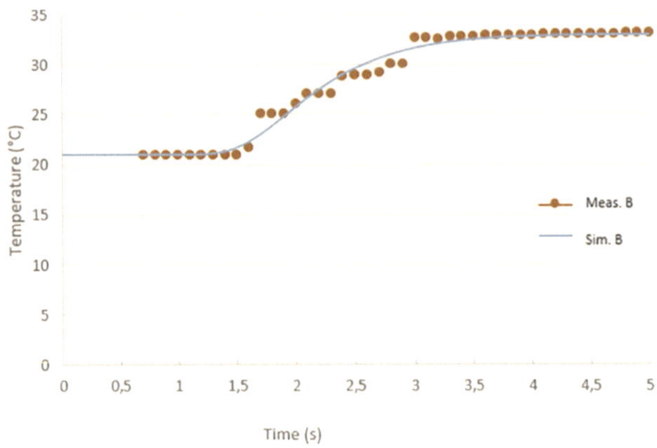

Figure 22.
Comparison B.

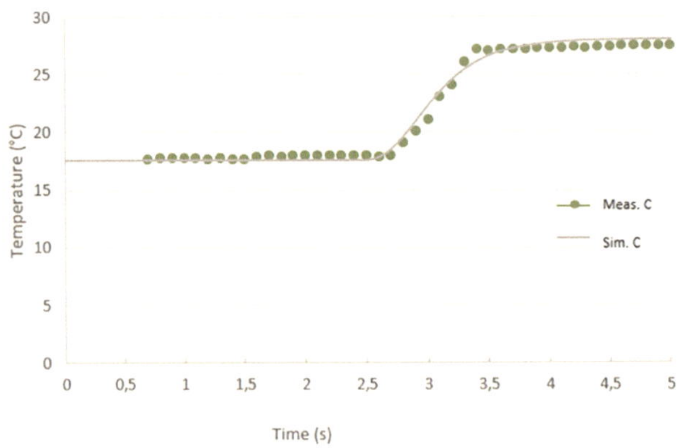

Figure 23.
Comparison C.

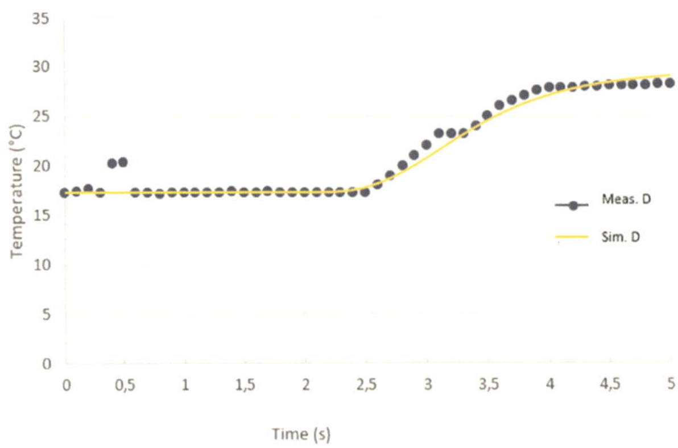

Figure 24.
Comparison D.

Figure 25.
Error between simulation and measurements.

The mathematical model can be simplified by ignoring the intermediate wall, complete cell mixing, and independence of material properties from temperature.

However, it can be improved by increasing the number of cells to improve the approximation of the real state or by considering the logarithmic distribution of the temperature in the cell.

Conflict of interest

"The authors declare no conflict of interest."

Nomenclature

A	m^2Area
C	-Constant
c_p	J/(kgK)Specific heat
G	Transfer function
h	JEnthalpy
I	J/kgReaction enthalpy
m	kgMass
$\dot{m}$	kg/sMass flow

Nu	-Nusselt number
o	mCircumference
P	WPower
Pr	-Prandtl number
p	PaPressure
Re	-Reynolds number
r	kmol/sVelocity of chemical reaction
s	Laplace operator
T	°C, KTemperature
t	sTime
$\boldsymbol{v_x}$	m/sShear velocity
v_T	m^3/kgSpecific volume
w	m/sFluid speed
x	mCoordinate
y	mCoordinate
α	W/(m^2K)Heat convection coefficient
β	m^3/KTemperature dilatation coefficient
γ	1/sHeat exchanger coefficient
δ	J/(m^2K)Heat exchanger coefficient
ε	N/mm^2Compression Module
ε	-Heat exchanger efficiency
η	Pa·sDynamic viscosity
ρ	kg/m^3Density
λ	W/(mK)Heat conduction coefficient
ν	kg/(ms)Kinematic viscosity
τ	N/m^2Shear stress
τ	sHeat exchanger coefficient
1	Primary side
11	Primary side inlet
12	Primary side outlet
+1	Next cell
−1	Previous cell
2	Secondary side
21	Secondary side inlet
22	Secondary side outlet
c	Cold side
eff	Effective
dej	Real
h	Hot side
i	Inlet
i	i-th element
j	j-th element
m	Exponent
n	Exponent
o	Outlet
p	Pressure
w	Wall

Author details

Miha Bobič
Danfoss Trata d.o.o, Ljubljana, Slovenia

*Address all correspondence to: miha.bobic@danfoss.com

References

[1] Abu-Khader MM. Plate heat exchangers: Recent advances. Renewable an Sustainable Energy Reviews. 2012;**16**: 1883-1891

[2] Aslam Bhutta MM, Hayat N, Bashir MH, Khan AR, Ahmad KN, Khan S. CFD applications in various heat exchangers design: A review. Applied Thermal Engineering. 2012;**32**:1-12

[3] Zhang J, Zhu X, Mondejar ME, Haglind F. A review of heat transfer enhancement techniques in plate exchangers. Renewable and Sustainable Energy Reviews. 2019;**101**:305-328

[4] Gullapalli VS, Sundén B. CFD simulation of heat transfer and pressure drop in compact brazed plate heat exchangers. Heat Transfer Engineering. 2014;**35**(4):358-366

[5] Freund S, Kabelac S. Investigation of local heat transfer coefficients in plate heat exchangers with temperature oscillation IR thermography and CFD. International Journal of Heat and Mass Transfer. 2010;**53**(19–20): 3764-3781

[6] Ayub ZH, Khan TS, Salam S, Nawaz K, Ayub AH, Khan MS. Literature survey and universal evaporation correlation for plate type heat exchangers. International Journal of Refrigeration. 2019;**99**: 408-418

[7] Guo Y, Wang F, Jia M, Zhang S. Modeling of plate heat exchanger based on sensitivity analysis and model updating. Chemical Engineering Research and Design. 2018;**138**:418-432

[8] Bobič M, Ogorevc T, Mazej M, Petkovšek J, Pribošek J, Sevčnikar J, et al. Vizualizacija delovanja ploščnega prenosnika toplote, Akademija strojništva 2014 in 4. In: Mednarodna konferenca strojnih inženirjev. Ljubljana, Slovenija: ZSIS; 2014

[9] Pribošek J, Bobič M, Golobič I, Diaci J. Correcting the periodic optical distortion for particle-tracking velocimetry in corrugated-plate heat exchangers. Strojniški Vestnik. 2016;**62**(1):3-10

[10] Jin S, Hrnjak P. A new method to simultaneously measure local heat transfer and visualize flow boiling in plate heat exchanger. International Journal of Heat and Mass Transfer. 2017; **113**:635-646

[11] Kabelac S. Analysis of local heat transfer in plate heat exchangers for flow pattern optimization. In: 14th International Heat Transfer Conference. Vol. 4. ASME; 2010. pp. 375-384

[12] Jin S, Hrnjak P. Effect of end plates on heat transfer of plate heat exchanger. International Journal of Heat and Mass Transfer. 2017;**108**:740-748

[13] Clarke DD, Vasquez VR, Whiting WB, Greiner M. Sensitivity and uncertainty analysis of heat exchanger designs to physical properties estimation. Applied Thermal Engineering. 2001;**21**:993-1017

[14] Thorsen JE, Iversen J. Impact of lowering dt for heat exchangers used in district heating systems. In: DHC 13 the 13th International Symposium on District Heating and Cooling. Copenhagen, Denmark; 3rd-4th 2012

[15] Thorsen JE, Gudmundsson O, Brand M. Performance Specifications for Heat Exchanger for District Heating Substations of the Future, the 14th International Symposium on District Heating and Cooling, 7th–9th Sep.

Stockholm, Sweden: Danish Technical University; 2014

[16] Ansari MS, Mishra P, Gaikwad K, Awachat PN. Heat transfer analysis of corrugated plate heat exchanger over coil type heat exchanger: A review. International Journal of Engineering Research and Technology. 2014;**3**(3): 446-449

[17] Carlomagno GM, Cardone G. Infrared thermography for convective heat transfer measurements. Experiments in Fluids. 2010;**49**:1187-1218

[18] Gustafsson J, Delsing J, Van Deventer J. Improved district heating substation efficiency with a new control strategy. Applied Energy. 2010;**87**:1996-2004

[19] Huang C, Cai W, Wang Y, Liu Y, Li Q, Li B. Review on the characteristics of flow and heat transfer in printed circuit heat exchangers. Applied Thermal Engineering. 2019;**153**:190-205

[20] Laszczyk P. Simplified modeling of liquid-liquid heat exchangers for use in control systems. Applied Thermal Engineering. 2017;**119**:140-155

[21] Piper M, Zibart A, Djakow E, Springer R, Homberg W, Kenig EY. Heat transfer enhancement in pillow-plate heat exchangers with dimpled surfaces: A numerical study. Applied Thermal Engineering. 2019;**153**:142-146

[22] Qian Z, Wang Q, Cheng J. Analysis of heat and resistance performance of fin-and-tube heat exchanger with rectangle-winglet vortex generator. International Journal of Heat and Mass Transfer. 2018;**124**:1198-1211

[23] Sharifi F, Glokar Narandji MR, Mehravaran K. Dynamic simulation of plate heat exchangers. International Communications in Heat and Mass Transfer. 1995;**22**:213-225

[24] Wang L, Sundén B, Manglik RM. Plate Heat Exchangers: Design, Applications and Performance. Southampton: WIT Press; 2007

[25] Whalley R, Ebrahimi KM. Heat exchanger dynamic analysis. Applied Mathematica Modeling. 2018;**62**:38-50

[26] Arsenyeva OP, Kapustenko PO, Tovazhnyanskyy LL, Khavin G. The influence of plate corrugations geometry on plate heat exchanger performance in specified process conditions. Energy. 2016;**57**:201-207

[27] Dović D, Palm B, Svaić S. Generalised correlations for predicting heat transfer and pressure drop in plate heat exchanger channels of arbitrary geometry. International Journal of Heat and Mass Transfer. 2009;**52**(19–20):4553-4563

[28] Muley A, Manglik RM. Experimental study of turbulent flow heat transfer and pressure drop in a plate heat exchanger with Chevron plates. Journal of Heat Transfer. 1999;**121**(1):110-117

[29] Arsenyeva OP, Tovazhnyanskyy LL, Kaputenko PO, Demirskiy OV. Generalized semi empirical correlation for heat transfer in channels of plate heat exchanger. Applied Thermal Engineering. 2014;**70**(2):1208-1215

[30] Khan TS, Khan MS, Chyu MC, Ayub ZH. Experimental investigation of single-phase convective heat transfer coefficient in a corrugated plate heat exchanger for multiple plate configurations. Applied Thermal Engineering. 2010;**30**(8–9):1058-1065

[31] Sidebotham G. Nusselt number correlations. In: Heat Transfer Modeling. Springer-Verlag; 2015

[32] Aprea C, Renno C. Experimental analysis of a transfer function for an air-cooled evaporator. Applied Thermal Engineering. 2001;**21**:481-493

[33] Maddali RK. Modeling ordinary differential equations in Mathlab Simulink. Indian Journal of Computer Science and Engineering. 2012;**3**(3): 406-410

[34] Badami M, Fonti A, Carpignano A, Grosso D. Design of district heating networks through an integrated thermo-fluid dynamics and reliability modelling approach. Energy. 2018;**144**(1):826-838

[35] Brand M, Thorsen JE, Svendsen S. Numerical modelling and experimental measurements for a low- temperature district heating substation for instantaneous preparation of DHW with respect to service pipes. Energy. 2012;**41** (1):392-400

[36] Das SK, Roetzel W. Dynamic analysis of plate heat exchangers with dispersion in both fluids. International Journal of Heat and Mass Transfer. 1995; **38**(6):1127-1140

[37] Hamze S, Witrant E, Bresch-Pietri D, Fauvel C. Estimating heat-transport and time-delays in a heat exchanger. In: 2018 IEEE Conference on Control Technology and Applications (CCTA). Copenhagen: Danish Technical University; 2018. pp. 1514-1519

[38] Michel A, Kugi A. Accurate low-order dynamic model of a compact plate heat exchanger. International Journal of Heat and Mass Transfer. 2013;**61**(33): 323-331

[39] Michel A, Kugi A. Model based control of compact heat exchangers independent of the heat transfer behavior II. Journal of Process Control. 2014;**50**(11–12):1859-1868

[40] Bobič M. Development of Measuring System of Coupled Pressure-Temperature Control System[Thesis]. Ljubljana, Slovenia: University of Ljubljana, Faculty for Mechanical Engineering; 2021

[41] Bobič M, Gjerek B, Golobič I, Bajsić I. Dynamic behaviour of a plate heat exchanger: Influence of temperature disturbances and flow configurations. International Journal of Heat and Mass Transfer. 2020;**163**:120439

[42] Bobič M. Dinamika preizkuševališča za razvoj in simuliranje delovanja toplotne podpostaje [magistrsko delo], mentor: Bajsić I., Ljubljana. Ljubljana, Magistrska dela: Fakulteta za strojništvo; 2000. p. 1079

[43] Bobič M., Muhič D., Zupnčič M., Sedmak I., Golobič I.: Dinamika ploščnega prenosnika toplote ogrevalnega sistema, Akademija strojništva 2017: inženirstvo - za kakovostnejše življenje, 6 mednarodna konferenca strojnih inženirjev 2017, Ljubljana, oktober 2017. Zveza strojnih inženirjev Slovenije, 2017. Vol. 6, No. 3/ 4, 78, Svet strojništva.

[44] Miao Q, You S, Zheng W, Zheng X, Zhang H, Wang Y. A Grey-box dynamic model of plate heat exchangers used in an urban heating system. Energies. 1398; **10**(9):2019

[45] Chapman AJ. Fundamentals of heat transfer. MacMilian Publishing Company; 1987

Section 4

Fouling and Its Mitigation in Heat Exchangers

Chapter 13

Fouling and Mechanism

Obaid Ur Rehman, Nor Erniza Mohammad Rozali and Marappa Gounder Ramasamy

Abstract

Fouling is the deposition of material on the heat transfer surface which reduces the film heat transfer coefficient. The impact of fouling on the heat exchanger is manifested as the reduction of thermal and hydraulic performance, in which the latter has a minor effect. This factor needs to be considered when calculating the effectiveness of the heat exchanger. During the design of heat exchangers, the fouling factor increases the required heat transfer area, which adds extra manufacturing costs. With less efficient heat exchangers, the economic cost of fouling is related to excess fuel consumption, loss of production, and maintenance or cleaning. The extra fuel consumption also damages the environment by increasing greenhouse gas production. Although much of the research work has been done on modeling and predicting fouling, it is still a poorly understood phenomenon representing the complexity of its mechanism. The common fouling mitigation action after the onset of fouling is to optimize the operating condition, e.g., increase the bulk flow velocity or decrease surface temperature. However, many quantitative and semi-empirical models have been developed to predict the fouling rate for preventive actions and optimizing cleaning schedules.

Keywords: heat exchanger, fouling, crude oil, heat transfer, thermohydraulic

1. Introduction

The deposition of foulant on the heat exchanger surface is usually in a solid or semi-solid state which gradually solidifies by the action of heat. This deposition reduces the capacity of exchangers to deliver the required outlet temperatures due to the low thermal conductivity of the deposit. The subsequent result of this reduction is the loss of thermal efficiency of the heat exchangers. Fouling is referred to the deposition of unwanted material on heat transfer surfaces resulting in the loss of thermohydraulic capacity of heat exchangers. The fouling usually occurs in most of the heat exchangers with varying severity. There are typically various sources that contribute to the development of foulants attached to the heat exchanger surface. The most common fouling precursors in crude oil are dirt particles, asphaltenes, salts, sulfur, and nitrogen compounds. It is crucial to consider the provision of fouling resistance in the heat exchanger design to compensate for the loss of efficiency due to fouling. At some point, it is desirable to take periodic shut down for mechanical cleaning of the heat transfer surface [1]. Usually, the outlet temperature of the heat exchanger is monitored, and required actions are taken to achieve the desired value.

Typically, the reduction of thermal efficiency is compensated by increasing the load on the furnace, which is manifested as an economic penalty in terms of extra fuel consumption. Van Nostrand et al. [2] divide the cost related to fouling into three categories, i.e., additional energy consumption, production loss, maintenance, and cleaning cost. They estimated the cost of fouling for a refinery processing 100,000 bbl per day at US$ 31.17 MM (extrapolated to 2022). The cost distribution for production loss, extra energy consumption, and maintenance was 40%, 10%, and <1%. Fouling mitigation strategies can reduce the cost of fouling. It is estimated that [3] 15% of energy savings can be achieved by applying fouling mitigation strategies in pre-heat trains and furnaces in oil refineries. Other options to mitigate the effect of fouling is to use retrofit of heat exchangers or to introduce tube inserts [4].

With the decrease in thermal efficiency, the flow area in the heat exchanger tube is also reduced, resulting in increased pressure drop and sometimes tube blockage. The monitoring of pressure drop and outlet temperature is therefore highly required to predict the extent of fouling and apply the mitigation actions by adjusting the operating conditions. It is also imperative to improve the heat exchanger design by incorporating the effect of operating conditions and the type of fluid used. The typical design strategy uses the fouling factor, which is merely a correction factor to compensate for the loss of heat transfer area due to fouling. The heat exchanger design also considers the ease of cleaning and elimination of hot spots to mitigate fouling. However, this scheme does not consider the dependence of fouling on the operating condition and fluid type.

In most cases, the mitigation strategy is adopted during the operation of heat exchangers. The feed is filtered to remove contaminants to minimize particle deposition, and chemical inhibitors are added. Different inhibitors used in the industry provide diverse functionality in preventing inorganic and organic fouling. Ethylenediamine tetraacetic acid (EDTA) inhibitors are commonly used against crystallization fouling. Similarly, some antioxidants and surfactants are also used to prevent fouling from polymerization and particulate matter in crude oil. An alternative to using chemical inhibitors is the application of tube inserts such as wire matrix, twisted coils, or hiTRAN. They serve as turbulence promoters to increase the removal rate of fouling by the action of wall shear forces [5].

During the operation of heat exchangers, it is sometimes desirable to promote foulant removal through operational parameters. The most common strategy is to increase the velocity to detach the foulant layer from the heat transfer surface by turbulent sweeps and bursts. These actions can work effectively against weak foulant deposits and are adopted for a longer duration, especially in refineries. For hardened deposits such as in particulate fouling, the gas rumbling is useful for promoting the dislodging of foulant. It is carried out with pressurized air or nitrogen stream introduced in the fluid for a short duration.

2. Fouling estimation

Heat exchangers are designed for the required heat duty, temperatures, and flow rates of the hot and cold streams. The overall heat transfer coefficient is estimated to predict the required surface area of the heat exchanger using the following equation:

$$Q = UA\Delta T \tag{1}$$

The overall heat transfer coefficient can be calculated as follows:

$$\frac{1}{U} = \frac{1}{U_0} + R_f \tag{2}$$

There are two approaches usually adopted during the operations to estimate the extent of fouling in heat exchangers, namely thermal and hydraulic approaches.

2.1 Thermal resistance

The estimation of the extent of fouling depends on the mode of operation of the heat exchanger. If the operation has a constant temperature difference, which is usually encountered in the operation of furnace and reactor [6], then the heat flux is decreased to maintain the constant temperature difference. The fouling resistance can be estimated as follows:

$$R_f = \left(\frac{\Delta T}{q_o}\right) - \left(\frac{\Delta T}{q_t}\right) \tag{3}$$

The second mode is with constant heat flux, which is predominantly applied in lab-scale fouling experimental units [7], in which the deposition on the surface is estimated with increasing temperature difference by the following equation:

$$R_f = \left(\frac{\Delta T_t}{q_o}\right) - \left(\frac{\Delta T_o}{q_o}\right) \tag{4}$$

2.2 Hydraulic resistance

The effect of pressure drop on heat exchanger performance is also very significant, manifested as decreases in the hydraulic efficiency of the exchangers. The pressure drop increases with the thickness foulant layer. It is crucial to estimate the pressure drop changes in tubes to predict the accurate thickness of the fouling layer. Sometimes the estimation of foulant thickness based on only the thermal resistance can lead to erroneous results due to the change in thermal conductivity of the foulant with aging [8]. Commonly thin-slab approximation [9] is employed to estimate the thickness of foulant as follows:

$$R_f = \frac{x}{\lambda_f} \tag{5}$$

The aging of foulant manifested in its increasing thermal conductivity. Thus, Eq. (5) can be employed by incorporating the foulant's aging to predict the fouling rate accurately. Usually, the heat exchangers are operated at constant mass flow rate operations in which pressure drop increases with the deposition. Usually, the pumping power is increased to overcome increasing pressure drops. The velocity increases with time, affecting the film heat transfer coefficient. The pressure drop for a clean tube is given by

$$\Delta P_o = 4f\frac{L}{d_i}\frac{\rho u_o^2}{2} \tag{6}$$

Similarly, for a fouled tube, the pressure drop can be represented as

$$\Delta P_f = 4f\frac{L}{d-2x}\frac{\rho u_f^2}{2} \tag{7}$$

If the friction factor is assumed constant, the pressure drop can be related to foulant thickness by Blasius correlation [10] as follows:

$$\Delta P_f = \Delta P_0\left(1-\frac{2x}{d}\right)^{-4.75} \tag{8}$$

2.3 Fouling rate

Kern and Seaton [11] considered the deposition of foulant due to two competing processes: deposition and removal. Some of the foulant material are removed from the layer by the action of turbulent eddies. The net deposition can be given as follows:

$$\frac{dR_f}{dt} = \dot{m}_D - \dot{m}_R \tag{9}$$

The fouling rate can be determined from the slope of fouling resistance-time curves, which changes typically during the process, as shown in **Figure 1**.

Initially, with the clean surface of the heat exchanger, the fouling resistance does not increase with time due to the low strength of the deposit to stick with the surface. This period of zero fouling rate is known as the induction period. Sometimes, during the induction period, the fouling rate becomes negative, representing the enhancement of heat transfer because of increased surface roughness contributed by the initial foulant layer. After the termination of the induction period, a constant initial fouling rate is usually attained. The fouling phenomenon determines the shape of the fouling resistance-time curve. As shown in **Figure 1**, an accelerating fouling curve represents a negligible removal rate due to hard deposits.

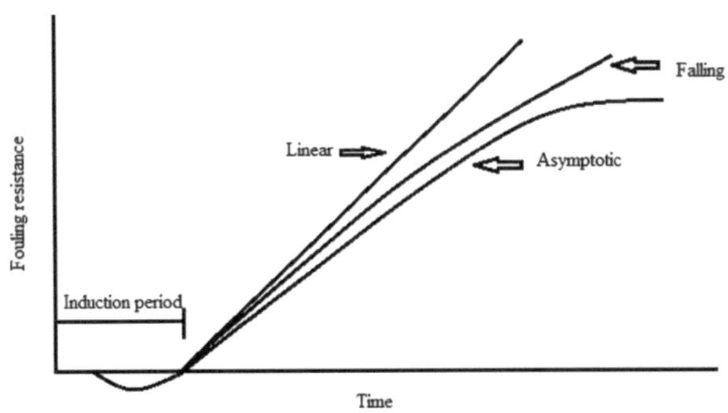

Figure 1.
Fouling resistance-time curves for different fouling mechanisms [11].

The linear fouling curve is the most common behavior in heat exchangers; it represents constant deposition and removal rate, while the deposition dominates over the removal. It is usually associated with crystallization fouling, but some studies also showed the asymptotic behavior for crystallization fouling [12, 13]. The falling and asymptotic curves depict the existence of the removal rate. As the deposit builds up, the flow area is reduced, and the velocity increases, allowing the foulant removal by shear forces. This retardation phenomenon is the cause of the asymptotic behavior of the fouling resistance-time curve and is proportional to the thickness of the foulant layer. This behavior is most familiar with particulate fouling and constant heat flux operations. The sawtooth behavior is typical in crude oil fouling, where the sudden removal of the deposits causes a negative fouling rate.

This initial fouling rate was represented by Ebert and Panchal [14] as a semi-empirical threshold model for crude oil fouling as:

$$\frac{dR_f}{dt} = \alpha \, Re^{-\beta} \exp\left(\frac{-E}{RT_f}\right) - \sigma\tau_w \tag{10}$$

They described the threshold point where the induction period is terminated. The above model was developed to predict the film temperature for a particular velocity to achieve a zero initial fouling rate, i.e., threshold point. They found that if this condition is maintained during the operation, there would be negligible or no fouling. If the left-hand side of Eq. (10) is taken to be zero, then a set of data appoints can be generated as shown in **Figure 2** for different values of film temperatures. The developed curve in **Figure 2** can be employed to estimate the corresponding film temperature in the non-fouling zone for a particular velocity.

The fouling prediction models are modified to produce a similar behavior caused by complex interaction of different mechanisms existing in the crude oil fouling process. The feedstock composition also plays a critical role in defining the fouling behavior. The presence of inorganic and asphaltenes molecules in crude oil makes it difficult to predict feedstock composition effects on fouling rate.

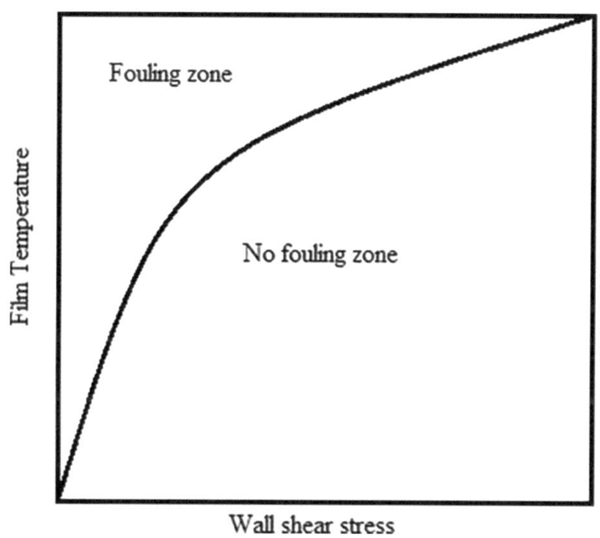

Figure 2.
Data curve to estimate film temperature for corresponding shear stress at zero fouling rate [14].

3. Types of fouling

Fouling phenomena depend on the type and concentration of foulant precursors in the bulk fluid entering the heat exchanger. There are several types of fouling based on the feedstock handled by the heat exchanger. Typically, foulant material consists of organic and inorganic constituents with deposition rates proportional to each other [15]. The following paragraphs are devoted to explaining different types of fouling.

3.1 Particulate fouling

The deposition of particles on the heat transfer surface is known as particulate fouling. These particles include originally suspended particles in the feedstock before entering the heat exchanger and particles formed within the heat exchanger by reaction or flocculation. The concentration of filterable solids in crude oil indicates the potential of particulate fouling. Wang and Watkinson [16] found that iron sulfide particles in crude oil were formed by the chemical reaction of hydrogen sulfide with iron oxides which resulted in particulate fouling. These particles can be inorganic, organic, or a mixture of both. For example, the salt contained by the boiler water and crude oils, corrosion products, and sand and dirt particles in cooling water is commonly referred to as inorganic particle fouling. Similarly, the deposition of organic particles produced by the aggregation and flocculation of asphaltenes in crude oils is known as organic particle fouling. The thermophoretic forces sometimes govern the particle deposition, which act on particles when there is a temperature difference between the hot surface and bulk fluid.

3.2 Chemical reaction fouling

Under a favorable condition, the soluble reactants present in the fluid tend to react to form insoluble products, which are then deposited on the heat transfer surface. The reaction in the bulk fluid is often polymerization of organic molecules, which results in products with high molecular weights. The presence of transition metals and heteroatoms such as nitrogen, sulfur, and oxygen tend to catalyze the reaction by polarizing other organic molecules, leading to lower activation energies for the reaction to occur. The oxygen reacts with organic molecules to form peroxides, increasing the polymerization reaction rate. The formation of highly viscous organic foulant from the deformation of organic molecules is also an example of chemical reaction fouling in refinery pre-heat train and furnace. Some reactions take place on the heated surface to transform them into harder products that are not easily removed, such as coke formation at high temperatures. The presence of unsaturated hydrocarbons, acidic compounds, mercaptans, and amines in crude oils induce polymerization reactions. Commonly, the acid number of crude oil is estimated to indicate its susceptibility to fouling. Temperature is also an important factor in chemical reaction fouling which promotes the fouling process. Therefore, it is desirable to operate the process at lower temperatures to mitigate the fouling phenomena.

3.3 Corrosion fouling

The reaction of metal surfaces with the fluid stream in contact to produce deposits is called corrosion fouling. This type of fouling is highly dependent on the surface roughness, composition, and impurities in the fluid stream. The most common impurity is sulfur, nitrogen, and chlorine compounds such as ammonia, hydrogen sulfide, and hydrogen chloride. The salt present in crude oils in the from sodium, calcium and magnesium chlorides sometimes enters the heat exchanger due to less efficient desalter process. When heated in the form of chlorides, these salts break down to hydrogen chloride, which is very toxic for steel. The hydrochloric acid deposits on the heat transfer surface and form corrosion product such as iron chloride. The reaction is incessant with the continuous production of hydrogen chloride.

$$CaCl_2 + H_2O \rightarrow CaO + 2HCl$$

$$2HCl + Fe \rightarrow FeCl_2 + H_2$$

$$FeCl_2 + H_2S \rightarrow FeS + 2HCl$$

The sulfur present in crude oils also reacts with iron to form iron sulfide, which is toxic to the metal surface. Mostly, the organic acids such as naphthenic acids present in crude oils tend to react with iron to form iron-organic acids, which then react with hydrogen sulfide to form iron sulfide.

$$Fe + 2RCOOH \rightarrow Fe(RCOO)_2 + H_2$$

$$Fe(RCOO)_2 + H_2S \rightarrow FeS + 2RCOOH$$

The amount of naphthenic acid present in crude oils is represented by total acid number (TAN), which indicates crude oil susceptibility to corrosion. The iron sulfide is insoluble in water and crude oil which act as protection from further corrosion at low shear stress regions. Sulfur compounds like mercaptans decompose to hydrogen sulfide in the presence of hydrogen gas and become highly corrosive [16]. Temperature is also plays a major role in the corrosion process. The decomposition of mercaptans into H_2S, called sulfidation, occur in the presence of hydrogen gas at temperatures higher than 230°C. If temperature exceeds 370°C the H_2S decompose into elemental sulfur which is the most toxic state of sulfur.

3.4 Biofouling

This type of fouling can be caused by microorganisms (bacteria, algae, or fungus) or macroorganisms (barnacles, mussels, or seaweeds). These organisms colonize the metal surface by taking the nutrition from the interface between biofilm and fluid. Initially, the aerobic species consume the available oxygen and lead to anaerobic conditions, which only require nutrition to thrive [17]. This condition results in the rapid growth of biofilm. Biofouling is commonly found in sea and wastewater systems. The biofilm deposition by the microorganisms and their products served as the resistance to heat transfer in recirculation systems. The flow reduction and blockage of fluid passage are directly related to macroorganisms' growth in once-through systems. These organisms can trap the silt and other suspended particles, which reduce the thermal efficiency of heat exchangers. Due to the depletion of oxygen near biofilm, the metal surface beneath biofilm becomes anodic, which results in an enhanced corrosion rate [18]. Biofouling can also result in corrosion by supplying ion-concentrated sites. The most common industries affected by biofouling are food, marine, water, medicine, and mining. Biofouling is highly concerned in power plants where condenser operations are highly susceptible to biofouling and corrosion. Chlorine is added to the water to overcome this difficulty.

3.5 Crystallization fouling

In this type of fouling, the inorganic salts become insoluble in the fluid at supersaturated state and form deposits on the heat transfer surface. These salts can behave with normal solubility or inverse solubility. With normal solubility behavior, the solubility increases with temperature and vice versa with inverse solubility. The nuclei of the salt crystal can be formed in the bulk fluid and on the heated surface with inverse solubility at high temperatures. These nucleation sites may be corrosion products, impurities like crystals, or irregular sites on the surface. The crystals formed in the bulk fluid are deposited on the heated surface. The fouling grows by the deposition of the crystal layer over another layer. This fouling is common in desalination systems, boilers, and evaporators [19]. The crystallization fouling from inorganic salts from water is known as scaling. These salts usually are sulfates and carbonates of calcium and magnesium.

3.6 Precipitation fouling

The deposition of asphaltenes separated from crude oils is categorized as precipitation fouling. Asphaltenes are considered the major foulant in crude oil fouling. When precipitated and flocculated from the crude oil, the asphaltenes molecules form large particles. These particles are aggregated and deposited on the surface of the heat exchanger. The chemical formula for asphaltenes is not specific; the number of atoms a molecule contains varies depending upon the source. Also, the molecular weight of asphaltenes differs in a wide range. The average molecular weight of asphaltene is 750 amu [20]. The structure of the asphaltene molecule contains polycyclic aromatic rings with alkyl side chains. The presence of heteroatoms, e.g., oxygen, nitrogen,

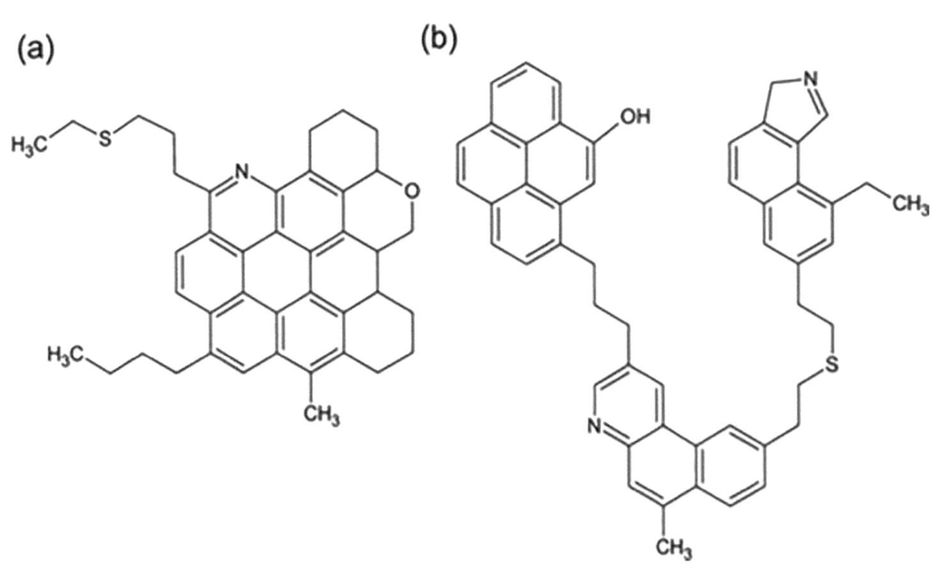

Figure 3.
Structure of asphaltenes molecule: Yen–Mullins island model (a) and archipelago model (b) [23].

sulfur, nickel, and vanadium, renders its polarity. There are two suggested structure types of asphaltene molecule, i.e., island and archipelago. The island structure is comprised of a monomer containing six or more rings, while the archipelago is a polymeric structure with polycondensed groups of rings linked by aliphatic side chains [21]. A rough estimate suggested about 7–8 fused rings in a single polycyclic aromatic hydrocarbon (PAH) [22] which favors the proposed island structure.

The concentration of peptizers such as aromatics and resins in the crude oil should be enough to hold the asphaltenes soluble in the crude oil (**Figure 3**). The saturates present in the crude oils are considered as flocculants for the asphaltenes [24]. The ability of crude oils to maintain asphaltenes soluble is called stability. Sometimes the combination of different crude oils induces the precipitation of asphaltenes which cause deposition in the heat exchangers. The Colloidal Instability Index (CII) is considered as a good indication of the fouling propensity by asphaltenes precipitation.

4. Fouling mechanism

Fouling is a complex phenomenon that involved multiple processes. These processes occur in a sequential manner, and sometimes one process dominates over another. The general sequence can be shown in the steps below:

Precursors → Foulant → Hard Deposit

Initially, the fouling precursors enter the heat exchanger or formed within the heat exchanger. These fouling precursors are then reacted to form insoluble deposits on the heat transfer surface. After the deposition, the foulant is transformed into a hard material by aging, which is difficult to be removed. It is still debatable whether the formation of precursor and foulant is in the bulk fluid or on the heat transfer surface. However, it is evident from the experimental work [25, 26] that the solubility of precursors and temperature have a high impact on fouling and both of these factors can affect fouling in the bulk fluid as well as on the heat transfer surface.

Epstein [27] categorized the fouling in the five stages known as the 5×5 matrix, as shown in **Figure 4**. This matrix relates the fouling stages with the type involved in the process.

The details of these main five events, i.e., initiation, transport, attachment, removal, and aging, are described in the subsequent paragraphs.

4.1 Initiation or formation of foulant material

The fouling phenomenon starts with a delay or induction period with no increase in the resistance to heat transfer. On the contrary, the fouling resistance sometimes becomes negative due to an increase in surface roughens, which enhances heat transfer by turbulence [28]. The surface roughness is caused by the initial deposition of particles on the clean surface present initially in the bulk fluid or formed by polymerization or corrosion reaction.

Only a few studies were done on predicting the length induction period. Yang et al. [29] developed a model for fractional surface coverage (θ) of the clean surface. The model relates the surface coverage with time using a sigmoidal function as:

$$\theta = \frac{k_1 - k_2}{k_1} \frac{1}{1 + we^{-(k_1 - k_2)t}} \tag{11}$$

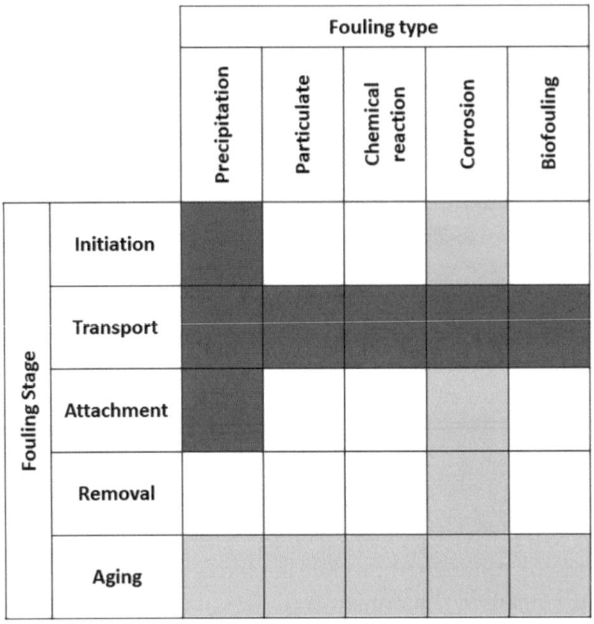

Figure 4.
The current standing of the 5 × 5 fouling matrix by Epstein [27]. The color change from lighter to darker indicates the increase in the degree of understanding.

To maintain a long induction period, the value of deposition rate constant (k_1) should be equal to removal rate constant (k_2). They defined the induction period length as the time to reach 50% of the maximum surface coverage. A good agreement of model found with experimental data of deposition from crude oil, protein, and salt. They found very short induction periods with initially not cleaned surfaces. The length of the induction period depends on various factors such as fluid composition, surface roughness, temperature, and velocity. However, the temperature is the most significant parameter that affects the induction period length because the other factors, such as solubility of precursors and reaction rate, are also temperature-dependent. High temperature and low velocity significantly reduce the length of the induction period.

4.2 Transport to the heat transfer surface

The transfer of fouling precursors from the body of the fluid to the heat transfer surface occurs in many ways, e.g., diffusion, sedimentation, thermophoresis, or impaction. The migration of the precursor from bulk fluid to the surface boundary layer due to the concentration difference is called diffusion [30]. The deposition rate by diffusion can be described as

$$\dot{m}_D = k_m(c_b - c_s) \tag{12}$$

where k_m is the deposition rate constant. The mass transfer coefficient, k_m, can be estimated using the correlations of Sherwood number.

The sedimentation process by the action of gravitational force is common for large deposit particles such as sand, clay, or rust material [31]. The drag force on the particle should be lower than the gravitational force to settle it down on the surface. Thermophoresis, also known as thermodiffusion or Soret effect, is caused by the diffusion due to temperature gradient [32]. The particles are transferred from the hot zone to the cold zone by the thermophoretic force. This phenomenon is common in gas phase deposition, where particles are derived towards the colder regions of the heat exchanger. The deposition of particles by impaction is also significant as some particles achieve high velocity with respect to the fluid and reach the hot surface by inertial force.

4.3 Attachment or formation of the deposit at the heat transfer surface

The attachment of particles transported to the surface is highly dependent on the condition of the surface. The rough surface contains the potential to bind with the particles due to the availability of nucleation sites. The protective coatings on the heat exchanger surface provide a low-energy surface, reducing its wettability and preventing fouling. The sticking probability of the particles, defined as the residence time of the particles on the heat transfer surface, is also a function of velocity and surface temperature. The experimental studies suggest increased deposition with low velocity and high surface temperature [33]. Considerable work has been performed to model foulant deposition on heat transfer surfaces [34]. Many empirical or semi-empirical models were proposed to describe the deposition mechanism. Kern and Seaton [35] were the first to propose the deposition model as a competing phenomenon between deposition and removal as follows:

$$dx/dt = \alpha c_b J - \beta \tau x \tag{13}$$

The first and second terms on the left-hand side represent the rate of deposition removal. They consider the removal rate to be the function of the thickness of the fouling layer. Paterson and Fryer [36] proposed a model for fouling by skimmed milk. They assumed the laminar sub-layer as a differential chemical reactor in which the foulant is produced, and the deposition rate is the product of the foulant production rate and sticking probability. The proposed model is given as

$$\frac{dR_f}{dt} = \alpha \exp\left(-E/RT\right)/u_o \tag{14}$$

Polley et al. [37] suggested a fouling prediction model which relates the sticking probability (S) with shear stress as:

$$\frac{dR_f}{dt} = \frac{A_o}{\alpha} e^{-E/RT_f} S \tag{15}$$

The sticking probability was defined as the function of shear stress

$$S = 1 - \left(\frac{\tau_w - 2}{98}\right)^{0.5} \tag{16}$$

Other forces can influence particle residence time on the heat transfer surface, such as London–van der Waals force, electric double-layer structure, and viscous hydrodynamic force [38].

4.4 Removal of material from the heat transfer surface

It is possible that the part of the deposited material loses its bonding with the surface or the deposition layer. The net deposition of the foulant is the result of the difference between the deposition rate and the removal rate. These two mechanisms occur simultaneously during the fouling process. Several factors affect the removal process, such as erosion, diffusion, or dissolution [39]. The shear forces generated by the interaction of moving fluid and the static foulant layer cause the breakup of the foulant particles from the deposited layer. These shear forces can also be generated from the turbulent sweeps and bursts. As the deposited layer becomes thicker, the cross-sectional area reduces, resulting in increased velocity and shear stress at the fluid-layer interface, referred to as the auto-retardation process [40]. This phenomenon leads to the removal of foulant particles by spalling or erosion. There is also a possibility that the deposit is removed by diffusion back to the bulk fluid. It is only possible if the foulant is a reaction product transformed into a soluble product or by the concentration gradient between the foulant layer and the bulk fluid. Kern and Seaton [35] related the deposition rate with the thickness of the foulant layer as:

$$\dot{\mathrm{m}}_R = \mathrm{k}_r \tau x \tag{17}$$

Polley et al. [41] modified the model by suggesting the controlling mechanism of removal as mass transfer dependent rather than shear stress. They replaced the shear stress parameter with $\mathrm{Re}^{0.8}$ as:

$$\dot{\mathrm{m}}_R = \mathrm{k}_r \, Re^{\,0.8} \tag{18}$$

The removal term is also referred to as entrainment because initially, there is no fouling layer on the clean surface to be removed [42].

4.5 Aging

After the foulant deposition, physical and chemical transformation occurs, which results in its mechanical and thermal properties known as the aging process. As time passes, the foulant becomes hard and thermally more conductive by the action of heat. In a few cases, such as in biofouling, the deposit becomes weaker in strength due to the corrosive decay of microorganisms. The organic deposit becomes harder in crude oil fouling as the coking process starts after the deposition. The thermal conductivity of the deposit also increases with aging, which also enhances the heat transfer resulting in a lower fouling rate. The change in thermal conductivity due to the conversion of organic deposit into a coke-like material was reported to be from 0.2 to 1 W/m K [43]. This mechanism is crucial in estimating the accurate thickness of the foulant layer. Therefore, aging plays a key role in describing the thermohydraulic behavior of heat exchangers. A long time is required for aging to occur, rendering its measurement impossible in laboratory experiments. The aging phenomenon was incorporated in the fouling prediction models by Ishiyama et al. [44]. They correlated the foulant thermal conductivity with a youth factor, y, which gradually decreases from 1 to 0 as aging takes place. The deposit thermal conductivity was represented as:

$$\lambda_f = \lambda_f^{\infty} + \left(\lambda_f^{o} - \lambda_f^{\infty}\right) y \tag{19}$$

The youth factor was modeled using a first-order kinetics as:

$$\frac{dy}{dt} = ky \tag{20}$$

where k is the rate constant for aging, which depends on temperature by Arrhenius law.

They assumed that the overall fouling resistance is the sum of resistance by individual sublayers with different aging histories. The thickness of each sublayer was assumed constant; however, the thermal conductivities were varied with time. This modeling scheme enabled to produce a close to the real behavior of the fouling process.

5. Conclusion

Fouling is still a long-standing problem in heat transfer research. It is related to the economic and environmental penalties in heat transfer operations. The complexity of the fouling mechanism entails difficulty in its understanding. However, recent development in fouling prediction models has enabled researchers to estimate the fouling rate accurately. A successful design of a heat exchanger requires a deep understanding of the fouling mechanism. This understanding is promising in improving the accuracy of mitigation actions to optimize the operation of the heat exchanger. Similarly, it also assures cost-effective cleaning and production planning.

Nomenclature

A_o	Arrhenius coefficient
A	Heat transfer area, m^2
c_b	Concentration of foulants in bulk fluid, mol/m^3
c_s	Concentration of foulants on surface, mol/m^3
d	diameter of tube, m
E	Activation energy, J/mol
f	fanning friction factor
J	mass flux, kg/s
k_1, k_2	lumped rate constants, s^{-1}
L	length of tube, m
$\dot{m}_D$	deposition flux, kg/m^2 s
$\dot{m}_R$	removal flux, kg/m^2 s
P	pressure, atm
Q	rate of heat transfer, J/s
q	heat flux, J/m^2s
q_o	initial heat flux, J/m^2s
q_t	heat flux after time t, J/m^2s
R	Gas constant, J/mol K
R_f	thermal resistance, m^2K/W
Re	Reynolds number
S	sticking probability

t	time, s
T	temperature, K
T_f	film temperature, K
U	overall heat transfer coefficient, W/m^2K
U_o	clean overall heat transfer coefficient, W/m^2K
u_o	initial flow velocity, m/s
u_f	flow velocity in fouled tube, m/s
w	wettability parameter
x	deposit thickness, m
y	youth factor
Greek	
$\alpha, \beta, \gamma, \sigma$	model fitting parameters.
λ_f	foulant thermal conductivity, W/m K
λ_f^o	fresh foulant thermal conductivity, W/m K
λ_f^∞	aged foulant thermal conductivity, W/m K
ρ	density, kg/m^3
τ	shear stress, N/m^2
τ_w	wall shear stress, N/m^2
μ	dynamic viscosity, Pa s
θ	fractional surface coverage

Author details

Obaid Ur Rehman[1*], Nor Erniza Mohammad Rozali[1] and Marappa Gounder Ramasamy[2]

1 Universiti Teknologi Petronas, Seri Iskandar, Malaysia

2 KPR Institute of Engineering and Technology, Coimbatore, India

*Address all correspondence to: obaid_19000977@utp.edu.my

References

[1] Watkinson AP. Deposition from crude oils in heat exchangers. Heat Transfer Engineering. 2007;**28**(3): 177-184

[2] Van Nostrand JWL, Leach SH, Haluska JL. Economic penalties associated with the fouling of refinery heat transfer equipment. In: Somerscales EFC, Knudsen JG, editor. Fouling of Heat Transfer Equipment. New York: Hemisphere Pub., 1981. pp. 619-643

[3] DOE, Energy Bandwidth for Petroleum Refining Processes. 2006

[4] Polley GT, Tamakloe E, Picon Nunez M, Ishiyama EM, Wilson DI. Applying thermo-hydraulic simulation and heat exchanger analysis to the retrofit of heat recovery systems. Applied Thermal Engineering. 2013;**51** (1–2):137-143

[5] Müller-Steinhagen H, Malayeri MR, Watkinson AP. Heat exchanger fouling: Mitigation and cleaning strategies. Heat Transfer Engineering. 2011;**32**(3–4): 189-196

[6] Ishiyama EM, Paterson WR, Ian Wilson D. Exploration of alternative models for the aging of fouling deposits. AICHE Journal. 2011;**57**(11): 3199-3209

[7] Wang Y, Yuan Z, Liang Y, Xie Y, Chen X, Li X. A review of experimental measurement and prediction models of crude oil fouling rate in crude refinery preheat trains. Asia-Pacific Journal of Chemical Engineering. 2015;**10**(4):607-625

[8] Diaz-Bejarano E, Coletti F, Macchietto S. Thermo-hydraulic analysis of refinery heat exchangers undergoing fouling. AICHE Journal. 2017;**63**(3): 984-1001

[9] Yeap BL, Wilson DI, Polley GT, Pugh SJ. Mitigation of crude oil refinery heat exchanger fouling through retrofits based on thermo-hydraulic fouling models. Chemical Engineering Research and Design. 2004;**82**(1):53-71

[10] Ishiyama EA, Coletti F, Macchietto S, Paterson WR, Wilson DI. Impact of deposit ageing on thermal fouling: Lumped parameter model. AICHE Journal. 2009;**56**(2):531-545

[11] Awad MM. Fouling of heat transfer surfaces. In: Heat Transfer. Rijeka: IntechOpen; 2011. pp. 505-542

[12] Ritter RB. Crystalline fouling studies. Journal of Heat Transfer. 1983;**105**(2): 374-378

[13] Watkinson AP, Martinez O. Scaling of heat exchanger tubes by calcium carbonate. Journal of Heat Transfer. 1975;**97**(4):504-508

[14] Ebert W, Panchal CB. Analysis of Exxon crude-oil-slip stream coking data. In: Fouling Mitigation of Industrial Exchange Equipment. New York: Begell House; 1995. pp. 451-460

[15] E. Diaz-Bejarano, F. Coletti, and S. Macchietto, "Impact of complex layering structures of organic and inorganic foulants on the thermohydraulic performance of a single heat exchanger tube: A simulation study," Industrial and Engineering Chemistry Research, vol. 55, no. 40, pp. 10718–10734, Oct. 2016.

[16] Wang W, Watkinson AP. Iron sulphide and coke fouling from sour oils: Review and initial experiments. In: International Conference on Heat Exchanger Fouling and Cleaning IX. Crete Island: HTRI; 2011. pp. 23-30

[17] Rao TS, Rani PG, Venugopalan VP, Nair KVK. Biofilm formation in a freshwater environment under photic and aphotic conditions. Biofouling. Dec 1997;**11**(4):265-282

[18] Rao TS. Chapter 6 - Biofouling in industrial water systems. Z. Amjad and K. D. B. T.-M. S. and D. Demadis, Eds. Amsterdam: Elsevier. 2015, pp. 123–140

[19] Klepetsanis PG, Dalas E, Koutsoukos PG. Role of temperature in the spontaneous precipitation of calcium sulfate dihydrate. Langmuir. 1999;**15**(4): 1534-1540

[20] Rashid Z, Wilfred CD, Gnanasundaram N, Arunagiri A, Murugesan T. A comprehensive review on the recent advances on the petroleum asphaltene aggregation. Journal of Petroleum Science and Engineering. 2019;**176**:249-268

[21] Silva FB, Guimarães MJOC, Seidl PR, Garcia MEF. Extraction and characterization (compositional and thermal) of asphaltenes from brazilian vacuum residues. Brazilian Journal of Petroleum and Gas. 2013;**7**(3): 107-118

[22] Andrews AB et al. Molecular orientation of asphaltenes and PAH model compounds in langmuir–blodgett films using sum frequency generation spectroscopy. Langmuir. 2011;**27**(10): 6049-6058

[23] Majumdar RD. A nuclear Magnetic Resonance Spectroscopic Investigation of the Molecular Structure and Aggregation Behavior of Asphaltenes. Canada: University of Lethbridge; 2016

[24] Hong E, Watkinson AP. Precipitation and fouling in heavy oil-diluent blends. Heat Transfer Engineering. 2009;**30**(10–11):786-793

[25] Rogel E, Hench K, Miao T, Lee E, Dickakian G. Evaluation of the compatibility of crude oil blends and its impact on fouling. Energy & Fuels. 2018; **32**(9):9233-9242

[26] Saleh ZS, Sheikholeslami R, Watkinson AP. Fouling characteristics of a light Australian crude oil. Heat Transfer Engineering. 2005;**26**(1):15-22

[27] Epstein N. Thinking about heat transfer fouling: A 5 × 5 matrix. Heat Transfer Engineering. 1983;**4**(1):43-56

[28] Kazi SN, Duffy GG, Chen XD. Fouling and fouling mitigation on heated metal surfaces. Desalination. 2012;**288**:126-134

[29] Yang M, Young A, Niyetkaliyev A, Crittenden B. Modelling fouling induction periods. International Journal of Thermal Sciences. 2012;**51**:175-183

[30] Epstein N. A Model of the Initial Chemical Reaction Fouling Rate for Flow within a Heated Tube and its Verfication. In: 10th International Heat Transfer Conference. Brighton: Begell House; 1994. pp. 225-229

[31] Bennett CA. A theory describing sedimentation particulate fouling thresholds inside heat exchanger tubes. Heat Transfer Engineering. 2016;**37**(5): 468-474

[32] Gounder RM, Emani S. CFD simulations of crude oil fouling on heat transfer surfaces. In: Ionescu SEE-A, editor. Computational Fluid Dynamics. Rijeka: IntechOpen; 2017. pp. 357-375

[33] Deshannavar UB, Marappagounder R. Revisiting threshold fouling models for crude oil fouling. Heat Transfer Engineering. 2021;**42**(17):1489-1505

[34] Ogbonnaya S, Ajayi O. Fouling phenomenon and its effect on heat

exchanger: A review. Frontiers in Heat and Mass Transfer. 2017;**9**

[35] Kern DQ, Seaton RE. A theoretical analysis of thermal surface fouling. British Chemical Engineering. 1959;**4**: 258-262

[36] Paterson WR, Fryer PJ. A reaction engineering approach to the analysis of fouling. Chemical Engineering Science. 1988;**43**(7):1714-1717

[37] Polley G, Tamakloe E, Wilson I, Macchietto S, Coletti F. Development of a model for the prediction of fouling in heat exchangers processing crude oil. In: 2010 Spring Meeting & 6th Global Congress on Process Safety. San Antonio: AIChE; 2010

[38] Emani S, Ramasamy M, Shaari KZK. Discrete phase-CFD simulations of asphaltenes particles deposition from crude oil in shell and tube heat exchangers. Applied Thermal Engineering. 2019;**149**:105-118

[39] Coletti F, Crittenden BD, Macchietto S. Basic Science of the Fouling Process. In: Coletti F, Hewitt GF, editor. Crude Oil Fouling. Waltham: Gulf Professional Pub; 2015. pp. 23-50

[40] Berce J, Zupančič M, Može M, Golobič I. A review of crystallization fouling in heat exchangers. PRO. 2021;**9** (8)

[41] Polley GT, Wilson DI, Yeap BL, Pugh SJ. Evaluation of laboratory crude oil threshold fouling data for application to refinery pre-heat trains. Applied Thermal Engineering. 2002;**22**(7): 777-788

[42] Wilson DI, Ishiyama EM, Polley GT. Twenty years of ebert and panchal— What next? Heat Transfer Engineering. 2017;**38**(7–8):669-680

[43] Diaz-Bejarano E, Coletti F, Macchietto S. Modeling and prediction of shell-side fouling in shell-and-tube heat exchangers. Heat Transfer Engineering. 2019;**40**(11):845-861

[44] Wilson DI, Ishiyama EM, Paterson WR, Watkinson AP. Ageing: Looking backward and looking forward. In: International Conference on Heat Exchanger Fouling and Cleaning VIII 2009. 2009. pp. 221-230

Chapter 14

Fouling in Industrial Heat Exchangers: Formation, Detection and Mitigation

Rania Jradi, Christophe Marvillet and Mohamed Razak Jeday

Abstract

Heating or cooling of a fluid by another is made in a heat exchanger with heat dissipation from surfaces of the equipment. Over time, the abundant quantity of impurities promotes fouling in a heat exchanger. This equipment has extensive domestic and industrial applications. The concepts of design, operation, and maintenance of heat exchanger are available in the form of complete technical literature, but this literature is extensively distributed throughout the industrial bulletins, industrial design codes and standard, technical journals, etc. Thus, the aim of this book chapter is to reveal the concepts design, operation, cleaning, and maintenance of heat exchanger closely related to the industrial practices.

Keywords: heat exchanger, fouling, industrial applications

1. Introduction

Heat exchanger has an important role in an industrial application whose aim is to heat and cool of large-scale industrial process fluids [1]. Due to their dynamic design, heat exchangers can be personalized to adapt to any industrial process depending on many thermodynamic properties, such as the temperature, pressure, type of fluid, phase flow, density, chemical composition, and viscosity [2, 3]. Efficient heat recovery or heat dissipation becomes a vital challenge for scientists and engineers because of the global energy crisis [2]. The optimization of the surface area of the wall between two fluids in order to maximize the efficiency while minimizing resistance to fluid flow across the exchangers within the limits of material cost is the main reason for the design of heat exchangers. The enhancement of heat exchanging surfaces performance could be done by the addition of corrugations or fins in the heat exchanger, which increase surface area and may channel fluid flow or induce turbulence [4]. Online monitoring of the industrial heat exchanger's efficiency is made by following up the overall heat transfer coefficient based on its temperatures, which tend to decline over time due to fouling [1].

Scale formation can cause potential damage to equipment. This deposit can be very costly if processed water is not treated correctly. In the industry, chemical products are commonly used to treat the water. It is estimated that a total of 7.3 billion dollar worth of chemical products per year in the U.S. is rejected into the air, dumped in streams, and buried in landfills every year. Industry purchased 40% of these chemical products, which represents more than 2 billion dollar of toxic waste which contribute to trillion of gallon contaminated water disposed annually into the earth, for control of scale in the cooling tower, boiler, and other heat transfer equipment. Several methods can be used to clean fouled tubular heat exchangers such as acid cleaning, sandblasting, high-pressure water jet, and bullet cleaning or drill rods [5]. Water treatment as purification, catalytic approach, addition of chemicals, etc., are used to minimize fouling of the heat exchanging equipment in large scale cooling water systems for heat exchangers [2] and other processes are also used in steam systems for power plants to minimize fouling of the heat exchanger and other equipment. For fouling mitigation, most of the chemicals and additives used are dangerous to the environment [6]. For that reason, it is necessary to apply chemical products that have a mild approach to the environment [1, 6].

2. Heat exchanger in industry

Heat exchanger in the industry is heat transfer equipment that uses a thermal energy exchange process between two or more medium available at a different temperature [7]. This equipment is applied in various industrial processes [8], for example in power plant generation, in petroleum oil and gas industry, chemical processing plants, transportation, alternate fuels, cryogenic, air conditioning, refrigeration, heat recovery [9], and in other industries and this device is always closely related to our daily life, such as evaporators, air preheaters, automobile radiators, condensers, and oil coolers. In general, a heat transfer surface separates the fluid for most heat exchangers. This surface incorporates a wide range of different flow configurations to achieve the desired performance in different applications [9, 10]. Heat exchangers can be classified in many different ways. For industrial heat exchangers, the classification is generally based on the construction, transfer processes, degrees of surface compactness, flow arrangements; pass arrangements, phase of the process fluids, and heat transfer mechanisms as is shown in **Figure 1** [9].

3. Design concepts for heat exchanger

Normal process requirements specified through service conditions for combinations of un-corroded and corroded conditions and the clean and fouled conditions must be taken into consideration in the design concepts of the heat exchanger. The most important of the critical criteria is the design of a heat exchanger for the facility of maintenance, which means parts cleaning or replacing (tubes, fittings, etc.) damaged by aging, vibration, corrosion, or erosion throughout the service period. Therefore, the design of the heat exchanger should be as simple as possible especially if significant fouling is expected. By minimizing temperature in conjunction with the choice of fluid velocity and by reducing the concentration of foulant precursors, this incidence of potential fouling can be reduced. Furthermore, the highest flowing velocity should be allowed under the constraints of pressure drop and erosion from the flow. In addition, material selection within a limited cost delays the accumulation of deposits and allows shorter residence time, which should also be compatible in terms of pH, corrosion, and not only just heat exchanger, but also in terms of heat equipment and transfer lines of the heat exchanger.

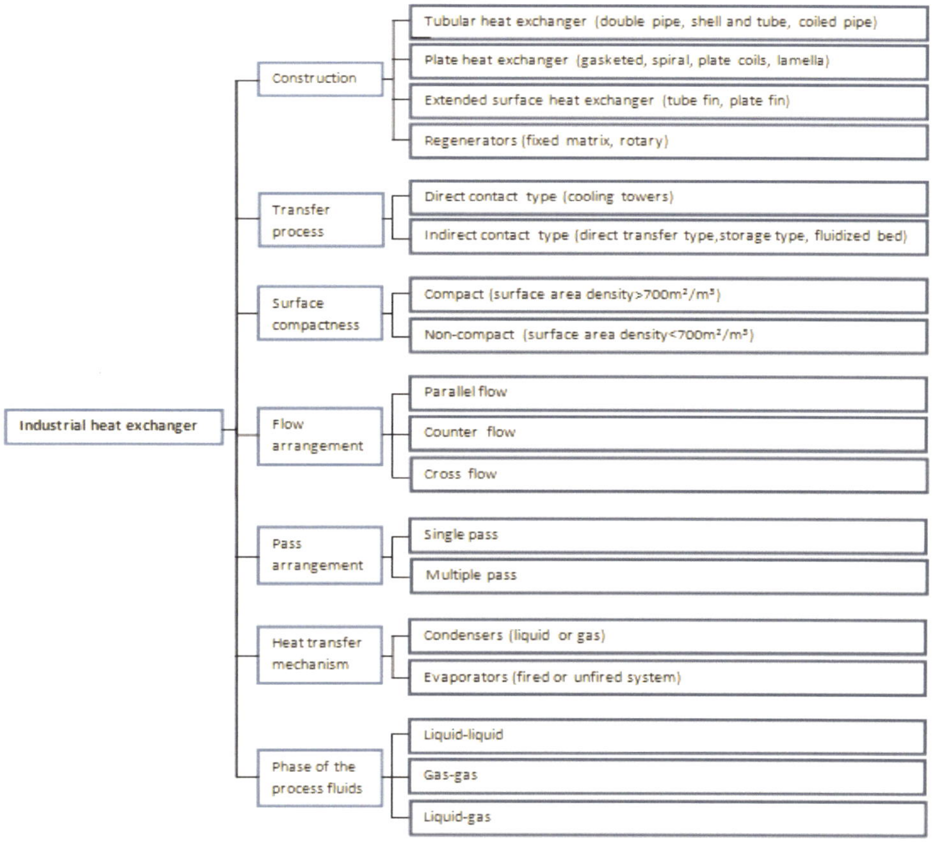

Figure 1.
Classification of industrial heat exchanger [9].

4. Fouling

4.1 Definition

The deposition and the accumulation of undesirable materials or substances on heat exchanger surfaces formed the fouling of thermal equipment [11]. This deposit, which forms on one or both sides of the heat exchanger surface has a lower thermal conductivity than that of the metal constituting the exchange surface, which creates a significant increase in overall resistance to heat transfer and therefore, a decrease in the performance of the heat exchanger. It also impacts the cross-section of the fluids, which causes an increase in the pressure drop.

It should be noted that the larger size of the heat exchanger, the greater the duration of the production shutdown.

By way of illustration, **Figure 2** [12] shows a bundle of heat exchanger tubes in a dirty (**Figure 2a**) and clean state (**Figure 2b**).

Other than its repercussions on energy performance, fouling in heat exchangers poses problems in terms of their exploitation and inevitably induces notable additional costs on the investment during the design of the devices but especially during their exploitation. These additional costs are mainly due to the growth of energy

Figure 2.
Fouled and clean heat exchanger tubes bundles [12].

consumption, production losses, and maintenance and cleaning costs associated with the capping and clogging of pipes [13].

4.2 Different types of fouling

It is possible to classify fouling according to the mechanism which controls the deposition rate, according to the conditions of use of the exchanger, or according to the dominant mechanism, even if it does not control the deposition velocity. In our work, we have adopted, like most authors [14], the classification which distinguishes four different types of fouling:

- Fouling
- Corrosion
- Scaling
- Mixed types

4.2.1 Fouling

Fouling appears following to the accumulation of finely divided solids suspended in the treatment fluid on the heat transfer surface. Three different forms of fouling can be distinguished [14]:

- Particulate fouling;
- Chemical reaction fouling;
- Biological fouling.

4.2.1.1 Particulate fouling

This type of fouling arises when the solid particles suspended in the process stream accumulate onto the heat transfer surface [13, 15]. The process is seen as "sedimentation" fouling if the settling occurs due to gravity as well as other deposition mechanisms. This also involves deposition of corrosion products dispersed in fluids as clay and mineral particles in river water, suspended solids in cooling water, soot particles of incomplete combustion, magnetic particles in economizers, deposition of salts in desalination systems, deposition of dust particles in air coolers, particulates partially present in fire-side (gas-side) fouling of boilers, etc.

The concentration of suspended particles, fluid flow velocity, temperature conditions on the fouled surface (heated or non-heated), and heat flux at the heat transfer surface are some factors that may influence the particulate fouling. **Figure 3** [16] shows a photograph of the particulate fouling.

4.2.1.2 Chemical reaction fouling

Result of the formation of deposits due to one or more chemical reactions between reactants contained in the flowing fluid in which the surface material itself is not a reagent or participant [13, 15]. In this case, the heat transfer surface may act as a catalyst as in cracking, coking, polymerization, and autoxidation. Thermal instabilities of chemical species can also induce fouling precursors, as asphaltenes and proteins. The occurrence of his types of fouling is for over a wide temperature range from ambient to over 1000°C but is more pronounced at higher temperatures. An unwanted chemical reaction occurring during the heat transfer process is the consequence of the mechanism of this fouling. Several applications in process industry promoted the formation of chemical reaction fouling, such as petrochemical industries, oil refining, vapor-phase pyrolysis, cooling of gas and oils, polymerization of process monomers, etc. In addition, fouling of heat transfer surface by biological fluids may involve complex heterogeneous chemical reactions and physicochemical processes. In the case of the formation of the protective oxide layer is inhibited, the deposits from chemical

Figure 3.
Accumulation of solid particles in the plate of a tubular heat exchanger [16].

Figure 4.
Residue of hydrocarbon in the tubes of a heat exchanger [16].

reaction fouling may promote corrosion at the surface. This type of fouling is often extremely tenacious that requires a special measure to clean off the deposits from heat exchanger surfaces to provide them with satisfactory operation life [6]. **Figure 4** [16] shows a photograph of the chemical reaction fouling.

4.2.1.3 Biological fouling

Biological fouling is due by the attachment and growth of macroorganisms and/or microorganisms and their products on the heat transfer surface, called usually "Bio-fouling", and it is generally a problem in water streams [13, 15]. Biological fouling generally can be divided into two main subtypes of fouling: microbial and macrobial. Microbial fouling can be defined as the accumulation of microorganisms such as algae, fungi, yeasts, bacteria, and molds, while macrobial fouling is the accumulation of macroorganisms, for example, clams, barnacles, mussels, and vegetation as found in seawater or estuarine cooling water. Generally, microbial fouling precedes macrobial deposition so it may be considered of primary interest. Biological fouling generally has the shape of a biofilm or a slime layer on the surface that is uneven, filamentous, and deformable but difficult to remove. Biological fouling is generally associated with open recirculation or once-through systems with cooling water despite it can occur in suitable liquid streams. Biological fouling may promote corrosion fouling under the slime layer. One of the common problems [16] in heat exchanger operation is the growth of attached organisms. Several processes may suffer biofouling such as food processing industries, power plant condensers using seawater, etc. A photograph illustrating biological fouling is given in **Figure 5** [16].

4.2.2 Corrosion

Corrosion involves a chemical or electrochemical reaction between the heat transfer surface itself and the fluid stream to produce corrosion products which, in turn, change the surface's thermal characteristics and foul it. This type of fouling is both a reactant and consumed. The reaction between the surface and the fluid allows creating

Figure 5.
Presence of biofouling in tubular bundle of a heat exchanger [16].

Figure 6.
Corroded tube of a heat exchanger [16].

a corroded surface [13, 15]. Corrosion fouling may be caused in two ways. Firstly, the accumulation and adhesion of corrosion products perform on the surface providing resistance to heat transfer, such as, the presence of sulfur in fuel can cause corrosion in gas and oil-fired boilers. Secondly, corrosion products may be transported from the corrosion site as particulate materials and be deposited on the heat transfer surface in another site of the system as particulate fouling, such as corrosion products originating in the condenser or feedtrain may cause fouling on the waterside of boilers. Corrosion is widespread in many applications where chemical reaction fouling takes place and the protective oxide layer is not formed on the surface that is of significant importance in the design of the boiler and condenser of fossil fuel-fired power plant. **Figure 6** [16] shows a corroded tube bundle.

4.2.3 Scaling

Scaling is defined as the formation of hard incrustations, adhering to heat exchange surfaces and thermally insulating [13, 15]. These incrustations are known as "scale" associated mainly with the production of crystalline solid from liquid solution. The type of scaling depends on the nature of the heated solution, the heating process used, and its operating conditions. This phenomenon reduces the efficiency of heat transfer and increases the effort deployed to clean the scaled surfaces. Scaling may occur in heat exchangers, in water-cooled exchangers, in

Figure 7.
Precipitated salts in the tubes of a heat exchanger [16].

seawater or brackish water desalination units, in boilers, etc. Crystal growth during precipitation, which requires the formation of a primary nucleus is the most important phenomenon involved in this type of fouling. Nucleation is the mechanism that controls the process, as a rule, heterogeneous in the presence of impurities and on the heat transfer surface. An example of scaling cases on an exchange surface is presented in **Figure** 7 [16].

4.2.4 Mixed types

Various mechanisms of fouling have been briefly described, in the practical case; it is rare that heat exchanger fouling is the result of a single mechanism. Two or probably more mechanisms are involved in most process streams where fouling occurs [15]. From a practical standpoint, one mechanism may be dominant, and the other mechanisms present can be ignored when remedial action is being considered. For instance, the circulating water in cooling water systems may contain dissolved solids, suspended particulate matter, and, perhaps, also aggressive chemicals in addition to microorganisms. Microorganisms, particles, scale, and products of corrosion maybe contained in the accumulated deposit on the equipment surfaces. The fouling on heat exchangers may be due to particle deposition, chemical reactions, and corrosion in fouling associated with combustion. From these two examples, it will be clear that the process of fouling may be extremely complex necessitating, a rather empirical approach to its understanding and investigation.

5. Fouling process

Generally, two simultaneous sub-processes formed the overall fouling process which is a deposition process and a removal process as illustrated in **Figure 8** [2]. It is important to note that, some of these sub-processes are not applicable in some fouling instances such as corrosion.

Three basic steps may be visualized concerning the deposition on surfaces from a moving fluid, which are:

1. The diffusional transport of the foulant or its precursors across the boundary layers adjacent to the solid surface within the flowing fluid.

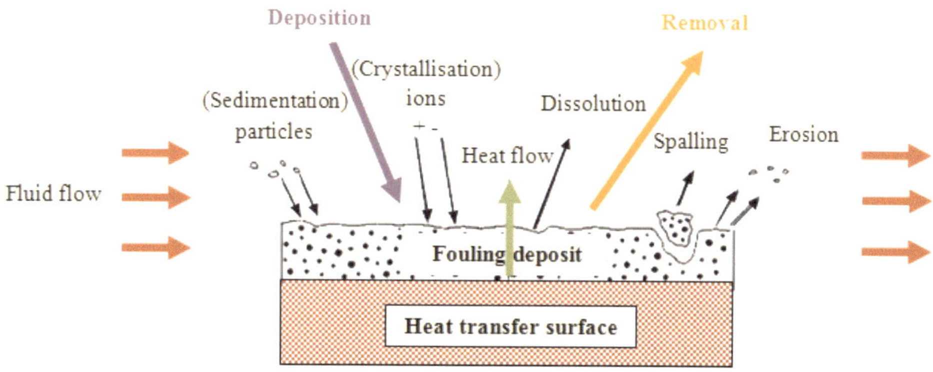

Figure 8.
Fouling process [2].

2. The adhesion of the deposit to the surface and to itself.

3. The transport of material away from the surface.

The sums of these three basic components correspond to the growth of the deposit on the surface. Mathematically, the rate of deposit growth (fouling resistance or fouling factor, *Rf*) is regarded as the difference between the deposition and removal rates [11]:

$$Rf = \dot{m}_d - \dot{m}_r \tag{1}$$

Where ($\dot{m}_d$) represents the deposition rate and ($\dot{m}_r$) represents the removal rate that can be expressed in the units of thermal resistance as m^2 K/W or in the units of the rate of thickness change as m/s or in the units of mass change as kg/m^2 s.

Five main stages are made successively to form the fouling and can be summarized as initiation of fouling, transport to the surface, attachment to the surface, removal from the surface, and aging at the surface [17], which can be summarized briefly as follows:

1. Initiation of fouling or formation of foulant materials in the bulk of the fluid is the first stage in the fouling process. This event is preceded by an induction period or a delay period, (t_d) as shown in **Figure 9** [11]. During this period, the basic mechanism involved is heterogeneous nucleation, and (t_d) is shorter with a higher nucleation rate. Several factors can affect (t_d), such as temperature, fluid velocity, composition of the fouling stream, and nature and condition of the heat exchanger surface. In the case of low-energy surfaces (unwettable), the induction periods are longer in comparison with those of high-energy surfaces (wettable). t_d appears to be decreased in crystallization fouling by the increasing degree of supersaturation, however, it tends to be decreased in chemical reaction fouling by increasing surface temperature. Due to available suitable sites for nucleation, adsorption, and adhesion, (t_d) decreases when the surface roughness increases in all fouling mechanisms.

2. Transport of species to the surface is the second stage in the fouling process. This event means the transfer of the fouling species itself from the bulk of the fluid to

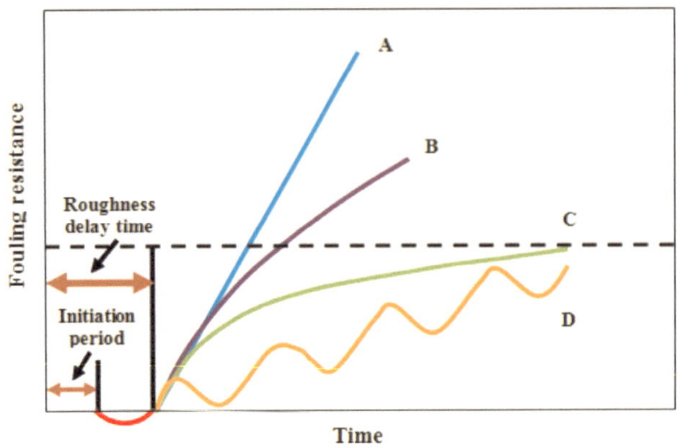

Figure 9.
Fouling curves [11].

the heat transfer surface. It is the best understood of all sequential events. Normally, this is done through the action of one or more of the following mechanisms:

- Diffusion: it involves the mass transfer of the fouling constituents from the circulating fluid toward the heat transfer surface by reason of the concentration difference between the bulk of the fluid and the fluid adjacent to the surface.

- Electrophoresis: fouling particles carrying an electric charge may move closer together or further from a charged surface according to the polarity of the surface and the particles due to the action of electric forces. Due to electrophoresis, deposition increases by decreasing the electrical conductivity of the fluid, by increasing fluid temperature, and by increasing fluid velocity and it also depends on the pH of the solution. London-van der Waals and electric double layer interaction forces are some of the surface forces, which are usually responsible for electrophoretic effects.

- Thermophoresis: it is about moving fine particles from a hot zone to a cold zone thanks to a "thermal force" in the direction of the negative temperature gradient. Near a hot wall, a high-temperature gradient will prevent particles from settling, while near a cold wall; the same absolute value of the gradient will promote particle deposition. The thermophoretic effect is more important for gases than liquids.

- Diffusiophoresis: it involves condensation of the gas stream onto a surface.

- Sedimentation: the particulate matter settles on the surface under gravity action as rust particles, clay, and dust. To occur sedimentation, the downward gravitational force must be greater than the upward drag force. Sedimentation is important for both large particles and low fluid velocities. In processes where rust and dust particles may act as catalysts and/or enter complex reactions as cooling tower waters processes, sedimentation is frequently observed.

- Inertial impaction: for not following fluid streamlines and as a result, depositing on the surface, the "coarse" particles should have sufficient inertia.

- Turbulent downsweeps: the fluid is being transported toward the surface by turbulent downsweeps if the viscous sublayer in a turbulent boundary layer is not truly steady, which can be thought of as suction areas of measurable strength distributed randomly all over the surface.

3. Attachment of the fouling species to the surface is the third stage of the fouling process and implies both physical and chemical processes. In this stage, three interlinked factors play crucial roles which are surface conditions, surface forces, and sticking probability. The combined and simultaneous actions of these factors are largely accounts responsible for the attachment event.

    - Surface properties: the most important properties of surface conditions for attachment events are the surface free energy, wettability (contact angle, spreadability), and heat of immersion. The difference between the surface free energy of the wall and the adjacent fluid layer increases by increasing wettability and heat of immersion. The induction period is longer for unwettable or low-energy surfaces which suffer less from deposition (such as polymer and ceramic coatings) in comparison with wettable or high-energy surfaces. The increase of the effective contact area of a surface and the provision of appropriate sites for nucleation and the promotion of the fouling initiation are the causes of surface roughness. This increases the wettability of wettable surfaces and decreases the unwettability of the unwettable ones.

    - Surface forces: among the most important surface forces is the London-van der Waals force. This force describes the intermolecular attraction between nonpolar molecules and it is always attractive, while the electric double layer interaction force can be attractive or repulsive. Whereas, viscous hydrodynamic force influences the attachment of a particle moving to the wall, which increases as it moves normal to the plain surface.

    - Sticking probability: the fraction of particles that reach the wall and stay there before any reentrainment occurs. A useful statistical concept conceived to analyze and explain the complicated event of attachment.

4. Removal of the fouling deposits from the surface: correspond to the fourth stage in the fouling process. This step may or may not occur simultaneously with deposition under the single or simultaneous action of the following mechanisms; shear forces, turbulent bursts, re-solution, and erosion.

    - Shear forces: the actions of the shear stress exerted by the circulating fluid on the depositing layer is the result of shear forces. The accumulation of deposit causes the cross-sectional area for flow decreases, thus increase in the average velocity of the fluid for a constant mass flow rate and increasing the shear stress. If the deposit bond resistance is greater than the prevailing shear forces at the solid-fluid interface, new deposits will form.

- Miniature tornadoes use to raise deposited material from the surface by periodic turbulent bursts randomly distributed (about less than 0.5% at any instant of time) and by continuity, these fluid bursts are offsetted by softer fluid return scans, which promote deposition.

- Re-solution: the solubility of the material deposited is related to the removal of the deposits by re-solution. The dissolution will occur only if there is a change in the properties of the deposit and in the circulating fluid, or in one of them, because of local changes of certain operational variables such as temperature, velocity, alkalinity, and this by considering that the fouling deposit is presumably insoluble at the time of its formation. For instance, sufficiently high or low temperatures could kill a biological deposit, thus weakening its attachment to a surface and causing sloughing or re-solution. The removal of corrosion deposits is done by re-solution at low alkalinity in power-generating systems and the resolution is associated with the removal of material in ionic or molecular form.

- Erosion: this process and the overall removal process are closely related. It is heavily dependent on the shear strength of the foulant and on the steepness and length of the sloping heat exchanger surfaces, where appropriate and it is associated with the removal of material in particulate form. If the fouling layer is composed of well-crystallized pure material (strong formations), the removal mechanism becomes largely ineffective; in contrast, if the fouling layer is composed of a wide variety of salts each having different crystal properties, the removal mechanism is highly effective.

5. Transport from the deposit-fluid interface to the bulk of the fluid: the destruction of deposits may or may not causes the transport from the deposit-fluid interface to the bulk of the fluid and this depends only on the mass and volume of the sloughed piece and on the hydrodynamic forces of the flowing fluid. In the case of the sloughed piece is sufficiently broad, it may move on the surface and settled on another site on the system as certain corrosion products. Because of the erosion effect, all deposits that are removed will be transported to the bulk of the fluid. Without this action, the removal process is not complete. The aging of deposits is the most important parameter, which affects deposit sloughing. It may strengthen or weaken the fouling deposits.

6. Fouling curves

Figure 9 [11] illustrates the initiation period or time delay in heat exchanger fouling. This period is considered the time when there is no deposition for some time during the commissioning of a clean heat exchanger [18, 19]. Due to changing flow characteristics near the wall, the initial growth of the deposit can cause an increase in heat transfer coefficient rather than a decrease, resulting in fouling resistance. By changing flow characteristics near the wall, the resulting turbulence increases the film heat transfer coefficient at the solid/liquid interface and this increase may overcome the thermal resistance offered by the deposits but the net heat transfer coefficient may increase.

The negative values of fouling resistance have been reported by several authors [20].

This process continues until the additional heat transfer resistance overcomes the advantage of increased turbulence. The time period from the beginning of the fouling process until the fouling resistance again becomes zero represents the roughness delay time [21]. The time period from the beginning of the induction period when the formation of stable crystalline nuclei and their concretion to a compact fouling layer takes place and ends up with the increase of fouling resistance above zero level represents the roughness delay time.

For particulate fouling, the initiation period and the roughness delay time are very small in comparison with scaling fouling where the delay time is fairly long [6].

According to the fouling mechanism and conditions and after the roughness delay time, the fouling curves can be classified into four categories: linear (A), falling (B), accelerating, asymptotic (C), or saw-tooth (D) as the case may be as illustrated in **Figure 9** [11].

Normally, the rate of fouling is defined as the average deposit surface loading per unit of surface area in a unit of time. Utilization of deposit thickness (μm) and porosity (%) are also frequently necessary for the description of the amount of fouling.

1. Linear fouling (curve A): it may be the most common type of fouling. The fouling rate can be steady with time with increasing fouling resistance and deposit thickness for this type of fouling. It occurs generally when the temperature of the deposit in contact with the flowing fluid remains constant.

   The fouling model developed by Ebert and Panchal has been reported in [11]. The average (linear) fouling rate under given conditions are expressed in the form of two competing terms. Mathematically, it is equal to the subtraction of the anti-deposition term from the deposition term.

$$\frac{dRf}{dt} = \alpha \,\mathrm{Re}^{\beta} \mathrm{Pr}^{\delta} \exp\left(\frac{-E}{RT_{film}}\right) - \gamma \tau_w \tag{2}$$

   α, β, γ, and δ represent the parameters determined by regression, τ_w represents the shear stress at the tube wall and T_{film} represents the fluid film temperature (average of the local bulk fluid and local wall temperatures). Depending on the relationship in Eq. (2), it is possible to identify a combination between the temperature and the velocity below which the fouling rates will be negligible. This is presented as the "threshold condition" by Ebert and Panchal. The developed model in Eq. (2) suggests that the heat exchanger geometry can be effectively applied to maintain the conditions below the "threshold conditions" in a given heat exchanger because it affects the surface and film temperatures, velocities, and shear stresses.

2. Falling fouling (curve B): the decreasing of the fouling rate with time, and the not mitigation of a constant value of the deposit thickness, despite the fouling rate never drops below a certain minimum value are the characteristics of the falling fouling. This type of fouling is generally due to an increase in removal rate with time.

3. Asymptotic fouling (curve C): this type of evolution is observed when the fouling rate decreases with time until it becomes negligible after a period of time when the deposition rate becomes equal to the deposit removal rate and the

deposit thickness remains constant. The asymptotic fouling occurs generally where the tube surface temperature remains constant while the temperature of the flowing fluid drops as a result of increased resistance of fouling material to heat transfer. It may also be the result of soft or poorly adherent suspended solid deposits upon heat transfer surfaces in areas of fast flow where they do not adhere strongly to the surface with the result that the thicker the deposit becomes, the more likely it is to wash off in patches and thus attain some average asymptotic value over a period of time.

By increasing particle concentration and by decreasing fluid bulk temperature, flow velocity, and particle diameter, the asymptotic fouling resistance increases. The first correlation describing the asymptotic fouling model was developed by Kern and Seaton. Note that no further increase in fouling occurs in this model beyond asymptotic fouling resistance. Based upon the asymptotic values, fouling factors for several fluids are suggested by the Tubular Heat Exchanger Manufacturers Association (TEMA). However, this approach does not address all fouling phenomena. For example, it does not address fouling at the "hot" end of a crude oil preheat train because fouling there does not exhibit asymptotic behavior.

4. Saw-tooth fouling (curve D): this type of fouling occurs when part of the deposit is detached after a critical residence time or once a critical deposit thickness has been reached. In that case, the fouling layer then builds up and breaks off again. Pressure pulses, spalling, trapping of air inside the surface deposits during shutdowns, or other reasons could cause the periodic variation which often corresponds to the moments of system shutdowns, startups, or other transients in operation.

7. Conditions affecting fouling

The most important conditions, which influence fouling is: operating parameters, heat exchanger parameters, and fluid properties [19, 22].

7.1 Operating parameters

Velocity, surface temperature, and bulk temperature are among the main important events of operating parameters that affect fouling at a significant level.

Velocity has a significant effect on fouling. The increase of fluid velocity in diffusion-controlled process causes more fouling. For high fluid velocities, the fouling decreases in most cases. By increasing flow velocity, the fluid shear stress increases, which causes more removal. This leads to lower fouling rates that causes lower fouling resistance. In the case of particulate fouling, for weak deposits, the increase of the flow velocity may completely eliminate fouling. In contrast, for stronger deposits, the increase of the flow velocity beyond a particular point may not decrease fouling significantly and for very strong deposits, the increase of the flow velocity may not have any effect at all.

Diverse behaviors, among which the increase of surface temperature may increase, decrease, or has no effect on the fouling rates [7]. The increase in temperature causes the increase in both the rates of the chemical reaction and inverse solubility crystallization. Because of higher concentration gradients and higher reaction rate constants, higher surface temperature increases fouling for inverse solubility salts. Cooling results in more fouling in the case of normal solubility salts.

The bulk temperature affected also the increase in the fouling rate. The increase of the temperature increases the rate of crystal formation and thus deposition and this is when precipitation happens in the fluid bulk in inverse crystallization. Thus, the bulk temperature has effects on chemical reaction rate and polymerization rate.

7.2 Heat exchanger parameters

Among the significant heat exchanger parameters that affect fouling are: surface material, surface structure (roughness), heat exchanger type, and geometry. Due to the potential to react and form corrosion products, the surface material is considered seriously for corrosion fouling. Various materials have different catalytic actions and may promote or reduce fouling for different processes. The surface roughness affects significantly on the initial fouling rate and scale formation. Theoretically, the free energy change associated with crystal nuclei formation is much less on a rough surface than on a smooth surface [6]. Because of protected zones in the cavities or pits where flow velocities are very low, rough surfaces result in the higher deposition.

The surface roughness strongly depends on the nuclei attachment and not on the rate of nucleation, according to Rankin and Adamson and Chandler [6]. Generally, for all types of fouling, the rough surface causes more fouling and that reduces the delay time. Surface roughness increases turbulence near the surface, which in turn increases the removal rate of fouling on the surface. The best performance corresponds to the increase of surface roughness with deposit formation [6]. The mirror-finished surfaces in heat exchangers are used to reduce fouling in practice which is reiterated by Marriott [6].

7.3 Fluid properties

Among the most known properties that affect fouling processes are the nature of the fluid and the species dissolved or entrained in the fluid. In cooling systems, for example, the quality of water has a significant impact on fouling mechanisms, which cause in crystallization of reverse solubility salts, particle deposition, corrosion, and biofouling [16].

The intrusion of small amounts of impurities into fluids can trigger or significantly increase fouling which can either deposit as a fouling layer or act as a catalyst for fouling processes. For instance, fouling by chemical reaction or the polymerization of hydrocarbons in refineries is due to the presence of oxygen and/or trace elements such as vanadium and molybdenum.

The presence of fine particles of impurities triggers the seeding deposition process in the case of crystallization fouling. The properties of impurities are the basis for the manufacture of many anti-fouling chemicals. These impurities as sand or other particles suspended in cooling water can sometimes have a deposit reduction or removal action [15].

Suspended solids promote the sedimentation of particles by gravity on the heat transfer surfaces. In this case, prevention is achieved by avoiding stagnant areas. Indeed, high water velocities (greater than 1 m/s) of the water help prevent the clogging of particles. It is often economical to install upstream filtration.

8. Fouling models

Various models have been proposed for different types of fouling but the analysis and the improvement of the model are still progressing because of the complex nature of deposit formation and the lack of reproducible measurement of fouling resistance. Several assumptions were assumed in order to simplify various models [23, 24], such as:

- Generally, only one type of fouling is considered.
- Homogeneity of the fouling layer.
- The form of deposits, such as crystals and particles, is ignored.
- In most cases, the changes in the physical properties of the fluids are neglected.
- In general, the changes in flow velocity with changing cross-sectional area because of fouling are overlooked.
- Surface roughness is neglected.
- Change in surface roughness with deposit formation is also neglected.

A few attempts have been made to model the initiation or roughness delay period and almost all the models predict fouling after the delay period. The majority of models usually take into account the main parameters related to fouling such as flow velocity, concentration, wall and bulk temperature, and time. In contrast, other some notable parameters that are effect of simultaneous action of different fouling mechanisms, equipment design, surface parameters such as surface material and surface roughness, increase in surface area with deposition, properties of foulant stream, nature of the process, and the fluctuations in operation are neglected in modeling [24].

Accumulation is usually considered to be the net result of two simultaneous processes: a deposition process and a removal process. Mathematically, the net rate of accumulation can be expressed as the difference between the deposition and removal rates which represent the equation of material balances. It should be noted that the semi-empirical modeling was developed on the basis of this equation whose general expression is:

$$\text{Rate of accumulation} = \text{Rate of deposition–Rate of removal}$$

$$\frac{dm_f}{dt} = \dot{m}_f = \dot{m}_d - \dot{m}_r \tag{3}$$

The deposition rate depends on the type and mechanism of fouling, while the removal rate depends on both the hardness and the adhesion strength of the deposit and the shear stress which results from the flow velocity.

Help the designer or indeed the operator of heat exchangers; to assess the impact of fouling on heat exchanger performance in certain operating conditions is the purpose of any fouling model. Providing a fouling model is based on its mathematical interpretation but the inclusion of an extensive set of conditions for such a model would be difficult and even impossible. Concerning the production of a mathematical model for the fouling process, the general material balance given in Eq. (3) is the basis of the modeling, which is centered on evaluating the functions $\dot{m}_d$ and $\dot{m}_r$ for specific fouling situations, some of these models are:

8.1 McCabe-Robinson model

This model developed in 1924 concerns the fouling of the surfaces of heat exchangers used as evaporators. The correlation proposed to evaluate the solid mass deposited is [24]:

$$\dot{m}_d = a\dot{q} \tag{4}$$

Note that no information has been specified concerning the physical meaning of the coefficient "a" and the deposit elimination mechanism.

8.2 Kern and Seaton model

In 1958, Kern and Seaton from some experimental fouling results managed to say that the rate of deposition mass, $\dot{m}_d$, remained constant with time t but the rate of removal mass, $\dot{m}_r$, is proportional to the accumulated mass, m_f, and therefore increased with time to approach $\dot{m}_d$ asymptotically.

The particle wall transport phase controls the deposition process whereas the shear stress controls the removal phase. Considering that ($\dot{m}_d$) is proportional to the deposited mass of particles, the expressions of ($\dot{m}_d$) and ($\dot{m}_r$) are the following [11]:

$$\dot{m}_d = k_p \times (C_b - C_w) \tag{5}$$

$$\dot{m}_r = C_1 \times \tau_w \times m_f \tag{6}$$

(k_p) represents the transport coefficient, (C_b) represents the particle concentration in the fluid, (C_w) represents the particle concentration at the wall, (C_1) represents a dimensional constant, and (τ_w) represents the shear stress exerted by the fluid on the deposit.

So, Eq. (3) becomes:

$$\frac{dm_f}{dt} = k_p \times (C_b - C_w) - C_1 \times \tau_w \times m_f \tag{7}$$

The integration of Eq. (7) from the initial condition (m_f = 0) at (t = 0) gives:

$$m_f = \frac{k_p \times (C_b - C_w)}{C_1 \times \tau_w} \times \left[1 - e^{(-C_1 \times \tau_w \times t)}\right] \tag{8}$$

Acknowledging that: $\tau = \frac{1}{C_1 \times \tau_w}$ and $m_f{}^* = \frac{k_p \times (C_b - C_w)}{C_1 \times \tau_w}$.

Eq. (8) is expressed as follows:

$$m_f = m_f{}^* \times \left(1 - e^{-\theta \times t}\right) \tag{9}$$

where (m_f^*) is the asymptotic value of m_f and $\theta = \frac{1}{t_c}$. (t_c) is the time constant which represents the average residence time for an element of fouling material at the heat transfer surface.

Assuming that the initial fouling flow is equal to the deposition flow and that the thermophysical properties of the deposit (conductivity and density) are constant, Eq. (9) can be written in the form of a thermal fouling resistance:

$$Rf(t) = Rf^* \times \left[1 - \exp\left(-\frac{t}{\tau}\right)\right] \tag{10}$$

($Rf\,(t)$) represents the evolution of the fouling resistance as a function of time (expressed in (m^2 K W^{-1})), (Rf^*) represents the asymptotic value of the fouling resistance (expressed in (m^2 K W^{-1})) (this value characterizes the situation where the deposition rate and the breakout speed are equals), (t) represents the time (expressed in (s)) and (τ) represents the characteristic time (expressed in (s)) and generally its value is attributed to the time required for the fouling resistance to reach its asymptotic value in case the evolution of this kinetics was linear.

The real solution would be to find expressions for (Rf^*) and (t_c) as a function of variables affecting the fouling process.

8.3 Taborek et al. model

Helalizadeh reported in [24] the water characterization factor introduced by Taborek et al in 1972 to the deposition term which is used to account for the effect of water quality. Diffusion of the potential depositing substance to the surface (1) and bonding at the surface (2) are the two processes of the deposition term. This deposition rate is expressed in an Arrhenius type equation as the following:

$$\dot{m}_d = k_1 P_d \Omega^n \exp\left(\frac{-Ea}{R_g T_s}\right) \tag{11}$$

(K_1) represents the deposition constant, (P_d) represents the deposition probability factor related to velocity and "Stickiness" or adhesion characteristics of the deposit, (n) represents an exponent, (Ω) represents the water characterization factor, $\left(\frac{-Ea}{R_g T_s}\right)$ represents the Arrhenius reaction rate function, (Ea) is the activation energy, (R_g) is the universal gas constant and (T_s) is the absolute surface temperature.

In the proposed model, the removal rate was assumed that is a function of shear stress, deposit thickness, and bonding strength of the deposit. The removal function was given as follows:

$$\dot{m}_r = k_2 \left(\frac{\tau}{\psi}\right) x_f \tag{12}$$

(k_2) represents the removal constant, (τ) represents the fluid shear stress exerted on the deposit surface, (Ψ) represents the strength or toughness of the deposited layer. In material balance (Eq. (1)), by replacing the deposition rate (Eq. (11)) and removal rate (Eq. (12)), the resulting equation is as follows;

$$Rf = \frac{x_f}{\lambda_f} = \frac{k_1 P_d \Omega^n e^{-Ea/R_g \times T_s}\left(1 - e - ^{-k_2 \lambda_f \tau t/\Psi}\right)}{\frac{k_2 \tau \lambda_f}{\Psi}} \tag{13}$$

And $Rf^* = \frac{k_1 P_d \Omega^n e^{-Ea/R_g \times T_s}}{\frac{k_2 \tau \lambda_f}{\Psi}}$, $\theta = \frac{k_2 \tau \lambda_f}{\Psi} = \frac{1}{t_c}$

8.4 Watkinson et de Martinez model

Watkinson and Martinez proposed in 1975 a model based on the models developed by Kern and Seaton (1958) and Reitzer (1964) taking into account the phase of elimination of the deposit. The general expression for this model is [24]:

$$Rf = Rf^* \times \left(1 - e^{-Bt}\right) \tag{14}$$

The parameter (B) is determined from the experimental data.

For crystallization fouling, the fouling resistance (Rf^*) is given by the following equation:

$$(Rf^*)^3 + \frac{2}{\alpha_i}(Rf^*)^2 + \frac{Rf^*}{\alpha_i^2} - \frac{0.0002751T - 0.08489}{(\alpha_i v)^2} = 0 \tag{15}$$

Where (α_i) is the inside heat transfer coefficient and (v) is the flow velocity.

8.5 Watkinson model

Based on the model developed by Watkinson and Martinez, Watkinson reported the effect of fluid velocity on the asymptotic fouling resistance in three cases as [15];

1. Calcium carbonate scaling (with constant surface temperature and constant composition)

$$Rf^* = \frac{0.101}{v^{1.33} \times D^{0.23}} \tag{16}$$

2. Gas oil fouling (with constant heat flux)

$$Rf^* = \frac{0.55}{v^2} \tag{17}$$

3. Sand deposition from water (with constant heat flux)

$$Rf^* = \frac{0.015}{v^{1.2}} \tag{18}$$

Where (Rf^*) is the asymptotic fouling resistance, (v) is the fluid velocity and (D) is the tube diameter.

8.6 Hasson et al. model

In 1981, Hasson et al. expressed the degree of fouling by crystallization due to calcium carbonate ($CaCO_3$) per interface unit by the following relation [24]:

$$\dot{m}_d = K_r\left(\left[Ca^{2+}\right]_i \left[CO_3^{2-}\right]_i - K_{sp}\right) \tag{19}$$

Where $[Ca^{2+}]$ is the concentration of calcium ion and $[CO_3^{2-}]$ is the concentration of carbonate ion.

By performing mathematical manipulations on the previous relationships, the fouling rate is obtained by the following relationship:

$$K_1\left(\frac{\dot{m}_d}{K_r}+K_{sp}\right)\left(\frac{\dot{m}_d}{\beta}+[CO_2]=4K_2\left[Ca^{2+}\right]^2\left(1-\frac{\dot{m}_d}{\beta\left[Ca^{2+}\right]}\right)\left(\frac{\left[HCO_3^-\right]}{2}-\frac{\dot{m}_d}{\beta}\right)^2\right. \quad (20)$$

Where (K_1) and (K_2) are the first and second dissociation constant, (K_{sp}) is the molar solubility product, and (k_r) is the reaction rate constant.

The expressions allowing having these constants are:

$$\log K_1 = \frac{-17052}{T} - 215.21\log T + 0.12675T + 545.56 \quad (21)$$

$$\log K_2 = \frac{-2902.39}{T} - 0.02379T + 6.498 \quad (22)$$

$$\log K_{sp} = -0.01183(T - 273.2) - 8.03 \quad (23)$$

$$K_r = \exp\left(41.04 - \frac{10417.7}{T}\right) \quad (24)$$

By doing mathematical operations, the expression of the Eq. (20) reduces to:

$$\dot{m}_d = \frac{0.5\beta b\left[Ca^{2+}\right]\left(\left(1+\frac{4ac}{b^2}\right)^{0.5}-1\right)}{a} \quad (25)$$

Where:

$$a = 1 - \frac{4K_2K_r\left[Ca^{2+}\right]}{K_1\beta} \quad (26)$$

$$b = \frac{[CO_2]}{\left[Ca^{2+}\right]} + \frac{4K_2K_r[HCO_3^-]}{K_1\beta} + \frac{K_{sp}K_r[CO_2]}{\beta\left[Ca^{2+}\right]} \quad (27)$$

$$c = \frac{K_2K_r[HCO_3^-]^2}{K_1\beta\left[Ca^{2+}\right]} - \frac{K_{sp}K_r[CO_2]}{\beta\left[Ca^{2+}\right]^2} \quad (28)$$

(β) is the mass transfer coefficient, calculated using the following relation, proposed by Hasson et al. and reported in [24]:

$$\beta = 0.023\,\mathrm{Re}^{0.85}Sc^{0.33}\frac{D}{d_{eq}} \quad (29)$$

8.7 Ritter model

In 1983, Ritter examined all possible parameters to correlate both the induction period and the fouling rate associated with calcium sulfate. He showed that the induction period and the fouling rate depend on the supersaturation of the solution and that the secondary correlation parameter is the mass transfer coefficient. The expressions proposed by Ritter are [24]:

For the induction period:

$$\theta = \frac{2.1 \times 10^{-4} \rho}{\beta \left(\frac{C_b - C^*}{C^*} \right)^2} \tag{30}$$

For the fouling rate:

$$\frac{dRf}{dt} = \frac{1.9 \times 10^{-9} \beta}{\rho} \left(\frac{C_b - C^*}{C^*} \right)^2 \tag{31}$$

Where (Θ) is the induction period and $\frac{dRf}{dt}$ is the fouling rate.

Note that for these results, Ritter did not specify the operating conditions.

8.8 Knudsen analysis

The fouling process is complicated and dynamic. Generally, the fouling resistance is not measured directly but must be determined from the degradation of the overall heat transfer coefficient [25], and the fouling factor, *Rf*, could be expressed as [15];

$$Rf = \frac{1}{U_f} - \frac{1}{U_c} \tag{32}$$

On the basis of the change in the overall heat transfer coefficient of the fouling test section experimental fouling data have been analyzed as in Eq. (34). It is assumed that the thermal-hydraulic condition in the test section remains reasonably constant for the duration of the fouling test. Through the use of the model of Taborek et al., the two parameters (Rf^*) and (t_c) can be determined for each fouling situation. The (Rf^*) represents the asymptotic fouling resistance contains all the factors that influence fouling and (t_c) represents the time constant of the fouling resistance exponential curve i.e. the time required for the fouling resistance to reach 63% of its asymptotic value (i.e. $(t_c) \approx 0.63 \times (t^*)$). This parameter (time constant) depends on the shear stress, the deposit strength factor, and the deposit thermal conductivity;

$$t_c = \frac{\psi}{\tau k_2 \lambda_f} \tag{33}$$

From the deposition-removal model, which was first presented by Kern and Seaton and reported in [15] (Eq. (10)) and from Eq. (32), the overall heat transfer coefficient of the fouled surface. U_f, may be given as;

$$U_f = \frac{U_c}{1 + U_c \times Rf} \tag{34}$$

Then

$$U_f = \frac{1}{\frac{1}{U_c} + Rf^* \times \left(1 - e^{\left(\frac{-t}{t_c}\right)}\right)} \tag{35}$$

In Eq. (35), if the two coefficients (Rf^*) and (t_c) can be obtained accurately either empirically or analytically, they will be useful for predicting the fouling factor which can be used in practical heat exchanger design.

9. Prediction of fouling factor

Fouling is generally defined as the formation of an essentially solid deposit of low thermal conductivity upon the heat transfer surface, through which heat must be transferred by conduction. But, the thermal conductivity of the fouling layer and its thickness is generally unknown, that's why the possible solution to the heat transfer problem is to introduce a fouling factor in order to take into account the additional resistance to heat transfer and make possible the calculation of the overall heat transfer coefficient [7]. The opposite value of the fouling factor represents the fouling coefficient.

It is necessary to be careful in selecting fouling factors in calculating heat transfer, particularly where fouling resistances completely dominate the thermal design. Generally, the uncertainties in design parameters such as fluid properties, flow rates, and temperatures have less effect than that of the uncertainty inherent in fouling factors [11, 26]. A significant fouling factor is considered as a safety margin to cover uncertainties in fluid properties and even in process knowledge, but the use of an excessively significant fouling factor will result in an oversized heat exchanger with two or three times more area than is really necessary.

Several experience-based tables available provide typical fouling factors such as TEMA Table RGP-T-2.4 [27]. For each particular application, an acceptable evaluation of the effects of fouling needs to be judged and evaluated. However, these tables are being considered as a guide in the absence of more specific information.

To predict the rate of fouling in heat exchangers or to estimate fouling factors to use in heat transfer calculations, several semi-empirical models have been developed over the years and some of them are presented in the previous section. Recent research work has recourse to the use of digital computers which are able to provide rapid means to perform calculations. The ANN method has been applied in many disciplines of engineering and has produced promising results. The ANN method becoming a powerful tool and progressing at an impressive rate due to its feature which is the ability to learn and generalize the relationships in a data set and to provide quick and satisfactory predictions. This technique became attractive for many different applications [28, 29].

The ANN method is an efficient and powerful non-linear computational structure derived from a biological neural system and composed of very simple and highly interconnected elementary units called neurons [30–32]. The neurons are organized according to architecture and the connection to each other is made by weighted links over which signals can pass. Each neuron treats the weighting factor, which is attached to the input (wX) and which is added to the bias coefficient (θ) with a suitable activation or transfer function (f). Thus, the output (Y) can be mathematically expressed as [30]:

$$Y = f\left(\sum wX + \theta\right) \tag{36}$$

Several used transfer functions, such as the hyberbolic tangent, the linear transfer function and the Gaussian function, and the tangent sigmoid function [33].

In engineering applications, the back propagation (BP) learning algorithm has become the most popular and exciting kinds of ANNs [34]. It is designed to solve the problem of determining weight values for a multi-layer ANN with feedforward connections from the input layer to the output layer through the hidden layer. The application of the ANN approach can be used to develop the best configuration in the training period.

The ANN approach has been widely used in different applications, especially in the field of heat exchangers such as the design and control of heat exchangers [35, 36], and in the simulation of heat exchanger performance [35–37] and estimation of heat exchanger parameters [7, 35].

As an alternative and practical technique, the ANN approach has also been used to evaluate the rate of heat exchange and heat transfer coefficient for a different type of heat exchangers [36] and in other applications including in particular the prediction of the rate of fouling and fouling factor in a shell-and-tube heat exchanger [7, 38].

To predict the fouling rate and fouling factor in heat exchangers, an ANN model can be developed and the available data set used for training the network and verifying its generalization capability. The fouling factor are then calculated and the input-output pairs are presented to the network, and the weights are adjusted to minimize the error between the network output and the actual value. Once training is complete, predictions from a new set of data may be done using the already trained network.

A few works relating to the application of artificial neural networks to predict the fouling rate or fouling factor are: in [39], the input layer of the artificial neural network comprised five parameters, namely: the inlet and outlet temperatures of the cold fluid, the inlet temperature of the hot fluid, mass flow rate of cold and hot fluids is used for predicting the fouling factor. Aminian and Shahhoseini [40] predict the fouling rate of crude oil pre-heaters using a four layers feedforward neural network model. The velocity of crude oil, tube surface temperature, and tube diameter were employed as independent variables of the ANN model. Fouling threshold in crude oil pre-heaters was predicted as a function of surface temperature, Reynolds, and Prandtl numbers using a neuro-based model [41]. Shell and tube side input temperatures and tube side crude oil flow rate was considered as the independent variables of the proposed model, while the output variable was chosen to calculate the fouling factor [42]. Davoudi and Vaferi [7] employed the ANN for predicting the fouling factor from some easily measured variables of the system which are density, velocity, temperature of the fluid, oxygen content, hydraulic diameter of the fluid passage, surface temperature, and time. And more recently, Jradi et al. [30] estimated the fouling resistance according to the inlet and outlet temperatures of the cold fluid, the temperature of the hot fluid, the density and the volume flow rate of the cold fluid, and the time for three types of heat exchanger by using three different methods which are: Kern and Seaton, Partial Least Squares (PLS) and Artificial Neural Networks (ANN). Results have shown that modeling by the use of ANN is very performing compared with modeling by PLS and Kern and Seaton.

10. Fouling control

To avoid technical problems associated with fouling of heat exchangers (plugging and clogging of pipes, corrosion of materials, drop in energy performance, loss of production due to the shutdown of installations for maintenance and cleaning), there are currently several methods for detecting and/or monitoring the evolution of fouling [13]:

- Predictive maintenance,
- Direct measurements at the terminals of the exchanger,
- Measurements using probes.

10.1 Predictive maintenance

Several techniques for developing a predictive maintenance program are available either by thermography or by visual inspection.

- Thermography makes it possible to measure the intensity of emissions of infrared rays, that is to say, heat, in order to determine the operating conditions of the installations.

Three types of infrared devices can be used in predictive maintenance:

Infrared thermometers: these are designed to measure the actual surface temperature at a single, relatively small point in an installation. As the measurement is restricted to a single point, this method offers limited possibilities.

Linear scanners: they take a temperature reading along a particular line, and therefore in one dimension. Although this method opens up a somewhat wider field of vision, it is also limited.

Imagers: they analyze all the infrared emissions from the exchanger. Most often these are infrared cameras.

These methods, although effective and new, make it possible to locate the problem of fouling, but remain qualitative due to the numerous errors in temperature measurement by radiometry.

- Visual inspection

Regular visual inspection of industrial facilities is an integral part of a predictive maintenance program. The user, guided by his own experience, can detect, through visual inspection, the level of fouling of the exchangers in his installation. Thus, this routine operation provides subjective but necessary information for setting up the maintenance program.

The principle common to these techniques lies in the regular monitoring of the condition of the installation. They provide more or less information on the state of fouling of the exchangers, but very little on heat transfer.

10.2 Direct measurements at the exchanger terminals

In industrial practice, there are different methods for detecting and monitoring the progress of fouling. Generally, three methods are used based on [13]:

- Monitoring the variation of the heat transfer coefficient or the fouling resistance.
- Simultaneous observations of pressure drop and flow rate indicators.
- Temperature measurements at the inlet and outlet of the heat exchanger.

10.3 Measurements using a probe

In this case, the fouling is monitored using a probe. The devices are classified into four main categories according to their mode of operation:

- Devices using optical methods: these devices use optical methods such as lasers or pyrometers to determine the thickness of the deposit. However, these techniques are restricted to laboratory tests under very controlled conditions and are therefore difficult to extrapolate to industrial conditions.

- Devices using a deposit measurement: these probes are used to collect deposits qualitatively under controlled conditions. With such devices, the mass of the deposit on the exchanger tube cannot be determined. These probes are used to characterize the thermal properties of the deposit.

- Devices allowing the measurement of the mass increase: such a device allows the quantitative measurement of the deposit. This technique requires the control of many parameters such as the wall temperature or the gas temperature.

- Devices using flowmeters: measurement by flux meter makes it possible to follow the decrease in heat flow as a function of the clogging time. It thus makes it possible to follow the evolution of fouling resistances. On the other hand, these probes do not make it possible to obtain either the thickness or the profile of the deposit. So far, this type of probe has mainly been studied in settings where radiation is the primary mode of heat transfer.

11. Economic aspect of fouling

A very few works have been reported to accurate determine economic penalties caused by fouling in spite of the high cost in the heat exchanger, which is attribute to the cost of the difference aspect of the design and operation of this equipment. In order to evaluate the cost-efficiency of various mitigation strategies, reliable knowledge of fouling economics is desirable [2].

Fouling of heat transfer equipment in industries imposes additional operating costs which are not assessed precisely. Several concordant studies have been undertaken to determine the fouling-related costs in the industry. Generally, fouling costs can be divided into three major categories as shown in **Figure 10** [43] which are: (1) cost related to the design of the heat exchanger, (2) cost related to the operation of heat exchanger, and (3) costs related to shutdown of operation.

The total fouling-related costs are as follows:

11.1 Costs related to the design of the heat exchanger

Generally, equipment manufacturers offer oversized heat exchangers, taking into account, among other things, an additional thermal resistance associated with the fouling phenomenon, the value of which is often of empirical origin. Thus, the additional surface area to the heat exchange surface required induces an additional cost of approximately 20% of the cost of acquisition of the heat exchanger.

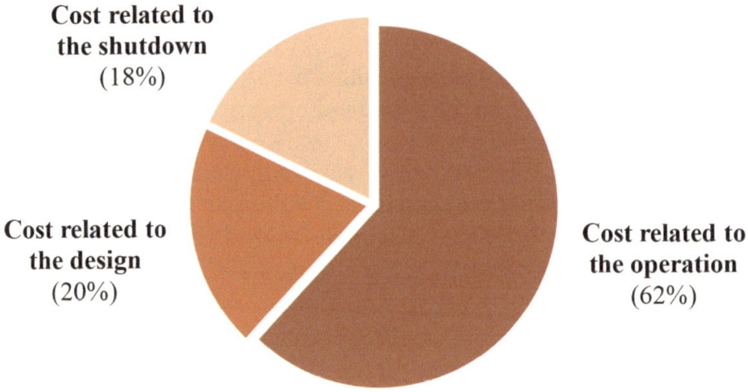

Figure 10.
Cost imposed due to fouling [43].

11.2 Costs related to the operation of the heat exchanger

Operators resort to very often costly actions such as the treatment of fluids before introduction into the devices, online control of the operating parameters, in particular, the temperatures and the inlet and outlet pressures of the two fluids, and the supply of additional energy to limit the negative repercussions of fouling on the operation of heat exchangers. These often-costly actions generate additional costs high to the total cost of operating the facilities in the order of 62%.

11.3 Costs related to the shutdown of the installation

Whatever the actions used by the operators; the fouling phenomenon of heat exchangers is inevitable. The planned or unplanned shutdowns of installations to carry out cleaning and maintenance operations have repercussions in particular on the nominal production rate. These equally costly operations contribute to an increase of around 18% in the costs associated with operating the facilities.

12. Fouling mitigation

To reduce the negative effects of fouling on energy performance and the operation of heat exchangers, several methods are envisaged at different levels of the design-construction, and operation of this type of device.

Table 1 [15] regroups the main actions to be respected during the four stages of life of a heat exchanger.

Several works have been done to reduce fouling formation when heat exchangers are functioning. In the operation stage, numerous methods have been developed in recent years to control fouling [2, 6]. These methods can be classified as chemical means (inhibitor), mechanical means, changing the phases of the solution, electromagnetic fields, electrostatic fields, acoustic fields, ultraviolet light, radiation or catalytic treatment, surface treatment, green additives, fiber as a suspension, etc. In the past, chromate was a successful chemical agent for crystal growth control until it was banned.

Phase	Actions to respect
Design	• Choice of heat exchanger type • Choice of geometry and configuration • Choice of operating conditions (temperature, flow, etc.).
Construction and assembly	• Quality assurance on manufacturing • Protection during transport and during storage • Cleaning and passivation operation.
Operation	• Maintaining operating conditions (temperatures, flow rates, etc.) • Fluid pretreatment • Use of additives (inhibitors, scale inhibitors, dispersants, acids).
Stop	• Disassembly and manual cleaning • Mechanical attacks on deposits (drilling, etc.) • Hydraulic cleaning with a lance (steam, air).

Table 1.
Actions to be respected to reduce the impact of heat exchangers fouling [15].

Then, chromate-based additives were been replaced with polyphosphate inhibitor which has a tendency to decompose the foulant in water containing high calcium hardness. Water fouling rich in calcium-containing phosphate fouling inhibitor has been studied by Knudsen et al. They demonstrated that acrylic acid/hydroxypropyl acrylate and acrylic acid/sulfonic acid were both very effective in inhibiting the deposition of calcium phosphate. Catalyst material composed of zinc and tourmaline was studied to mitigate fouling. To this end, Tijing et al. reported that this material potentially reduces calcium carbonate fouling formation. Teng et al. reported a similar finding but for calcium sulfate mitigation [44]. Furthermore, Tijing et al. further extended the research by using the same catalyst material to mitigate fouling on carbon steel piping. Most of the methods used in the past for fouling mitigation are dangerous to the environment. Thus, it is necessary to apply green technology methods and chemicals approaches benign to the environment [45].

One of safe and efficient nonchemical fouling mitigation method is physical water treatment (PWT) which is a good alternative. The PWT includes permanent magnets, solenoid coil devices, green additive and catalytic materials, and alloys [2]. Chemical additives are often used to mitigate scaling on heat transfer surfaces, but chemical products are expensive and pose a threat to the environment and health. Kazi [6] carried out the mitigation of calcium sulfate dehydrates scale formation on heat exchanger surfaces by using natural wood pulp fiber.

13. Cleaning of heat exchanger

It is often necessary to clean the heat exchangers to maintain or restore the efficiency of the heat exchanger. The cleaning operation of the heat exchanger may be classified into two groups: online and offline cleaning [46]. The cleaning can be done online during the phase of operation of the heat exchanger to maintain acceptable performance of the equipment without interruption of operation allowing thus to increase the service time between two maintenance shutdowns. In many other cases, it is necessary to proceed for complete cleaning of the equipment. The offline cleaning carries out during the stops phases of the installation.

13.1 Online cleaning

Generally, mechanical and/or chemical methods are used in online cleaning. These techniques of cleaning are designed for the tube side and do not require disassembly [13]. Among the main advantage of this type of cleaning is the continuity of service of the heat exchanger in the hope that no cleaning shutdown will occur. Abruptly stopping adds an extra cost of a new heat exchanger installation or a high cost of renovations and no guarantee that all the tubes would be cleaned sufficiently.

13.1.1 Online mechanical cleaning

I. Cleaning of the internal surface of tubes is made by means of circulation of sponge rubber balls [13], brushes, or a stretched spring. These techniques are able to prevent the accumulation of particulate matter, biofilm formation and scale, and corrosion product deposition. They are only applicable to flow through the inside of tubes.

II. Accoustic blow-molding [13]: this technique is based on the use of vibrations in the form of sound waves in high and low frequencies to relieve fouling problems on heat exchangers. It is a preventive technique and not curative. It has the advantage of not being aggressive for surfaces of exchange but it is much less effective in sticky and tenacious deposits that are generally associated with slagging.

III. Flow reversal: this technique allows for eliminating trapped particles in the entry collectors.

IV. Overspeed: this technique allows finding operating conditions close to clean conditions. This type of cleaning removes the surface layer of the deposit, but leashes germs on the wall for deposit growth, which accelerates the deposit growth phase.

V. Thermal shock [13]: this technique is characterized by a quick change of temperature causing thus cracking of the foulant layer with the possibility of flaking.

VI. Radiation [13]: the use of ultraviolet light and gamma rays has been considered for a long time for radiation sterilization of microbial-laden water.

13.1.2 Online chemical cleaning

I. Online chemical cleaning [13]: injection of chemical solutions into the process streams for cleaning purposes.

II. Scale inhibitors [13]

III. Chlorination used for combat bio-fouling [13]

13.1.3 Other online cleaning

It can be magnetic or electrical devices [13].

13.2 Offline cleaning

Offline cleaning is based on the stop operation and cleaning of the heat exchanger, which can be done by chemically or mechanical ways. This technique is used without needing to dismantle the heat exchangers, but generally, it is necessary to have access to the inside surfaces. It would be wise to consider the installation of a "standby" heat exchanger that provides the opportunity to clean the fouled heat exchanger while maintaining the production.

13.2.1 Offline mechanical cleaning

13.2.1.1 Pressurized water

This technique is proven and used in the most diverse industries. It is effective to remove deposits inside the tubes of the heat exchanger. However, there is a risk of surface erosion and especially of the destruction of the anti-corrosion protective film. The water flushing carries away the dislodged material and it is repeated until clean surfaces are obtained.

13.2.1.2 Using tools

This technique is used to remove even very hard deposits inside the tubes of heat exchangers. The removal devices are the pneumatic or electric motor, hole-punches, and hydraulic gun. The application of these devices may be to the rotating shaft including drills, cutting and buffing tools, and brushes that can be manufactured from different materials such as steels or nylon, brasses according to the tube material and the nature of the deposit.

13.2.2 Offline chemical cleaning

a. Acid solutions such as hydrofluoric, hydrochloric, citric, sulfuric, or complexing agents as EDTA are inhibitors for iron oxides, calcium/magnesium scales, etc. [13].

b. Chlorinated or aromatic solvents are inhibitors for heavy organic deposits such as tars and polymers [13].

c. Oxidant solutions of potassium permanganate [13] are inhibitors of carbon deposition.

14. Conclusion

Fouling is the main unresolved problem in heat exchanger operation and still remains a phenomenon misunderstood. Among the most serious concern of heat exchanger users is the problem of fouling deposition and its impact on the economy because there is a lack of awareness among relevant authorities. There are numerous and varied penalties for fouling and their effects on the effective, reliable, and safe functioning of equipment or structures are often very serious. Consequently, this present paper will encourage relevant organizations in different countries, seriousness of this problem, and application of possible mitigation approach. Fouling control and appropriate cleaning play an important role to reduce the production costs in the industry. Because of chemical usage, maintenance work and downtime loss, and water wastage, the production cost significantly increases. For that reason, the relevant authorities must understand the importance of control and cleaning of fouling and apply a specific standard of cleaning procedure in the industries.

Nomenclature

C	concentration (kg/m^3)
D	diameter (m)
deq	equivalent diameter (m)
E_a	activation energy (J/mol)
f	transfer function
K_{sp}	solubility product
k_1	first dissociation constant
k_2	second dissociation constant
k_r	reaction rate constant
$\dot{m}$	rate (kg/m^2 s)
R	perfect gas constant (J/mol K)
Rf	fouling resistance (m^2 K/W)
T	temperature (°C or K)
t	time (s)
U	global heat transfer coefficient (W/m^2 K)
v	flow velocity (m/s)

Greek letter

α	heat transfer coefficient (W/m^2 K)
β	mass transfer coefficient
δ	coefficient of linear expansion
ρ	density (kg/m^3)
γ	concentration effect
σ	shear resistance of the fouling layer (N/m^2)
τ	characteristic time or shear stress (s or N/m^2)
Θ	induction period (s)
Ψ	empirical structure parameter
Ω	water quality

Indices and exhibitors

d	relative to the deposit
f	fouling state
c	clean state
r	relative to removal
w	relative to the wall
*	relative to the asymptotic value

Dimensionless numbers

Pr	Prandtl number
Re	Reynolds number
Sc	Schmidt number

Author details

Rania Jradi[1]*, Christophe Marvillet[2] and Mohamed Razak Jeday[1]

1 Research Laboratory "Process, Energy, Environment & Electrical Systems", National Engineering School of Gabès, Gabès, Tunisia

2 CMGPCE Laboratory, French Institute of Refrigeration (IFFI), National Conservatory of Arts and Crafts (CNAM), Paris, France

*Address all correspondence to: raniajradi@yahoo.fr

References

[1] Ogbonnaya SK, Ajaji OO. Fouling phenomenon and its effect on heat exchanger: A review. Frontiers in Heat and Mass Transfer. 2017;**9**:1-12

[2] Hou TK, Kazi SN, Mahat AB, Teng CB, Al-Shamma'a A, Shaw A. Industrial heat exchanger: Operation and maintenance to minimize fouling and corrosion. In: Heat Exchangers-advanced Features and Applications. London, UK: Intechopen Edition; 2017

[3] Coquelet C, Valtz A, Théveneau P. Experimental determination of thermophysical properties of working fluids for ORC applications. In: Organic Rankine Cycles for Waste Heat Recovery-analysis and Applications. London, UK: Intechopen Edition; 2019

[4] Kareem ZS, Mohd Jaafar MN, Lazim TM, Abdullah S, Abdulwahid AF. Heat transfer enhancement in two-start spirally corrugated tube. Alexandria Engineering Journal. 2015;**54**:415-422

[5] Barker G.B. The engineer's guide to plant layout and piping design for the oil and gas industries. 1st ed. Kindle Edition; 2017

[6] Kazi SN. Fouling and fouling mitigation of calcium compounds on heat exchangers by novel colloids and surface modifications. Reviews in Chemical Engineering. 2020;**36**:653-685

[7] Davoudi E, Vaferi B. Applying artificial neural networks for systematic estimation of degree of fouling in heat exchangers. Chemical Engineering Research and Design. 2017;**130**:138-153

[8] Zohuri B. Heat exchanger types and classifications. In: Compact Heat Exchangers. Cham: Springer; 2017. pp. 19-56

[9] Thulukkanam K. Heat Exchanger Design Handbook. Madras, India: Marcel Dekker, Inc; 2000

[10] Magazoni FC, Cabezas-Gomez L, Alvarino PF, Saiz-Jabardo JM. Thermal performance of one pass shell and tube heat exchangers in counter flow. Brazilian Journal of Chemical Engineering. 2019;**36**:869-883

[11] Jradi R, Fguiri A, Marvillet C, Jeday MR. Tubular heat exchanger fouling in phosphoric acid concentration process. In: Inverse Heat Conduction and Heat Exchangers. London: Intechopen Edition; 2019. pp. 47-62

[12] Shell and tube heat exchanger cleaning. Apex Engineering Products. Available from: https://www.apexengineeringproducts.com/shell-and-tube-heat-exchanger-cleaning/ [Accessed: May 31, 2021]

[13] Demasles H, Mercier P, Tochon P, Thonon B. Guide de l'encrassement des échangeurs de chaleur. France: Editions GRETh; 2007

[14] Nadjet F. Etude de l'influence de la temperature du fluide sur la performance d'un échangeur de chaleur de la raffinerie d'Alger. Algérie: Université des sciences et de la technologie Houari Boumédienne; 2010

[15] Awad MM. Fouling of heat transfer surfaces. In: Heat Transfer-theoretical Analysis, Experimental Investigations and Industrial Systems. London, UK: Intechopen Edition; 2011

[16] Garcia S, Trueba A. Fouling in heat exchangers. In: Inverse Heat Conduction and Heat Exchangers. London, UK: Intechopen Edition; 2019. pp. 1-27

[17] Magens OM. Mitigation Fouling of Heat Exchangers with Fluoropolymer Coatings. England: University of Cambridge; 2018

[18] Deshannavar UB, Rafeen MS, Ramasamy M, Subbarao D. Crude oil fouling: A review. Journal of Applied Sciences. 2010;**10**:3167-3174

[19] Hong DY. Composite Fouling on Heat Exchanger Surfaces. New York: Nova Science Books Edition; 2007

[20] Majdi HS, Alabdly HA, Hasan BO, Hathal MM. Oil fouling in double-pipe heat exchanger under liquid-liquid dispersion and the influence of copper oxide nanofluid. Heat Transfer Asian Research. 2019;**48**:1963-1981

[21] Walker P. CFD Modeling of Heat Exchanger Fouling. Sydney, Australia: University of New South Wales; 2005

[22] Thulukkanam DK. Heat Exchanger Design Handbook. 2nd ed. United States: Boca Raton; 2013

[23] Shailaja M. A critical review on effects of fouling and fouling mitigation on heat exchanger surfaces. International Journal of Advance Research in Science and Engineering. 2016;**5**:338-343

[24] Helalizadeh A. Mixed Salt Crystallization Fouling. England: University of Surrey; 2002

[25] Jradi R, Fguiri A, Marvillet C, Jeday MR. Experimental analysis of heat transfer coefficients in phosphoric acid concentration process. Journal of Statistical Mechanics: Theory and Experiments. 2019;**2019**(8):15

[26] AlHaj-Ibrahim H. Fouling in heat exchangers. In: Matla, A Fundamental Tool for Scientific Computing and Engineering Applications. London, UK: Intechopen Edition, 2012;

[27] Standards of the tubular exchanger manufacturers association. 9th ed. Tarrytown, New York: 2007

[28] Ahangar RG, Yahyazadehfar M, Pournaghshband H. The comparison of methods artificial neural network with linear regression using specific variables for prediction stock price in Tehran stock exchange. International Journal of Computer Science and Information Security. 2010;**7**:38-46

[29] Pasini A. Artificial neural networks for small dataset analysis. Journal of Thoracic Disease. 2015;**7**:953-960

[30] Jradi R, Marvillet C, Jeday MR. Modeling and comparative study of heat exchangers fouling in phosphoric acid concentration plant using experimental data. Journal of Heat and Mass Transfer. 2020;**56**:2653-2666

[31] Trzcinski P, Markowski M. Diagnosis of the fouling effects in a shell and tube heat exchanger using artificial neural network. Chemical Engineering Transactions. 2018;**70**:355-360

[32] Hussami N, Al-Haj-Ibrahim H, Safwat A. Modelling of fouling in heat exchangers using the artificial neural network approach. International Journal of Chemistry, Mathematics and Physics. 2018;**2**:33-52

[33] Priddy K, Keller PE. Artificial Neural Networks: An Introduction. Vol. TT68. Washington, USA: SPIE Press; 2005

[34] Prevolnik M, Skorjanc D. Application of artificial neural networks in meat production and technology. In: Artificial Neural Networks: Industrial and Control Engineering. London, UK: Intechopen Edition; 2011

[35] Mohanraj M, Jayaraj S, Muraleedharan C. Applications of artificial neural networks for thermal analysis of heat exchangers—A review. International Journal of Thermal Sciences. 2014;**90**:150-172

[36] Tuntas R. Aneural network based controller design for temperature control in heat exchanger. Indian Journal of Chemical Technology. 2019;**26**: 342-346

[37] Gupta AK, Kumar P, Sahoo RK, Sahu AK, Sarangi SK. Performance measurement of plate fin heat exchanger by exploration: ANN, ANFIS, GA and SA. Journal of Computational Design and Engineering. 2017;**4**:60-68

[38] Wang J, Lv Z, Liang Y, Deng L, Li Z. Fouling resistance prediction based on Ga-Elman neural network for circulating cooling water with electromagnetic anti-fouling treatment. Journal of the Energy Institute. 2019;**92**:1519-1526

[39] Lalot S, Palson H. Detection of fouling in a cross flow heat exchanger using a neural network based technique. International Journal of Thermal Sciences. 2010;**49**:675-679

[40] Aminian J, Shahhosseini S. Evaluation of ANN modeling for prediction of crude oil fouling behavior. Applied Thermal Engineering. 2008;**28**: 668-674

[41] Aminian J, Shehhosseini S. Neuro-based formulation to predict fouling threshold in crude pre heaters. International Communications in Heat and Mass Transfer. 2009;**36**:525-531

[42] Navvab Kashani M, Aminian J, Shahhosseini S, Farrokhi M. Dynamic crude oil fouling prediction in industrial preheaters using optimized ANN based moving window technique. Chemical Engineering Research and Design. 2012; **90**:938-949

[43] Sadouk HC. Modélisation de l'encrassement en régime turbulent dans un échangeur de chaleur à plaques avec un revêtement fibreux sur les parois. France: Université de Paris Est; 2009

[44] Teng A, Baharudin RH, Bakar MA, Kazi SN. Mitigation of heat exchanger fouling in industry using catalytic materials. Desalination and Water Treatment. 2015;**57**:1-6

[45] Patil KD. Review of green chemical technologies for sustainable developments in chemical process industries. Journal of Current Trends in Chemical Engineering. 2014;**2**:1-7

[46] Muller-Steinhagen H, Malayeri MR, Watkinson P. Heat exchanger fouling: Mitigation and cleaning strategies. Heat Transfer Engineering. 2001;**32**:189-196